R. L. Pickholtz

Springer Texts in
Electrical Engineering

Springer Texts in Electrical Engineering

Multivariable Feedback Systems
F. M. Callier and C. A. Desoer

Linear Programming
M. Sakarovitch

Introduction to Random Processes
E. Wong

Stochastic Processes in Engineering Systems
E. Wong and B. Hajek

EUGENE WONG
BRUCE HAJEK

Stochastic Processes in Engineering Systems

Springer-Verlag
New York Berlin Heidelberg Tokyo

Eugene Wong
Department of Electrical Engineering
 and Computer Sciences
University of California
Berkeley, California 94720 U.S.A.

Bruce Hajek
Department of Electrical and
 Computer Engineering
University of Illinois, Urbana-
 Champaign
Urbana, Illinois 61801 U.S.A.

Library of Congress Cataloging in Publication Data
Wong, Eugene, 1934-
  Stochastic processes in engineering systems.
  (Springer texts in electrical engineering)
  Bibliography: p.
  Includes index.
  1. Electric engineering—Mathematics. 2. Stochastic
processes. I. Hajek, Bruce. II. Title. III. Series.
TK153.W66   1984   519.2′024′62   84-20257

Previous edition, *Stochastic Processes in Information and Dynamical Systems* by E. Wong, was published by McGraw-Hill, Inc. in 1971.

© 1971, 1985 by Springer-Verlag New York Inc.
All rights reserved. No part of this book may be translated or reproduced in any form without written permission from Springer-Verlag, 175 Fifth Avenue, New York, New York 10010, U.S.A.
The use of general descriptive names, trade names, trademarks, etc., in this publication, even if the former are not especially identified, is not to be taken as a sign that such names, as understood by the Trade Marks and Merchandise Marks Act, may accordingly be used freely by anyone.

Typeset by Science Typographers, Inc., Medford, New York.
Printed and bound by R. R. Donnelley, Harrisonburg, Virginia.
Printed in the United States of America.

9 8 7 6 5 4 3 2 1

ISBN 0-387-96061-9 Springer-Verlag New York Berlin Heidelberg Tokyo
ISBN 3-540-96061-9 Springer-Verlag Berlin Heidelberg New York Tokyo

Preface

This book is a revision of *Stochastic Processes in Information and Dynamical Systems* written by the first author (E.W.) and published in 1971. The book was originally written, and revised, to provide a graduate level text in stochastic processes for students whose primary interest is its applications. It treats both the traditional topic of stationary processes in linear time-invariant systems as well as the more modern theory of stochastic systems in which dynamic structure plays a profound role. Our aim is to provide a high-level, yet readily accessible, treatment of those topics in the theory of continuous-parameter stochastic processes that are important in the analysis of information and dynamical systems.

The theory of stochastic processes can easily become abstract. In dealing with it from an applied point of view, we have found it difficult to decide on the appropriate level of rigor. We intend to provide just enough mathematical machinery so that important results can be stated with precision and clarity; so much of the theory of stochastic processes is inherently simple if the suitable framework is provided. The price of providing this framework seems worth paying even though the ultimate goal is in applications and not the mathematics per se.

There are two main topics in the book: second-order properties of stochastic processes (Chapter 3) and stochastic calculus in dynamical systems with applications to white noise in nonlinear systems, nonlinear filtering and detection (Chapters 4, 6 and 7). Each topic provides a convenient core for a one-semester or one-quarter graduate course. This material has been used for several years in two such courses in Berkeley, and the second topic has been used in advanced graduate courses at Urbana. The level of sophistication required for the first course is considerably lower than that required by the second. For the course centered on second-order properties, a good undergraduate background in probability theory and linear systems analysis is adequate; for the course centered on stochastic calculus, an aquaintance with measure theory is almost necessary. At Berkeley we have required a prerequisite of an undergraduate course on integration. However, a generally high level of mathematical sophistication is probably more important than formal prerequisites.

This revised text differs from the original in that the original Chapter 6 was replaced by two new chapters, Chapters 6 and 7, which provide an introduction to a more recently developed calculus of random processes in dynamical systems. This approach allows simultaneous treatment of jump processes (found in queueing and optical communication systems, for example) and sample-continuous processes arising as integrals of white Gaussian noise. This revision also contains a brief introduction to the two-parameter martingale calculus which has been developed during the last ten years, and it contains an introduction to a related, recently developed theory of stochastic differential forms.

The supporting material in Chapters 1 and 2 should be used as the instructor sees fit. Much of this material may be inappropriate for one or the other of the two courses and should be omitted.

We have included a fair number of problems and exercises. Solutions have been provided for most of them in order to facilitate self study and to provide a pool of examples and supplementary material for the main text.

We have kept the list of references short and specific. In cases where alternative references exist, we have chosen those which are most familiar to us. The necessary incompleteness of such a list is perhaps compensated by the high degree of relevance that each reference bears to the text. On basic points of probability theory, we have relied heavily on the three well-known books by Doob, Loève, and

Neveau. On a more elementary level we have found the text by Thomasian particularly comprehensive and lucid.

This book could not have been written without direct and indirect assistance from many sources. First, it is obviously a direct result of our teaching and research on stochastic processes. We are grateful to the Department of Electrical Engineering and Computer Sciences at the University of California, Berkeley and the Department of Electrical and Computer Engineering at the University of Illinois, Urbana, for their continuing support. The Army Research Office (Durham), the National Science Foundation, the Office of Naval Research, and the Department of Defense through its Joint Services Electronics Program have supported our research over the years. The organization of the original book and some of its initial writing were done during E. Wong's very pleasant sabbatical year at the University of Cambridge with the support of a John Simon Guggenheim fellowship. He is indebted to his Cambridge host, Professor J. F. Coales, F. R. S., for his kindness.

Many colleagues and friends have generously provided suggestions and criticisms. It is a pleasure to acknowledge the help given by Pierre Brémaud, Dominique Decavele, Tyrone Duncan, Terence Fine, Larry Shepp, Pravin Varaiya, and Moshe Zakai. We are very grateful to Joe Doob for his many helpful suggestions which were for the most part incorporated into the new chapters, and we are especially indebted to Bill Root for a painstaking review of a major portion of the original manuscript. The number of errors found so far makes us wonder how many still remain, and we still dare not emulate our friend Elwyn Berlekamp in offering a cash reward to readers for correction of errors! Nevertheless, any report of errors will be appreciated.

Miss Bonnie Bullivant typed the manuscript and its many revisions with both skill and patience.

Eugene Wong
Bruce Hajek

Contents

PREFACE		v
1 ELEMENTS OF PROBABILITY THEORY		1
1.	Events and probability	1
2.	Measures on finite-dimensional spaces	5
3.	Measurable functions and random variables	7
4.	Sequences of events and random variables	11
5.	Expectation of random variables	15
6.	Convergence concepts	18
7.	Independence and conditional expectation	24
2 STOCHASTIC PROCESSES		37
1.	Definition and preliminary considerations	37

2.	Separability and measurability	41
3.	Gaussian processes and Brownian motion	46
4.	Continuity	55
5.	Markov processes	61
6.	Stationarity and ergodicity	66

3 SECOND-ORDER PROCESSES — 74

1.	Introduction	74
2.	Second-order continuity	77
3.	Linear operations and second-order calculus	78
4.	Orthogonal expansions	81
5.	Wide-sense stationary processes	88
6.	Spectral representation	97
7.	Lowpass and bandpass processes	105
8.	White noise and white-noise integrals	109
9.	Linear prediction and filtering	116

4 STOCHASTIC INTEGRALS AND STOCHASTIC DIFFERENTIAL EQUATIONS — 139

1.	Introduction	139
2.	Stochastic integrals	141
3.	Processes defined by stochastic integrals	145
4.	Stochastic differential equations	149
5.	White noise and stochastic calculus	155
6.	Generalizations of the stochastic integral	163
7.	Diffusion equations	169

5 ONE-DIMENSIONAL DIFFUSIONS — 180

1.	Introduction	180
2.	The Markov semigroup	182
3.	Strong Markov processes	190
4.	Characteristic operators	193
5.	Diffusion processes	198

6 MARTINGALE CALCULUS — 209

1.	Martingales	209
2.	Sample-path integrals	217
3.	Predictable processes	222
4.	Isometric integrals	227
5.	Semimartingale integrals	233

6.	Quadratic variation and the change of variable formula	236
7.	Semimartingale exponentials and applications	241

7 DETECTION AND FILTERING 250

1.	Introduction	250
2.	Likelihood ratio representation	254
3.	Filter representation—change of measure derivation	257
4.	Filter representation—innovations derivation	262
5.	Recursive estimation	269

8 RANDOM FIELDS 279

1.	Introduction	279
2.	Homogenous random fields	280
3.	Spherical harmonics and isotropic random fields	285
4.	Markovian random fields	292
5.	Multiparameter martingales	296
6.	Stochastic differential forms	303

REFERENCES	311
SOLUTIONS TO EXERCISES	316
INDEX	355

1
Elements of Probability Theory

1. EVENTS AND PROBABILITY

The simplest situation in which probability can be considered involves a random experiment with a finite number of outcomes. Let $\Omega = \{\omega_1, \omega_2, \ldots, \omega_N\}$ denote the set of all outcomes of the experiment. In this case a probability p_i can be assigned to each outcome ω_i. The only restrictions that we place on these probabilities are: $p_i \geq 0$ and $\sum_{i=1}^{N} p_i = 1$. Every subset A of Ω in this case also has a well-defined probability which is equal to the sum of the probabilities of the outcomes contained in A.

If the number of outcomes in Ω is countably infinite, the situation is quite similar to the finite case. However, if Ω contains an uncountable number of points, then not more than a countable number of them can have nonzero probabilities, since the number of outcomes with probability $\geq 1/n$ must be less than or equal to n. In the general case, it is necessary to consider probabilities to be defined on subsets of Ω, rather than on points of Ω. A subset A on which a probability is defined is called an **event**. Let $\mathcal{C}$ denote the class of all events. A satisfactory theory of probability demands that the complement of an event should again be an event. Indeed, the occurrence of the complement A^c is just the non-occurrence of A. Similarly, if A and B are two events, then the simultaneous occurrence of both A and B should have a well-defined probability, i.e., the intersection of two events should again be an event. Thus, $\mathcal{C}$ should be closed under complementation and pairwise intersection. This immediately implies that $\mathcal{C}$ is closed under all finite Boolean set operations.[1] A class of sets which is closed under all finite Boolean set operations is called a **Boolean algebra** or simply an **algebra**. An **elementary probability measure** $\mathcal{P}$ is a function defined on an algebra $\mathcal{C}$ such that

$$0 \leq \mathcal{P}(A) \leq 1 \quad \text{and} \quad \mathcal{P}(\Omega) = 1 \qquad (1.1a)$$
$$\mathcal{P}(A \cup B) = \mathcal{P}(A) + \mathcal{P}(B) \quad \text{whenever } A \text{ and } B \text{ are disjoint (additivity)} \qquad (1.1b)$$

Both (1.1a) and (1.1b) are clearly natural conditions to be required of a probability measure.

In order that we can consider sequences of events and possible convergence of such sequences, it is necessary that not only finite, but all countable set operations on events again yield events. A class of sets is called a (Boolean) **σ algebra** if it is closed under all countable set operations.[2] It is easy to verify that the intersection of arbitrarily many σ algebras (of subsets of the same Ω) is again a σ algebra (of subsets of Ω). Therefore, given an arbitrary class $\mathcal{C}$ of subsets of Ω, there is a **smallest σ algebra** $\mathcal{C}(\mathcal{C})$ which contains $\mathcal{C}$. This is because we can define $\mathcal{C}(\mathcal{C})$ to be the intersection of all σ algebras containing $\mathcal{C}$, and there is at least one such σ algebra, viz., the collection of all subsets of Ω. We shall say that $\mathcal{C}$ **generates** its minimal σ algebra $\mathcal{C}(\mathcal{C})$. Now, consider a Boolean algebra $\mathcal{B}$ together with an elementary probability measure $\mathcal{P}$. Suppose that in addition to (1.1a) and (1.1b), $\mathcal{P}$ also satisfies

$$\text{Whenever } \{A_n\} \text{ is a sequence of sets in } \mathcal{B} \text{ such that } A_n \supset A_{n+1}$$
$$\text{and } \bigcap_{n=1}^{\infty} A_n = \emptyset \text{ then } \lim_{n \to \infty} \mathcal{P}(A_n) = 0 \text{ (monotone sequential continuity at } \emptyset) \qquad (1.1c)$$

---

[1] Complementation, union, and intersection are the most familiar Boolean set operations. Only complementation and either union or intersection need be defined. All other set operations are then expressible in terms of the two basic operations.

[2] Since all set operations are expressible in terms of complementation and union, to verify that a class is a σ algebra, we only need to verify that it is closed under complementation and countable union.

1. EVENTS AND PROBABILITY

where $\emptyset$ denotes empty set. Conditions (1.1a) to (1.1c) are equivalent to (1.1a) and the following condition taken together:

Whenever $\{A_n\}$ is a sequence of pairwise disjoint sets in $\mathcal{B}$ such that $\bigcup_{n=1}^{\infty} A_n$ is also in $\mathcal{B}$, then $\mathcal{P}\left(\bigcup_{n=1}^{\infty} A_n\right) = \sum_{n=1}^{\infty} \mathcal{P}(A_n)$

(σ additivity) (1.1d)

A set function $\mathcal{P}$ defined on an algebra $\mathcal{B}$ satisfying (1.1a) to (1.1c) is called a **probability measure**. It is a fundamental result of probability theory that a probability measure $\mathcal{P}$ on an algebra $\mathcal{B}$ extends uniquely to a probability measure on the σ algebra generated by $\mathcal{B}$.

Proposition 1.1 (Extension Theorem). Let $\mathcal{B}$ be an algebra and let $\mathcal{A}(\mathcal{B})$ be its generated σ algebra. If $\mathcal{P}$ is a probability measure defined on $\mathcal{B}$, then there is one and only one probability measure $\tilde{\mathcal{P}}$ defined on $\mathcal{A}(\mathcal{B})$ such that the restriction of $\tilde{\mathcal{P}}$ to $\mathcal{B}$ is $\mathcal{P}$ [Neveu, 1965, p. 23].

Thus, we have arrived at the basic concept of a probability space. A **probability space** is a triplet $(\Omega, \mathcal{A}, \mathcal{P})$ where Ω is a nonempty set whose elements are usually interpreted as outcomes of a random experiment, $\mathcal{A}$ is a σ algebra of subsets of Ω, and $\mathcal{P}$ is a probability measure defined on $\mathcal{A}$. The set Ω will be called the **basic space,** and its elements are called points. Elements of $\mathcal{A}$ are called **events**.

A subset of an event of zero probability is called a **null set**. Note that a null set need not be an event. A probability space $(\Omega, \mathcal{A}, \mathcal{P})$ is said to be **complete** if every null set is an event (necessarily of zero probability). If $(\Omega, \mathcal{A}, \mathcal{P})$ is not already complete, $\mathcal{P}$ can be uniquely extended to the σ algebra $\bar{\mathcal{A}}$ generated by $\mathcal{A}$ and its null sets. This procedure is called **completion**. The process of completion is equivalent to the following: For a given probability space $(\Omega, \mathcal{A}, \mathcal{P})$, define for every subset A of Ω an outer probability $\mathcal{P}^*(A)$ and an inner probability $\mathcal{P}_*(A)$ by

$$\mathcal{P}^*(A) = \inf_{C} \{\mathcal{P}(C): C \supset A, C \in \mathcal{A}\}$$
$$\mathcal{P}_*(A) = \sup_{C} \{\mathcal{P}(C): C \subset A, C \in \mathcal{A}\}$$
(1.2)

Obviously, on $\mathcal{A}$ we have $\mathcal{P}_* = \mathcal{P} = \mathcal{P}^*$. Thus, $\mathcal{P}$ can be uniquely extended to the class of all sets whose inner and outer probabilities are equal. The additional sets for which $\mathcal{P}$ can be so defined are exactly the same as those gotten by completing $(\Omega, \mathcal{A}, \mathcal{P})$ [Neveu, 1965, p. 17].

An example might best illustrate the preceding discussions. Set $\Omega = [0,1)$, and let $\mathcal{B}$ be the class consisting of $[0,1)$, $\emptyset$, all semiopen intervals $[a,b)$ with $0 \leq a < b \leq 1$, and all finite unions of disjoint semiopen intervals. The class $\mathcal{B}$ is an algebra, but not a σ algebra. If $A = [a,b)$ is a semiopen interval, we set $\mathcal{P}(A) = b - a$. If $A = \bigcup_{i=1}^{m} A_i$ is a union of disjoint intervals A_i, we set $\mathcal{P}(A) = \sum_{i=1}^{m} \mathcal{P}(A_i)$. Clearly, $\mathcal{P}$ satisfies conditions (1.1a) and (1.1b). We shall show that $\mathcal{P}$ is, in fact, σ additive. Suppose that $A_1, A_2, \ldots$ are disjoint sets in $\mathcal{B}$ such that $A = \bigcup_{i=1}^{\infty} A_i$ is also in $\mathcal{B}$. Then A is a finite union of disjoint semiopen intervals $I_1, \ldots, I_m$, and each $I_k \cap A_i$ is again a finite union of disjoint semi-open intervals. Therefore, to prove that $\mathcal{P}$ is σ additive, it is enough to show that if

$$[a,b) = \bigcup_{i=1}^{\infty} [a_i, b_i) \qquad a \leq a_i \leq b_i \leq a_{i+1} \leq b$$

$$i = 1, 2, \ldots$$

then

$$\mathcal{P}([a,b)) = b - a = \sum_{i=1}^{\infty} \mathcal{P}([a_i, b_i)) = \sum_{i=1}^{\infty} (b_i - a_i) \quad (1.3)$$

First, we note that

$$\sum_{i=1}^{n} (b_i - a_i) \leq \sum_{i=1}^{n} (b_i - a_i) + \sum_{i=1}^{n-1} (a_{i+1} - b_i) = b_n - a_1 \leq b - a$$

Hence, $b - a \geq \sum_{i=1}^{\infty} (b_i - a_i)$. Next, we note that for any $\delta > 0$,

$$[a, b - \delta] \subset \bigcup_{i=1}^{\infty} (a_i - \delta/2^i, b_i)$$

The Heine-Borel theorem [see, e.g., Rudin, 1966, p. 36] then states that there is a finite N such that

$$[a, b - \delta] \subset \bigcup_{i=1}^{N} (a_i - \delta/2^i, b_i)$$

It follows that

$$[a,b) \subset [b - \delta, b) \cup \bigcup_{i=1}^{N} [a_i - \delta/2^i, b_i)$$

$$b - a \leq \delta + \sum_{i=1}^{N} (b_i - a_i + \delta/2^i)$$

$$\leq 2\delta + \sum_{i=1}^{\infty} (b_i - a_i)$$

Therefore, for every $\delta > 0$, we have $0 \leq (b - a) - \sum_{i=1}^{\infty} (b_i - a_i) \leq 2\delta$, and (1.3) is proved. Hence, $\mathcal{P}$ is σ additive and is a probability measure. It can be uniquely extended to the σ algebra $\mathcal{A}$ generated by the semiopen intervals. Sets in $\mathcal{A}$ are called Borel sets of [0,1). $\mathcal{P}$ can be further extended by completion to $\bar{\mathcal{A}}$. Sets in $\bar{\mathcal{A}}$ are called Lebesgue measurable sets of [0,1). We note that $\mathcal{A}$ and $\bar{\mathcal{A}}$ include all intervals in [0,1) and not just semiopen ones. In particular, a point x in [0,1) can be considered to be a degenerate interval $[x,x]$ and is in $\mathcal{A}$.

Let Ω be a basic space, and let $\mathcal{A}$ be a σ algebra of subsets of Ω. The pair $(\Omega, \mathcal{A})$ is called a **measurable space** and is sometimes referred to as a **preprobability space**. A nonnegative σ additive set function μ defined on $\mathcal{A}$ is called a **measure**. Thus, a probability measure is a measure satisfying $\mathcal{P}(\Omega) = 1$. More generally, μ is said to be a **finite measure** if $\mu(\Omega) < \infty$. Even more generally, if Ω is a countable union of sets $A_1, A_2, \ldots$ in $\mathcal{A}$ such that $\mu(A_i)$ is finite for each i, then μ is said to be a σ-**finite measure**. For example, let $\Omega = R$ be the real line, and let $\mathcal{A}$ be the σ algebra generated by the class of all intervals, and for any finite interval A define $\mu(A) = $ length of A. Then, μ can be extended to a unique σ-finite measure on $(\Omega, \mathcal{A})$, and upon completion it is just the Lebesgue measure of the real line.

2. MEASURES ON FINITE-DIMENSIONAL SPACES

Let R denote the real line $(-\infty, \infty)$, and let R^n denote the collection of all ordered n-tuples $\mathbf{x} = (x_1, x_2, \ldots, x_n)$, $x_i \in R$. A subset of R^n of the form $\{\mathbf{x}: x_i \in A_i, i = 1, 2, \ldots, n\}$, where $A_1, \ldots, A_n$ are intervals, will be called a **rectangle**. We denote by $\mathcal{R}^n$ the smallest σ algebra of subsets of R^n containing all rectangles. The σ algebra $\mathcal{R}^n$ is called the n-dimensional **Borel σ algebra**, and sets in $\mathcal{R}^n$ are called **Borel sets**.

We note that $\mathcal{R}^n$ can be generated in many ways. $\mathcal{R}^n$ is the smallest σ algebra containing every set of the form $\{\mathbf{x}: -\infty < x_i < a\}$ where i is one of the integers $1, 2, \ldots, n$ and a is a real number. Thus, the class of sets generating $\mathcal{R}^n$ can be smaller than the class of all rectangles. Of course, the class of sets generating $\mathcal{R}^n$ can also be larger. For example, $\mathcal{R}^n$ is also the smallest σ algebra containing every set of the form $\{\mathbf{x}: x_i \in A_i, i = 1, \ldots, n\}$ where A_i are one-dimensional Borel sets. We note that not every set in $\mathcal{R}^n$ can be written in this way, for example, $\{(x_1, x_2): x_1 + x_2 = 2\}$. A set of the form $\{\mathbf{x}: x_i \in A_i, i = 1, \ldots, n\}$ will be called a **product** and will be denoted by $\prod_{i=1}^{n} A_i$.

The pair $(R^n, \mathcal{R}^n)$ is obviously a measurable space. Measures defined on $(R^n, \mathcal{R}^n)$ are called **Borel measures**, and probability measures defined on $(R^n, \mathcal{R}^n)$ are called **Borel probability measures**. Let μ be a finite Borel measure defined on $(R^n, \mathcal{R}^n)$, and define a function M on R^n by

$$M(a_1, a_2, \ldots, a_n) = \mu(\{\mathbf{x}, -\infty < x_i < a_i, i = 1, 2, \ldots, n\}) \quad (2.1)$$

The function M satisfies the following rather obvious conditions:

M is nonnegative and bounded by $\mu(R^n)$ (2.2a)
For each i $\lim_{a_i \to -\infty} M(\mathbf{a}) = 0$ (2.2b)
$M(\mathbf{a}) \xrightarrow[\mathbf{a} \to (\infty,\infty,\ldots,\infty)]{} \mu(R^n)$ (2.2c)

Condition (2.2a) is a simple property of measures, and conditions (2.2b) and (2.2c) follow from monotone sequential continuity of μ.

Let A be any rectangle of the form

$$A = \{\mathbf{x}: a_i \leq x_i < b_i, i = 1, \ldots, n\}$$
$$= \bigcap_{i=1}^{n} \{\mathbf{x}: a_i \leq x_i < b_i\}$$

Denoting $A_i = \{\mathbf{x}: a_i \leq x_i < b_i\}$, we have (because of additivity)

$$\mu\left(\bigcap_{i=1}^{n-1} A_i \cap (-\infty, b_n)\right) = \mu\left(\bigcap_{i=1}^{n-1} A_i \cap (-\infty, a_n)\right) + \mu\left(\bigcap_{i=1}^{n} A_i\right)$$

Hence,

$$\mu(A) = \mu\left(\bigcap_{i=1}^{n-1} A_i \cap (-\infty, b_n)\right) - \mu\left(\bigcap_{i=1}^{n-1} A_i \cap (-\infty, a_n)\right)$$

Continuing in this way, we find

$$\mu(A) = \sum_{\mathbf{x} \in S(A)} (-1)^{k(\mathbf{x})} M(\mathbf{x})$$

where M is defined by (2.1), $S(A)$ denotes the collection of 2^n vertices $\{\mathbf{x}: x_i = a_i, b_i\}$ of A, and $k(\mathbf{x})$ is the number of a's in $\mathbf{x}$. It follows that M satisfies two additional conditions:

$\sum_{\mathbf{x} \in S(A)} (-1)^{k(\mathbf{x})} M(\mathbf{x}) \geq 0$ for all rectangles A of the form

$$A = \bigcap_{i=1}^{n} \{\mathbf{x}: a_i \leq x_i < b_i\} \quad (2.3)$$

For each i $\lim_{a_i \uparrow b_i} \sum_{\mathbf{x} \in S(A)} (-1)^{k(\mathbf{x})} M(\mathbf{x}) = 0$ (2.4)

Condition (2.3) might be referred to as the **monotonicity condition**. In one dimension, it reduces to simply

$$M(b) - M(a) \geq 0 \quad b \geq a$$

Condition (2.4) is a continuity condition. In one dimension, it reduces to left continuity

$$\lim_{a \uparrow b} M(a) = M(b)$$

3. MEASURABLE FUNCTIONS AND RANDOM VARIABLES

A function satisfying (2.2) to (2.4) will be called a **distribution function.** Equation (2.1) defines a distribution function in terms of a finite Borel measure.

Conversely, every distribution function defines a finite Borel measure. To verify this, let C be the class of all rectangles of the form

$$A = \bigcap_{i=1}^{n} \{\mathbf{x}: a_i \leq x_i < b_i\} \tag{2.5}$$

Let $\mathcal{B}(C)$ be the smallest algebra (but not σ algebra) containing C. Then, every set in $\mathcal{B}(C)$ is a finite union of disjoint rectangles of the form (2.5). Given a distribution function M, we can define a nonnegative set function μ on C by

$$\mu(A) = \sum_{\mathbf{x} \in S(A)} (-1)^{k(\mathbf{x})} M(\mathbf{x})$$

and extend μ to $\mathcal{B}(C)$ by additivity. Monotone sequential continuity of μ follows from conditions (2.4) and (2.2). Hence, μ can be extended to a σ-additive set function on $\mathcal{A}(C)$, which is just $\mathcal{R}^n$. If M satisfies

$$M(\mathbf{a}) \xrightarrow[\mathbf{a} \to (\infty, \infty, \ldots, \infty)]{} 1 \tag{2.6}$$

μ is a probability measure. Distribution functions satisfying (2.6) are called **probability distribution functions** for obvious reasons.

We have established a one-to-one relationship between distribution functions and finite Borel measures. Distribution functions are useful characterizations of finite Borel measures, because being point functions rather than set functions, they are easier to specify. However, when they are used as integrators, distribution functions usually have to be interpreted as measures (cf. Sec. 1.5).

3. MEASURABLE FUNCTIONS AND RANDOM VARIABLES

Suppose that $(\Omega_1, \mathcal{A}_1)$ and $(\Omega_2, \mathcal{A}_2)$ are two measurable spaces, and f is a function with domain Ω_1 and range in Ω_2. The function f is said to be a **measurable function** or a **measurable mapping** of $(\Omega_1, \mathcal{A}_1)$ into $(\Omega_2, \mathcal{A}_2)$ if for every set A in $\mathcal{A}_2$, the set

$$f^{-1}(A) = \{\omega: f(\omega) \in A\} \tag{3.1}$$

is in $\mathcal{A}_1$. The set $f^{-1}(A)$ is called the **inverse image** of A.

If f is a measurable function $(R^n, \mathcal{R}^n) \to (R^m, \mathcal{R}^m)$, then f is called a **Borel function.** All continuous functions are Borel functions. The pointwise limit of a sequence of Borel functions is again a Borel function. Indeed, the class of Borel functions is the smallest class of functions, which contains all continuous functions and is closed under pointwise limit.

If $(\Omega,\mathcal{A})$ is a measurable space, and X is a measurable mapping of $(\Omega,\mathcal{A})$ into $(R,\mathcal{R})$, then X is usually referred to as a real-valued $\mathcal{A}$-**measurable** function. If a probability measure $\mathcal{P}$ on $(\Omega,\mathcal{A})$ is defined, then X is called a **real random variable** defined on the probability space $(\Omega,\mathcal{A},\mathcal{P})$. We note that if X is $\mathcal{A}$ measurable, and $\mathcal{A}'$ is a σ algebra containing $\mathcal{A}$, then X is also $\mathcal{A}'$ measurable.

Because $\mathcal{R}$ is generated by the collection of half lines $\{x: -\infty < x < a\}$, for X to be a real random variable, it is sufficient for every set for the form $\{\omega: X(\omega) < a\}$ to be an event. The function P_X defined by

$$P_X(a) = \mathcal{P}(\{\omega: X(\omega) < a\}) \qquad (3.2)$$

is called the **probability distribution function** of X. The relationship

$$\mathbf{P}_X(A) = \mathcal{P}(\{\omega: X(\omega) \in A\}) \qquad (3.3)$$

defines a probability measure $\mathbf{P}_X$ on $(R,\mathcal{R})$. Our discussion in Sec. 2 indicated that the distribution function P_X and the Borel measure $\mathbf{P}_X$ are mutually uniquely determined.

Suppose that $X_1, X_2, \ldots, X_n$ are n real random variables and $\mathbf{X} = (X_1, X_2, \ldots, X_n)$. Then, $\mathbf{X}$ is a measurable function mapping $(\Omega,\mathcal{A})$ into $(R^n,\mathcal{R}^n)$. The function

$$\begin{aligned}P_X(\mathbf{a}) &= P_X(a_1, \ldots, a_n) \\ &= \mathcal{P}(\{\omega: X_i(\omega) < a_i, i = 1, \ldots, n\}) \qquad \mathbf{a} \in R^n\end{aligned} \qquad (3.4)$$

is called the **joint probability distribution function** of $\mathbf{X}$. At the same time, the relationship

$$\mathbf{P}_X(A) = \mathcal{P}(\{\omega: \mathbf{X}(\omega) \in A\}) \qquad A \in \mathcal{R}^n \qquad (3.5)$$

defines a Borel probability measure.

A real random variable X is said to be **discrete** (with probability 1) if there exists a countable set $S = \{x_i\}$ such that

$$\sum_{x_i \in S} \mathcal{P}(\{\omega: X(\omega) = x_i\}) = 1 \qquad (3.6)$$

If X is discrete, then its distribution function P_X is a function which is constant except for jumps at x_i, $i = 1, 2, \ldots$, the size of the jump at x_i being $\mathcal{P}(\{\omega: X(\omega) = x_i\})$. The probability measure $\mathbf{P}_X$ is concentrated on the points in S. For an arbitrary Borel set A, we have

$$\mathbf{P}_X(A) = \sum_{x_i \in A \cap S} \mathcal{P}(\{\omega: X(\omega) = x_i\}) \qquad (3.7)$$

3. MEASURABLE FUNCTIONS AND RANDOM VARIABLES

Let $\mathbf{P}$ be a probability measure defined on $(R^n, \mathcal{R}^n)$. It is said to be **singular** (with respect to the Lebesgue measure) if there exists a set S in $\mathcal{R}^n$ such that $\mathbf{P}(S) = 1$ and the Lebesgue measure of S is zero. On the other hand, $\mathbf{P}$ is said to be **absolutely continuous** (with respect to the Lebesgue measure) if Lebesgue measure $(A) = 0$ implies $\mathbf{P}(A) = 0$. It is clear that if $X_1, X_2, \ldots, X_n$ are discrete random variables, then $\mathbf{P}_X$ is singular. However, $\mathbf{P}_X$ being singular does not imply that $X_1, X_2, \ldots, X_n$ are necessarily discrete, since the set S having simultaneous zero Lebesgue measure and $\mathbf{P}_X$ measure 1 need not be countable.

If $X_1, \ldots, X_n$ are such that $\mathbf{P}_X$ is absolutely continuous, then there exists a nonnegative Borel function $p_X(\mathbf{x})$, $\mathbf{x} \in R^n$ such that

$$\mathbf{P}_X(A) = \int_A p_X(\mathbf{x}) \, d\mathbf{x} \qquad A \in \mathcal{R}^n \tag{3.8}$$

The function p_X is called the **probability density function** for the random variables $X_1, \ldots, X_n$. Representation (3.8) results from an application of the Radon-Nikodym theorem [Loève, 1963, p. 132]. In terms of the distribution function, (3.8) takes on the more familiar form

$$P_X(a_1, a_2, \ldots, a_n) = \int_{-\infty}^{a_1} \cdots \int_{-\infty}^{a_n} p_X(x_1, x_2, \ldots, x_n) \, dx_1 \, dx_2 \cdots dx_n \tag{3.9}$$

In general, $\mathbf{P}_X$ is neither absolutely continuous nor singular, but we can always write

$$\mathbf{P}_X = \alpha \mathbf{P}_X^{(1)} + (1 - \alpha) \mathbf{P}_X^{(2)} \tag{3.10}$$

where $0 \leq \alpha \leq 1$, and $\mathbf{P}_X^{(1)}$ and $\mathbf{P}_X^{(2)}$ are, respectively, absolutely continuous and singular. The decomposition (3.10) is known as the Lebesgue decomposition [Neveu, 1965, p. 108].

Given a probability distribution function $P(\mathbf{x})$, $\mathbf{x} \in R^n$, a set of random variables $X_1, X_2, \ldots, X_n$ having P as their joint distribution function is called a **realization** of P. Every probability distribution function, i.e., every nonnegative function satisfying (2.3) to (2.6), has at least one realization. Indeed, it has numerous realizations. One standard realization can be constructed as follows: An n-dimensional probability distribution function P defines a probability measure $\mathbf{P}$ on $(R^n, \mathcal{R}^n)$. Let $\mathbf{X}: R^n \to R^n$ be the **coordinate function** defined by

$$X_j(\mathbf{x}) = x_j \tag{3.11}$$

Then $\mathbf{X} = (X_1, \ldots, X_n)$ as random variables on $(R^n, \mathcal{R}^n, \mathbf{P})$ will have P as the joint distribution function because

$$\mathbf{P}(\{\mathbf{x}: X_1(\mathbf{x}) < a_1, X_2(\mathbf{x}) < a_2, \ldots, X_n(\mathbf{x}) < a_n\})$$
$$= \mathbf{P}(\{\mathbf{x}: x_i < a_i, i = 1, 2, \ldots, n\}) = P(a_1, a_2, \ldots, a_n) \tag{3.12}$$

If $\mathbf{X} = (X_1, X_2, \ldots, X_n)$ are real random variables defined on a probability space $(\Omega, \mathcal{A}, \mathcal{P})$, and $f: R^n \to R^m$ is a Borel function, then

$$\mathbf{Y} = (Y_1, Y_2, \ldots, Y_m) = f(\mathbf{X}) \tag{3.13}$$

are again random variables defined on $(\Omega, \mathcal{A}, \mathcal{P})$. If we denote the Borel probability measure of $\mathbf{X}$ by $\mathbf{P}_X$, then for $A \in \mathcal{R}^m$,

$$\begin{aligned}\mathbf{P}_Y(A) &= \mathcal{P}(\{\omega: \mathbf{Y}(\omega) \in A\}) \\ &= \mathcal{P}(\{\omega: f(\mathbf{X}(\omega)) \in A\}) \\ &= \mathbf{P}_X(\{\mathbf{x}: f(\mathbf{x}) \in A\}) \\ &= \mathbf{P}_X(f^{-1}(A))\end{aligned} \tag{3.14}$$

Equation (3.14) expresses the transformation rule of Borel probability measures (equivalently, distribution functions). We note that (3.14) would be somewhat awkward to state in terms of distribution functions directly, without introducing the Borel probability measures.

Suppose that $f: R^n \to R^n$ is a one-to-one and invertible mapping such that both f and f^{-1} have continuous partial derivatives with respect to the coordinates. Further, suppose that $\mathbf{X} = (X_1, X_2, \ldots, X_n)$ are random variables with a density function p_X. Then $\mathbf{Y} = f(\mathbf{X})$ also has a density function, which is given by [see Thomasian, 1969, pp. 362–363]

$$p_Y(\mathbf{y}) = p_X(f^{-1}(\mathbf{y}))|J(\mathbf{y})| \tag{3.15}$$

where J is the matrix with elements $\partial f_i^{-1}(\mathbf{y})/\partial y_j$, and $|J(\mathbf{y})|$ stands for the absolute value of the determinant. As an example of (3.15), suppose that (X_1, X_2) has a joint density function

$$p_X(x_1, x_2) = \frac{1}{2\pi} e^{-\frac{1}{2}(x_1^2 + x_2^2)}$$

and $\begin{bmatrix} Y_1 \\ Y_2 \end{bmatrix} = \mathbf{A} \begin{bmatrix} X_1 \\ X_2 \end{bmatrix}$, where $\mathbf{A}$ is a constant nonsingular matrix. Denoting by $\mathbf{x}'$ the transpose of $\mathbf{x}$, we can write $(x_1^2 + x_2^2) = \mathbf{x}'\mathbf{x}$. Hence,

$$\begin{aligned}p_X(f^{-1}(\mathbf{y})) &= p_X(\mathbf{A}^{-1}\mathbf{y}) \\ &= \frac{1}{2\pi} e^{-\frac{1}{2}\mathbf{y}'(\mathbf{A}\mathbf{A}')^{-1}\mathbf{y}}\end{aligned}$$

and

$$p_Y(\mathbf{y}) = \frac{1}{2\pi|\mathbf{A}|} e^{-\frac{1}{2}\mathbf{y}'(\mathbf{A}\mathbf{A}')^{-1}\mathbf{y}}$$

Other examples can be found in the exercise section.

4. SEQUENCES OF EVENTS AND RANDOM VARIABLES

Finally, we note that the notation

$$\mathcal{P}(\{\omega: \mathbf{X}(\omega) \in A\})$$

is unduly clumsy. We shall use the simpler, but less exact, notation

$$\mathcal{P}(\mathbf{X} \in A)$$

instead.

4. SEQUENCES OF EVENTS AND RANDOM VARIABLES

For a given probability space $(\Omega, \mathcal{A}, \mathcal{P})$, let $\{A_n, n = 1, 2, \ldots\}$ be a sequence of events. The sequence $\{A_n\}$ is said to be **increasing** if $A_{n+1} \supset A_n$ for every n and **decreasing** if $A_{n+1} \subset A_n$ for every n. A sequence which is either increasing or decreasing is said to be **monotone**. The limit $\lim_{n \to \infty} A_n$ is defined for every monotone sequence as $\bigcup_{n=1}^{\infty} A_n$ or $\bigcap_{n=1}^{\infty} A_n$ according as $\{A_n\}$ is increasing or decreasing. For a general sequence $\{A_n\}$ (not necessarily monotone), we define superior and inferior limits as follows:

$$\limsup_{n \to \infty} A_n = \bigcap_{n=1}^{\infty} \bigcup_{k \geq n} A_k \tag{4.1}$$

$$\liminf_{n \to \infty} A_n = \bigcup_{n=1}^{\infty} \bigcap_{k \geq n} A_k \tag{4.2}$$

The superior limit is the set of all points which occur in an infinite number of A_n's, while the inferior limit is the set of all points which occur in all but a finite number of A_n's. Hence, $\limsup A_n \supset \liminf A_n$. If the superior limit and the inferior limit coincide, then we say that $\{A_n\}$ is a convergent sequence, and we set

$$\lim A_n = \limsup A_n = \liminf A_n \tag{4.3}$$

We note that lim sup, lim inf, and lim all involve only countable set operations. Hence all such sequential limits of events are again events.

Suppose that $\{A_n\}$ is a convergent sequence of events, and A is its limit. Then A is an event, and the following proposition relates $\mathcal{P}(A)$ to $\mathcal{P}(A_n)$.

Proposition 4.1. Every probability measure $\mathcal{P}$ is **sequentially continuous** in the sense that if $\{A_n\}$ is a convergent sequence of events, then

$$\lim_{n \to \infty} \mathcal{P}(A_n) = \mathcal{P}(\lim_{n \to \infty} A_n) \tag{4.4}$$

Proof: First, if $\{A_n\}$ decreases to the empty set $\emptyset$, then (4.4) is simply (1.1c) (sequential monotone continuity at $\emptyset$). If $\{A_n\}$ is a decreasing sequence with a nonempty limit A, then

$$\mathcal{P}(A_n) = \mathcal{P}(A_n - A) + \mathcal{P}(A)$$

and $\{A_n - A\}$ decreases to $\emptyset$. If $\{A_n\}$ is increasing with limit A, then

$$\mathcal{P}(A_n) = \mathcal{P}(A) - \mathcal{P}(A - A_n)$$

and $\{A - A_n\}$ decreases to $\emptyset$. In either case, (4.4) follows once again from (1.1c).

Now suppose $\{A_n\}$ is convergent, but not necessarily monotone. Set $B_n = \bigcap_{k \geq n} A_k$ and $C_n = \bigcup_{k \geq n} A_k$. Then, $B_n \subset A_n \subset C_n$, and $\{B_n\}$ and $\{C_n\}$ are monotone sequences. Therefore,

$$\mathcal{P}(\lim B_n) = \lim \mathcal{P}(B_n) \leq \lim \mathcal{P}(A_n)$$
$$\mathcal{P}(\lim C_n) = \lim \mathcal{P}(C_n) \geq \lim \mathcal{P}(A_n)$$

Since by definition $\lim A_n = \lim B_n = \lim C_n$, we have

$$\mathcal{P}(\lim A_n) \leq \lim \mathcal{P}(A_n) \leq \mathcal{P}(\lim A_n)$$

so that (4.4) is proved. ∎

The following rather obvious consequence of Proposition 4.1 is known as the **Borel-Cantelli lemma**. For such an obvious result, it is amazingly powerful and is a standard tool in proving properties which are true with probability 1.

Proposition 4.2. For an arbitrary sequence of events $\{A_n\}$, $\sum_{n=1}^{\infty} \mathcal{P}(A_n) < \infty$ implies $\mathcal{P}(\limsup A_n) = 0$.

Proof: From Proposition 4.1 we have

$$\mathcal{P}(\limsup A_n) = \mathcal{P}\left(\lim_{n \to \infty} \bigcup_{k \geq n} A_k\right)$$
$$= \lim_{n \to \infty} \mathcal{P}\left(\bigcup_{k \geq n} A_k\right) \leq \lim_{n \to \infty} \sum_{k \geq n}^{\infty} \mathcal{P}(A_k)$$

Because $\sum_{n=1}^{\infty} \mathcal{P}(A_n) < \infty$,

$$\lim_{n \to \infty} \sum_{k \geq n}^{\infty} \mathcal{P}(A_k) = 0$$

which proves the proposition. ∎

4. SEQUENCES OF EVENTS AND RANDOM VARIABLES

Before proceeding to the consideration of sequences of random variables, we first recall the definitions for infimum and supremum. Let S be a set of numbers in $[-\infty, \infty]$. A lower bound b (upper bound) of S is defined by the property $b \leq x (b \geq x)$ for all x in S. The infimum of S, denoted by inf S, is the greatest lower bound of S, and the supremum of S, denoted by sup S, is the least upper bound of S. Suppose that $\{X_n\}$ is a sequence of random variables defined on a common probability space $(\Omega, \mathcal{C}, \mathcal{P})$. If for every $\omega \in \Omega$ and every n, $X_{n+1}(\omega) \geq X_n(\omega)$, then $\{X_n\}$ is said to be an **increasing** sequence. The sequence $\{X_n\}$ is said to be **decreasing** if $\{-X_n\}$ is increasing. If a sequence $\{X_n\}$ is either increasing or decreasing, then it is said to be **monotone.**

If $\{X_n\}$ is a monotone sequence, then we define its limit as

$$\lim_{n\to\infty} X_n = \sup_n X_n \quad \text{or} \quad \lim_{n\to\infty} X_n = \inf_n X_n$$

according as $\{X_n\}$ is increasing or decreasing. We note that such limits may assume values $+\infty$ and $-\infty$. If $\{X_n\}$ is an arbitrary sequence, not necessarily monotone, we define inferior and superior limits as

$$\liminf_{n\to\infty} X_n = \lim_{n\to\infty} \inf_{k\geq n} X_k \tag{4.5}$$

and

$$\limsup_{n\to\infty} X_n = \lim_{n\to\infty} \sup_{k\geq n} X_k \tag{4.6}$$

If the inferior and superior limits of $\{X_n\}$ agree, we say that the sequence $\{X_n\}$ converges and set

$$\lim_{n\to\infty} X_n = \liminf_{n\to\infty} X_n = \limsup_{n\to\infty} X_n \tag{4.7}$$

Again, we note that even though X_n is finitely valued for each n, the limit $\lim_{n\to\infty} X_n$ may assume values $\pm \infty$.

If $\{X_n\}$ is a sequence of random variables, then both $\liminf X_n$ and $\limsup X_n$ are random variables *provided that they are finite at every $\omega \in \Omega$*. This is because

$$\{\omega: \liminf X_n(\omega) < a\} = \{\omega: \sup_n \inf_{k\geq n} X_n(\omega) < a\}$$
$$= \bigcap_n \bigcup_{k\geq n} \{\omega: X_k(\omega) < a\} \tag{4.8}$$

and

$$\{\omega: \limsup X_n(\omega) < a\} = \{\omega: \inf_n \sup_{k\geq n} X_n(\omega) < a\}$$
$$= \bigcup_n \bigcap_{k\geq n} \{\omega: X_k(\omega) < a\} \tag{4.9}$$

so that every set of the form

$$\{\omega: \liminf X_n(\omega) < a\} \quad \text{or} \quad \{\omega: \limsup X_n(\omega) < a\}$$

is an event. Indeed, if we had defined random variables to be *extended* real-valued functions, as is often done, then inferior and superior limits of a sequence of random variables are always random variables. The following is an immediate consequence of (4.8) and (4.9).

Proposition 4.3. Let $\{X_n\}$ be a sequence of random variables converging to a limit X. Suppose that $X(\omega)$ is finite at every $\omega \in \Omega$, then X is a random variable.

Let A be an event. Define the indicator function I_A as follows:

$$I_A(\omega) = \begin{cases} 1 & \omega \in A \\ 0 & \omega \in \Omega - A \end{cases} \tag{4.10}$$

It is obvious that I_A is a random variable. Suppose that X is a function that can be written as

$$X = \sum_{\nu=1}^{n} \alpha_\nu I_{A_\nu}$$

where $\alpha_1, \alpha_2, \ldots, \alpha_n$, are real numbers, and $A_1, A_2, \ldots, A_n$, are events. Then, X is called a **simple** random variable.

Proposition 4.4. Every random variable is the limit of a sequence of simple random variables. Every nonnegative random variable is the limit of an increasing sequence of nonnegative simple random variables.

Proof: We define $\{X_n\}$ as follows:

$$X_n(\omega) = \begin{cases} -2^n & \text{if } X(\omega) < -2^n \\ k/2^n & \text{if } X(\omega) \in [k/2^n, k+1/2^n) \\ & \quad -2^{2n} \leq k \leq 2^{2n} - 1 \\ 2^n & \text{if } X(\omega) \geq 2^n \end{cases} \tag{4.11}$$

For a fixed ω and for $n \geq \log_2 |X(\omega)|$, we have

$$\sup_{k \geq n} |X(\omega) - X_k(\omega)| \leq 2^{-n} \xrightarrow[n \to \infty]{} 0$$

so that $\{X_n\}$ converges to X at every ω. If X is nonnegative, then $\{X_n\}$ as defined by (4.11) is an increasing sequence of nonnegative random variables. ∎

5. EXPECTATION OF RANDOM VARIABLES

The expectation of a random variable X can be defined as the Stieltjes integral

$$EX = \int_{-\infty}^{\infty} x\, dP(x) \tag{5.1}$$

where $P(\cdot)$ is the probability distribution function for X. Provided that at least one of the two integrals $\int_{0}^{\infty} x\, dP(x)$ and $-\int_{-\infty}^{0} x\, dP(x)$ is less than ∞, the integral in (5.1) is well defined. As a definition for the expectation, (5.1) is less than satisfactory on two counts. First, the definition is not elementary in that it depends on the definition of a Stieltjes integral. Secondly, it obscures the fact that EX is really the integral of X on Ω with respect to the probability measure $\mathcal{P}$. In other words,

$$EX = \int_{\Omega} X(\omega)\, \mathcal{P}(d\omega) \tag{5.2}$$

where the integral needs to be defined. For these reasons, we shall give a definition for EX by defining the integral in (5.2) for a random variable X.

First, we define EX when X is a simple random variable. By definition, this means that X has the form

$$X = \sum_{i=1}^{n} x_i I_{A_i} \tag{5.3}$$

where A_i are events and x_i are real constants. For such X we define

$$EX = \sum_{i=1}^{n} x_i \mathcal{P}(A_i) \tag{5.4}$$

Thus defined, EX satisfies the following properties:

Additivity:	$E(X + Y) = EX + EY$	(5.5a)
Homogeneity:	$E(cX) = cEX$	(5.5b)
Order preservation:	$X \geq Y \Rightarrow EX \geq EY$	(5.5c)

In addition, EX also satisfies the following important property.

Lemma. Let $\{X_n\}$ be a *monotone* sequence of simple random variables converging to a *simple random variable* X. Then

$$\lim_{n \to \infty} EX_n = EX$$

Proof: If $\{X_n\}$ is decreasing, then $\{X_n - X\}$ decreases to zero. If $\{X_n\}$ is increasing, then $\{X - X_n\}$ decreases to zero. Since expectation is

additive, we only need to prove that if $\{X_n\}$ decreases to zero, then
$$\lim_{n\to\infty} EX_n = 0$$
To do this, we note that for every $\epsilon > 0$,
$$\begin{aligned} 0 \leq EX_n &\leq (\max X_n)\mathcal{P}(X_n > \epsilon) + \epsilon \\ &\leq (\max X_1)\mathcal{P}(X_n > \epsilon) + \epsilon \\ &\xrightarrow[n\to\infty]{} \epsilon \end{aligned}$$
which completes the proof of the lemma. ∎

Next, let X be a nonnegative random variable, and let $\{X_n\}$ be an increasing sequence of nonnegative simple random variables converging to X. Since $\{EX_n\}$ is a nondecreasing sequence of nonnegative numbers, $\lim_{n\to\infty} EX_n = \sup_n EX_n$ always exists, but may be infinite. We shall show that two such sequences $\{X_n\}$, $\{Y_n\}$, both converging to X, have the same limit $\lim_{n\to\infty} EX_n = \lim_{n\to\infty} EY_n$. Hence, we can unambiguously define
$$EX = \lim_{n\to\infty} EX_n \tag{5.6}$$
for nonnegative random variables. For a general X, we write
$$X = X_+ - X_- \tag{5.7}$$
where X_+ and X_- are both nonnegative and define
$$EX = EX_+ - EX_- \tag{5.8}$$
provided that the right-hand side is not of the form $\infty - \infty$. It is easy to verify that, thus defined, EX satisfies properties (5.5). The only thing that remains to be shown is the uniqueness of the right-hand limit in (5.6).

Proposition 5.1. Let $\{X_n\}$ and $\{Y_n\}$ be two increasing sequences of nonnegative simple random variables converging to the same limit X. Then
$$\lim_{n\to\infty} EX_n = \lim_{n\to\infty} EY_n$$

Proof: Let p be a fixed integer and set
$$Z_n = \min(X_n, Y_p)$$
It is clear that $\{Z_n\}$ is an increasing sequence of simple random variables. Since $X_n \xrightarrow[n\to\infty]{} X \geq Y_p$, we have
$$\lim_{n\to\infty} Z_n = Y_p$$

5. EXPECTATION OF RANDOM VARIABLES

Using the lemma, we get

$$\lim_{n\to\infty} EX_n \geq \lim_{n\to\infty} EZ_n = EY_p$$

Hence, $\lim_{n\to\infty} EX_n \geq \lim_{p\to\infty} EY_p$. Reversing the roles of $\{X_n\}$ and $\{Y_n\}$ yields $\lim_{n\to\infty} EY_n \geq \lim_{n\to\infty} EX_n$ and proves the theorem. ∎

The basic properties (5.5) of EX are precisely those of an integral. When we want to emphasize the nature of EX as an integral, we shall write

$$EX = \int_\Omega X(\omega)\mathcal{P}(d\omega)$$

We shall also use the abbreviation $\int_\Omega X\, d\mathcal{P}$. A random variable X is said to be **integrable** if $E|X| < \infty$. The following results on sequences of integrable random variables are counterparts of standard results on sequences of integrable functions in integration theory. Proofs will be omitted [Loève, 1963, pp. 124–125].

Proposition 5.2 (Monotone Convergence). Let $\{X_n\}$ be an increasing sequence of nonnegative random variables converging to X. Then,

$$EX = \lim_{n\to\infty} EX_n \tag{5.9}$$

Proposition 5.3 (Fatou's Lemma). Let $\{X_n\}$ be a sequence of random variables. Suppose that there exists an integrable random variable X such that $X_n(\omega) \geq X(\omega)$ for all n and ω. Then,

$$\liminf_{n\to\infty} EX_n \geq E \liminf_{n\to\infty} X_n \tag{5.10}$$

Proposition 5.4 (Dominated Convergence). Let $\{X_n\}$ be a sequence of random variables converging to X. Suppose that there exists an integrable random variable Y such that $|X_n(\omega)| \leq Y(\omega)$ for all n and ω. Then,

$$\lim_{n\to\infty} EX_n = EX \tag{5.11}$$

Suppose that $\mathbf{X} = (X_1, \ldots, X_n)$ are random variables defined on $(\Omega, \mathcal{A}, \mathcal{P})$ with a joint distribution function P_X. Let f be a nonnegative Borel function. Then $f(\mathbf{X}(\omega))$, $\omega \in \Omega$, is a nonnegative random variable defined on $(\Omega, \mathcal{A}, \mathcal{P})$, and $f(\mathbf{x})$, $\mathbf{x} \in R^n$, is a random variable defined on

$(R^n, \mathfrak{R}^n, \mathbf{P}_X)$. Both of the integrals $\int_\Omega f(\mathbf{X}(\omega)) \, \mathcal{P}(d\omega)$ and $\int_{R^n} f(\mathbf{x}) \, \mathbf{P}_X(d\mathbf{x})$ are well defined, and we have

$$\int_\Omega f(\mathbf{X}(\omega)) \mathcal{P}(d\omega) = \int_{R^n} f(\mathbf{x}) \mathbf{P}_X(d\mathbf{x}) \tag{5.12}$$

We leave the proof of (5.12) to Exercise 1.3. If f is not nonnegative, then we write $f = f^+ - f^-$ as before, and (5.12) still holds provided that one of the pair $Ef^+(\mathbf{X})$ and $Ef^-(\mathbf{X})$ is finite. The integral $\int_{R^n} f(\mathbf{x}) \, \mathbf{P}_X(d\mathbf{x})$ is called a **Lebesgue-Stieltjes integral** and is often written as $\int_{R^n} f(\mathbf{x}) \, dP_X(\mathbf{x})$ to emphasize the role of P_X as a distribution function. However, for the definition of the integral, it is the role of $\mathbf{P}_X$ as a measure which is crucial.

If $\mathbf{X} = (X_1, \ldots, X_n)$ are real random variables, the function $F_X(\mathbf{u})$, $\mathbf{u} \in R^n$, defined by

$$F_X(\mathbf{u}) = E \exp\left(i \sum_{k=1}^n u_k X_k\right) = E \cos \sum_{k=1}^n u_k X_k + iE \sin \sum_{k=1}^n u_k X_k \tag{5.13}$$

is called the **characteristic function** for $\mathbf{X}$. Expressed in terms of $\mathbf{P}_X$, the characteristic function is given by

$$F_X(\mathbf{u}) = \int_{R^n} \exp\left(i \sum_{k=1}^n u_k x_k\right) \mathbf{P}_X(d\mathbf{x}) \tag{5.14}$$

Every distribution function is uniquely determined by its corresponding characteristic function [compare with Eqs. (3.5.21) and (3.5.32)]. This is easy to see when P_X is absolutely continuous with density p_X. In that case,

$$F_X(\mathbf{u}) = \int_{R^n} \exp\left(i \sum_{k=1}^n u_k x_k\right) p_X(\mathbf{x}) \, d\mathbf{x} \tag{5.15}$$

and p_X can be obtained from F_X by the inversion formula of the Fourier integral, viz.,

$$p_X(\mathbf{x}) = \frac{1}{(2\pi)^n} \int_{R^n} \exp\left(-i \sum_{k=1}^n u_k x_k\right) F_X(\mathbf{u}) \, d\mathbf{u} \tag{5.16}$$

6. CONVERGENCE CONCEPTS

Thus far, when we speak of the convergence of a sequence of random variables, we have been referring to pointwise convergence at every point

6. CONVERGENCE CONCEPTS

in Ω. For this type of convergence, the probability measure $\mathcal{P}$ plays no role. In probability theory, convergence concepts which depend on $\mathcal{P}$ are of greater interest. In this section, we shall define some of these convergence concepts and explore their interrelation.

Definition. A sequence of random variables $\{X_n\}$ is said to **converge almost surely** to X if there exists a set A such that $\mathcal{P}(A) = 0$, and for every $\omega \notin A$

$$\lim_{n \to \infty} |X_n(\omega) - X(\omega)| = 0 \qquad (6.1)$$

We abbreviate the words "almost sure" and "almost surely" to a.s., and we adopt the equivalent notations lim a.s. $X_n = X$ and $X_n \xrightarrow[n \to \infty]{\text{a.s.}} X$ to denote the a.s. convergence of $\{X_n\}$ to X. We observe that as we have defined it, the limit a.s. convergent sequence of random variables is always finite except on a set of probability zero. Furthermore, two functions which are both a.s. limits of the same sequence are equal except on a set of probability zero. With these considerations, we see that we can always take the limit X of an a.s. convergent sequence of random variables $\{X_n\}$ to be a random variable, i.e., it is finite and measurable.

We note that the convergence theorems for expectation, Propositions 5.2 to 5.4, remain valid if in the statements of these theorems "convergence at every point" is replaced by "convergence almost surely." This follows simply from the fact that if $\mathcal{P}(A) = 0$, then $\int_A X \, d\mathcal{P} = 0$. Thus, if $X_n \xrightarrow[n \to \infty]{\text{a.s.}} X$, and we denote by A the set on which convergence does not take place, then $\int_{\Omega - A} X_n \, d\mathcal{P} \to \int_{\Omega - A} X \, d\mathcal{P}$ is the same as $\int_\Omega X_n \, d\mathcal{P} \to \int_\Omega X \, d\mathcal{P}$.

Given a sequence $\{X_n\}$ which may or may not be almost surely convergent, it is difficult to apply the definition of a.s. convergence to it, because we have no candidate for the limit X. For this reason, the concept of **mutual convergence** is useful. We say that a sequence $\{X_n\}$ converges mutually a.s. if $\sup_{m \geq n} |X_m - X_n| \xrightarrow[n \to \infty]{\text{a.s.}} 0$. By virtue of the Cauchy criterion for the convergence of a sequence of real numbers, for each ω, we have $\sup_{m \geq n} |X_m(\omega) - X_n(\omega)| \xrightarrow[n \to \infty]{} 0$, if and only if the sequence of real numbers $\{X_n(\omega)\}$ converges. Therefore, mutual a.s. convergence and a.s. convergence are equivalent, and we have the following.

Proposition 6.1. A sequence of random variables $\{X_n\}$ converges a.s. if and only if it converges mutually a.s.

Definition. A sequence of random variables $\{X_n\}$ is said to **converge mutually in probability** if for every $\epsilon > 0$,

$$\sup_{m \geq n} \mathcal{P}(|X_m - X_n| \geq \epsilon) \xrightarrow[n \to \infty]{} 0 \qquad (6.2)$$

A sequence $\{X_n\}$ is said to **converge in probability,** if there exists X such that for every $\epsilon > 0$,

$$\mathcal{P}(|X_n - X| \geq \epsilon) \xrightarrow[n \to \infty]{} 0 \qquad (6.3)$$

We use the notations $\lim$ in p. $X_n = X$ or $X_n \xrightarrow[n \to \infty]{\text{in p.}} X$ to denote that $\{X_n\}$ converges in probability to X.

Proposition 6.2. Let $\{X_n\}$ be a sequence of random variables:
 (a) If $\{X_n\}$ converges a.s., then it converges in probability to the same limit.
 (b) $\{X_n\}$ converges in probability if and only if it converges mutually in probability.
 (c) If $\{X_n\}$ converges in probability, then there is a subsequence converging a.s. to the same limit.

Remark: The only part of Proposition 6.2 that is easy to prove is (a). If $X_n \xrightarrow[n \to \infty]{\text{a.s.}} X$, then

$$\mathcal{P}(\sup_{m \geq n} |X_m - X| \geq \epsilon) \xrightarrow[n \to \infty]{} 0 \qquad (6.4)$$

for every $\epsilon > 0$, which certainly implies that

$$\mathcal{P}(|X_n - X| \geq \epsilon) \xrightarrow[n \to \infty]{} 0 \qquad (6.5)$$

for every $\epsilon > 0$. The proof for parts (b) and (c) will be omitted.

There is a third type of convergence which is important to us. We define it as follows.

Definition. A sequence of random variables $\{X_n\}$ is said to **converge in νth mean** ($\nu > 0$) to X if

$$E|X_n - X|^\nu \xrightarrow[n \to \infty]{} 0 \qquad (6.6)$$

We use the notation $\lim_{n \to \infty} \nu.\text{m}. \ X_n = X$ or $X_n \xrightarrow[n \to \infty]{\nu.\text{m}.} X$.

Remark: The case $\nu = 2$ is of particular importance and is known as convergence in **quadratic mean.** We abbreviate quadratic mean by q.m.

6. CONVERGENCE CONCEPTS

Proposition 6.3.

(a) If $\{X_n\}$ converges in νth mean, then it converges in probability to the same limit.

(b) $\{X_n\}$ converges in νth mean if and only if

$$\sup_{m \geq n} E|X_m - X_n|^\nu \xrightarrow[n \to \infty]{} 0 \tag{6.7}$$

Proof:

(a) We make use of the following rather obvious inequality known as the **Markov inequality**:

$$E|Z|^\nu = \int_\Omega |Z|^\nu \, d\mathcal{P} \geq \int_{|Z| \geq \epsilon} |Z|^\nu \, d\mathcal{P}$$
$$\geq \epsilon^\nu \mathcal{P}(|Z| \geq \epsilon) \tag{6.8}$$

Therefore, if $\{X_n\}$ converges in νth mean to X, then

$$\mathcal{P}(|X_n - X| \geq \epsilon) \leq \frac{1}{\epsilon^\nu} E|X_n - X|^\nu \xrightarrow[n \to \infty]{} 0$$

which proves that $X_n \xrightarrow[n \to \infty]{\text{in p.}} X$.

(b) We first suppose that $\{X_n\}$ converges in νth mean to X. Then,

$$\sup_{m \geq n} E|X_m - X_n|^\nu \leq 2^\nu \{E|X_n - X|^\nu + \sup_{m \geq n} E|X_m - X|^\nu\} \xrightarrow[n \to \infty]{} 0$$

Conversely, suppose that (6.7) holds. Then, by virtue of the Markov inequality (6.8), $\{X_n\}$ also converges mutually in probability. It follows from Proposition 6.2 that there is a subsequence $\{X'_n\}$ converging almost surely, and we denote the limit by X. Using Fatou's lemma, we find

$$E|X_n - X|^\nu \leq \liminf_{m \to \infty} E|X_n - X'_m|^\nu$$

Since $\{X'_n\}$ is a subsequence of $\{X_n\}$, we have

$$\lim_{n \to \infty} \liminf_{m \to \infty} E|X_n - X'_m| = \lim_{m,n \to \infty} E|X_m - X_n|^\nu$$

which is zero, because $\{X_n\}$ converges mutually in ν.m. Hence,

$$E|X_n - X|^\nu \xrightarrow[n \to \infty]{} 0$$

and $\{X_n\}$ converges in ν.m. to X. ∎

If we are given the pairwise joint distributions of a sequence of random variables $\{X_n\}$, we can determine whether $\{X_n\}$ converges in probability and whether it converges in νth mean. This is simply because if we know the joint distribution P_{mn} of X_m and X_n, we can compute $E|X_m - X_n|^\nu$ and $\mathcal{P}(|X_m - X_n| \geq \epsilon)$. Thus, mutual convergence in ν.m.

and in probability can be determined, hence, also convergence. On the other hand, to determine whether $\{X_n\}$ converges a.s. generally requires that we know the joint distribution of all finite subsets of random variables from the sequence $\{X_n\}$. There are, however, sufficient conditions on pairwise distributions which ensure a.s. convergence. We state one of the simplest and most useful of such conditions as follows.

Proposition 6.4. Suppose that for every $\epsilon > 0$,

$$\sum_n \sup_{m \geq n} \mathcal{P}(|X_m - X_n| \geq \epsilon) < \infty \tag{6.9}$$

Then $\{X_n\}$ converges almost surely.

Proof: Since (6.9) implies $\sup_{m \geq n} \mathcal{P}(|X_m - X_n| \geq \epsilon) \xrightarrow[n \to \infty]{} 0$, the sequence $\{X_n\}$ converges in probability. Let X denote the limit and define

$$A_n{}^\epsilon = \{\omega : |X(\omega) - X_n(\omega)| \geq 2\epsilon\}$$

Since $A_n{}^\epsilon \subset \{\omega : \max(|X(\omega) - X_m(\omega)|, |X_m(\omega) - X_n(\omega)|) \geq \epsilon\}$, we have

$$\mathcal{P}(A_n{}^\epsilon) \leq \mathcal{P}(|X - X_m| \geq \epsilon) + \mathcal{P}(|X_m - X_n| \geq \epsilon)$$

By letting $m \to \infty$, we get

$$\mathcal{P}(A_n{}^\epsilon) \leq \sup_{m \geq n} \mathcal{P}(|X_m - X_n| \geq \epsilon)$$

It follows from (6.9) that $\sum_n \mathcal{P}(A_n{}^\epsilon) < \infty$, and it follows from the Borel-Cantelli lemma (Proposition 4.2) that

$$\mathcal{P}(\limsup_n A_n{}^\epsilon) = 0$$

so that for every $\epsilon > 0$, $|X_n - X| \geq 2\epsilon$ for, at most, a finite number of values of n with probability 1. If we take $A = \bigcup_{k \geq 1} \limsup_n A_n{}^{1/k}$, then $\mathcal{P}(A) = 0$ and $\omega \notin A$ implies that

$$\lim |X_n(\omega) - X(\omega)| = 0$$

proving the theorem. ∎

The following example illustrates some of the ideas introduced in this section. Suppose $\{X_n\}$ is a sequence of random variables such that

$$\mathcal{P}[(X_m - X_n) < a] = \frac{1}{\sqrt{2\pi\sigma_{mn}{}^2}} \int_{-\infty}^{a} \exp\left(-\frac{x^2}{2\sigma_{mn}{}^2}\right) dx \tag{6.10}$$

6. CONVERGENCE CONCEPTS

and $\sigma_{mn}^2 = |m - n|/mn$. We want to investigate the convergence of $\{X_n\}$. First, we compute $E|X_m - X_n|^2$ and find

$$E|X_m - X_n|^2 = \frac{1}{\sqrt{2\pi\sigma_{mn}^2}} \int_{-\infty}^{\infty} x^2 \exp\left(-\frac{x^2}{2\sigma_{mn}^2}\right) dx$$

$$= \sigma_{mn}^2 = \frac{|m - n|}{mn} \qquad (6.11)$$

Therefore,

$$\limsup_{n \to \infty} E|X_m - X_n|^2 = \lim_{n \to \infty} \frac{1}{n} = 0$$

and we have shown that $\{X_n\}$ converges in quadratic mean. It follows that it also converges in probability.

To prove a.s. convergence is more difficult. First, we note the formula

$$\int_{-\infty}^{\infty} x^{2k+2} \exp\left(-\tfrac{1}{2}x^2\right) dx = -\int_{-\infty}^{\infty} x^{2k+1} \frac{d}{dx}[\exp\left(-\tfrac{1}{2}x^2\right)] dx$$

$$= (2k+1) \int_{-\infty}^{\infty} x^{2k} \exp\left(-\tfrac{1}{2}x^2\right) dx$$

Therefore,

$$E|X_m - X_n|^4 = 3\left(\frac{|m-n|}{mn}\right)^2$$

$$E|X_m - X_n|^6 = 15\left(\frac{|m-n|}{mn}\right)^3 \qquad (6.12)$$

. . . .

Using the Markov inequality, we have

$$\sup_{m \geq n} \mathcal{P}(|X_m - X_n| \geq \epsilon) \leq \sup_{m \geq n} \frac{E|X_m - X_n|^4}{\epsilon^4} = \frac{3}{n^2 \epsilon^4}$$

Therefore, for every $\epsilon > 0$,

$$\sum_n \sup_{m \geq n} \mathcal{P}(|X_m - X_n| \geq \epsilon) \leq \frac{3}{\epsilon^4} \sum_n \frac{1}{n^2} < \infty$$

and it follows from Proposition 6.4 that $\{X_n\}$ converges almost surely.

As the final topic in this section, we consider the distribution function of the limit of a convergent sequence of random variables.

Proposition 6.5. Let $\{X_n\}$ converge in probability to X. Let $P_n(\cdot)$ denote the distribution function of X_n, and let $P(\cdot)$ denote the distribution function of X. Then, at every continuity point x of $P(\cdot)$,

$$\lim_{n \to \infty} P_n(x) = P(x) \qquad (6.13)$$

Proof: We write

$$P(x - \epsilon) = \mathcal{P}(X < x - \epsilon, X_n < x) + \mathcal{P}(X < x - \epsilon, X_n \geq x)$$
$$P_n(x) = \mathcal{P}(X < x - \epsilon, X_n < x) + \mathcal{P}(X \geq x - \epsilon, X_n < x)$$

Therefore,

$$P(x - \epsilon) \leq P_n(x) + \mathcal{P}(|X - X_n| > \epsilon)$$

Similarly, we find

$$P_n(x) \leq P(x + \epsilon) + \mathcal{P}(|X - X_n| > \epsilon)$$

Because $\{X_n\}$ converges to X in probability, we have

$$P(x - \epsilon) \leq \lim_{n \to \infty} P_n(x) \leq P(x + \epsilon)$$

Because $P(\cdot)$ is continuous at x, we get (6.13) by letting $\epsilon \to 0$. ∎

Remark: We note that a distribution function is completely determined by its values at points of continuity. Because of this, $P(\cdot)$ is determined everywhere by the sequence $\{P_n(\cdot)\}$. The proof that a distribution function $P(\cdot)$ is completely determined by its values at points of continuity can be outlined as follows: Because $P(\cdot)$ is nondecreasing and bounded by 1, the number of jumps of size $1/n$ or greater must be less than n. Hence, the points of discontinuity are, at most, countable. This means that points of continuity are dense in $(-\infty, \infty)$, that is, every $x \epsilon (-\infty, \infty)$ is the limit of a sequence of continuity points. Because $P(\cdot)$ is left continuous at every x, it is completely determined by its values at points of continuity.

7. INDEPENDENCE AND CONDITIONAL EXPECTATION

Two events A and B are said to be **independent** if

$$\mathcal{P}(A \cap B) = \mathcal{P}(A)\mathcal{P}(B) \tag{7.1}$$

N events $A_1, A_2, \ldots, A_N$ are said to be independent if for any subset $\{k_1, \ldots, k_r\}$ of $\{1, 2, \ldots, N\}$,

$$\mathcal{P}\left(\bigcap_{i=1}^{r} A_{k_i}\right) = \prod_{i=1}^{r} \mathcal{P}(A_{k_i}) \tag{7.2}$$

More generally, we say that a family (countable or not) of events is a family of independent events if every finite subfamily is a set of indepen-

7. INDEPENDENCE AND CONDITIONAL EXPECTATION

dent events. If A and B are events such that $\mathcal{P}(B) > 0$, then we define the conditional probability of A given B by

$$\mathcal{P}(A|B) = \frac{\mathcal{P}(A \cap B)}{\mathcal{P}(B)} \tag{7.3}$$

If A and B are independent, then clearly

$$\mathcal{P}(A|B) = \mathcal{P}(A) \tag{7.4}$$

We say that the random variables $X_1, \ldots, X_N$, are independent if for arbitrary constants $x_1, \ldots, x_N$, the events $\{X_1 < x_1\}, \ldots, \{X_N < x_N\}$ are independent. If we denote the distribution function of X_i by $P_i(\cdot)$ and the joint distribution function of $X_1, \ldots, X_N$ by $P(\cdot)$, then $X_1, \ldots, X_N$ are independent if and only if

$$P(x_1, \ldots, x_N) = \prod_{i=1}^{N} P_i(x_i) \tag{7.5}$$

We leave the proof of this fact as an exercise. More generally, we define an infinite and possibly uncountable collection of random variables to be independent if every finite subset is independent.

Let $X_1, X_2, \ldots, X_N$ and Y be random variables. We define the conditional distribution function of Y relative to $X_1, \ldots, X_N$ as follows:

1. If $x_1, x_2, \ldots, x_N$ are such that $\mathcal{P}(X_1 = x_1, \ldots, X_N = x_N) > 0$, we set

$$P(y|x_1, \ldots, x_N) = \mathcal{P}(Y < y | X_1 = x_1, \ldots, X_N = x_N) \tag{7.6}$$

2. If $X_1, X_2, \ldots, X_N$ and Y have a joint density function p, we set

$$P(y|x_1, \ldots, x_N) = \frac{\int_{-\infty}^{y} p(x_1, x_2, \ldots, x_N, \eta) \, d\eta}{\int_{-\infty}^{\infty} p(x_1, x_2, \ldots, x_N, \eta) \, d\eta} \tag{7.7}$$

for all $(x_1, \ldots, x_N)$ such that the denominator is positive.
3. If neither of the above conditions is satisfied, $P(y|x_1, \ldots, x_N)$ remains undefined for the present.

If the numerator and the denominator in (7.7) are continuous at $(x_1, \ldots, x_N)$ then it is not hard to see that

$$P(y|x_1, \ldots, x_N) = \lim_{\substack{\epsilon_i \downarrow 0 \\ i=1,2,\ldots,N}} \mathcal{P}(Y < y | x_i \leq X_i < x_i + \epsilon_i, i = 1, \ldots, N) \tag{7.8}$$

The function defined by

$$p(y|x_1, \ldots, x_N) = \frac{p(x_1, \ldots, x_N, y)}{\int_{-\infty}^{\infty} p(x_1, \ldots, x_N, \eta) \, d\eta} \quad (7.9)$$

is called the conditional density function of Y relative to $X_1, \ldots, X_N$, and it has the interpretation

$$p(y|x_1, \ldots, x_N) \, dy = \mathcal{P}(y \leq Y < y + dy | X_1 = x_1, \ldots, X_N = x_N) \quad (7.10)$$

Conditional distributions involve rather cumbersome notations, and their use is limited to situations involving only a finite number of random variables. In more advanced applications of probability theory, they are all but abandoned in favor of the simpler yet more powerful concept of conditional expectation.

Before defining the conditional expectation, we shall try to get an intuitive picture of what it is. Suppose that $\mathcal{C}$ is a σ algebra of subsets of some basic space Ω. We call A an **atom** of $\mathcal{C}$ if $A \in \mathcal{C}$, and no subset of A belongs to $\mathcal{C}$ other than A itself and the empty set $\emptyset$. Atoms are a kind of irreducible units of a σ algebra. It is obvious that two distinct atoms must be disjoint. Suppose that there is a *countable* collection of sets $B_1, B_2, \ldots$, which generates $\mathcal{C}$, that is, $\mathcal{C}$ is the minimal σ algebra containing all the B_j's. Then $\mathcal{C}$ is said to be **separable**. Atoms of a separable σ algebra have some important properties that we summarize as follows.

Proposition 7.1. Let $\mathcal{C}$ be a separable σ algebra generated by a countable collection of sets $B_1, B_2, \ldots$.
(a) The collection of all atoms of $\mathcal{C}$ can be indexed by a subset of the real line. We denote the index set by $T(T \subset R)$ and denote the collection of atoms by $\{A_t, t \in T\}$.
(b) Every set B in $\mathcal{C}$ is a union of atoms.
(c) $\Omega = \bigcup_{t \in T} A_t$.

Proof: Take the generating sets $B_1, B_2, \ldots$ and define $\tilde{B}_i$ to be B_i or its complement. Now, the atoms of $\mathcal{C}$ can be constructed by forming the countable intersections $\bigcap_i \tilde{B}_i$, because for any atom A, the intersection $A \cap B_i$ must be either A or empty. Since each atom, thus constructed, is indexed by a sequence (possibly infinite) of binary numbers, the index set T can be taken to be a subset of the real line. Each ω belongs to either B_i or its complement for each i. Hence, each ω belongs to one and only one A_t so that $\Omega = \bigcup_{t \in T} A_t$. Each set B in $\mathcal{C}$ is the union of all

7. INDEPENDENCE AND CONDITIONAL EXPECTATION

intersections $\bigcap_i \tilde{B}_i$ such that $B \cap \tilde{B}_i$ is nonempty for every i. Hence, B is a union of atoms. ∎

On an atom of $\mathcal{A}$, an $\mathcal{A}$-measurable function X can only assume a single value. Let A be an atom, and let x be a real number, then $\{\omega: X(\omega) = x\}$ must be in $\mathcal{A}$ and so must $A \cap \{\omega: X(\omega) = x\}$. By definition, A being an atom implies

$$A \cap \{\omega: X(\omega) = x\} = \emptyset \quad \text{or} \quad A$$

Now suppose that $\mathcal{A}_1$ and $\mathcal{A}_2$ are two separable σ algebras such that $\mathcal{A}_1 \supset \mathcal{A}_2$. Since every atom of $\mathcal{A}_2$ is a set in $\mathcal{A}_1$, every atom of $\mathcal{A}_2$ is a union of atoms of $\mathcal{A}_1$. Therefore, the atoms of $\mathcal{A}_2$ are bigger than the atoms of $\mathcal{A}_1$, and the collection of atoms of $\mathcal{A}_2$ gives a coarser partition of Ω than the corresponding collection from $\mathcal{A}_1$. A function measurable with respect to $\mathcal{A}_2$ is necessarily measurable with respect to $\mathcal{A}_1$, but not conversely, because an $\mathcal{A}_1$-measurable function may take on more than one value on an atom of $\mathcal{A}_2$.

Let X be a random variable defined on a probability space $(\Omega, \mathcal{A}, \mathcal{P})$. We can write X as the difference $X^+ - X^-$ of a pair of nonnegative random variables X^+ and X^-. We assume that at least one of the pair EX^+ and EX^- is finite so that EX is well defined. Let $\mathcal{A}'$ be a sub-σ algebra of $\mathcal{A}$. Roughly speaking, the **conditional expectation** of X with respect to $\mathcal{A}'$ is an $\mathcal{A}'$-measurable random variable obtained by averaging X on the atoms of $\mathcal{A}'$. The precise definition is as follows.

Definition. $E^{\mathcal{A}'}X$ is uniquely defined up to sets in $\mathcal{A}'$ of probability zero by the following two conditions:
 (a) $E^{\mathcal{A}'}X$ is measurable with respect to $\mathcal{A}'$.
 (b) Let I_A denote the indicator function of A. Then

$$EI_A E^{\mathcal{A}'}X = EI_A X \quad \text{for all } A \text{ in } \mathcal{A}' \tag{7.11}$$

The existence of $E^{\mathcal{A}'}X$ as a possibly extended real-valued function is guaranteed by the Radon-Nikodym theorem, one version of which we state as follows.

Proposition 7.2. Let $(\Omega, \mathcal{A}, \mathcal{P})$ be a probability space, and let μ be a nonnegative σ-additive set function on $(\Omega, \mathcal{A})$, that is, μ is a measure. Suppose that for every A in $\mathcal{A}$ such that $\mathcal{P}(A) = 0$, we also have $\mu(A) = 0$. Then, there exists a nonnegative, $\mathcal{A}$-measurable, extended real-valued function φ such that for every $A \in \mathcal{A}$,

$$\mu(A) = \int_A \varphi \, d\mathcal{P} \tag{7.12}$$

Furthermore, φ is unique up to sets of $\mathcal{P}$-measure zero.

Remark: The function φ is called the **Radon-Nikodym derivative** of μ with respect to $\mathcal{P}$ and is sometimes denoted by $d\mu/d\mathcal{P}$.

We now apply the Radon-Nikodym theorem to the problem of defining conditional expectation. Suppose that X is a random variable on $(\Omega, \mathcal{A}, \mathcal{P})$, and $\mathcal{A}'$ is a sub-σ algebra of $\mathcal{A}$. We write $X = X^+ - X^-$ as usual and assume that at least one of the pair EX^+ and EX^- is finite. Now, define measures μ^+ and μ^- on $(\Omega, \mathcal{A}')$ by

$$\begin{aligned} \mu^+(B) &= EI_B X^+ \\ \mu^-(B) &= EI_B X^- \end{aligned} \quad B \in \mathcal{A}' \tag{7.13}$$

If $B \in \mathcal{B}$ and $\mathcal{P}(B) = 0$, then

$$\mu^+(B) = \int_B X^+ \, d\mathcal{P} = 0$$
$$\mu^-(B) = \int_B X^- \, d\mathcal{P} = 0$$

Therefore, there exist $\mathcal{A}'$-measurable functions φ^+ and φ^- such that for all B in $\mathcal{A}'$,

$$\mu^{\pm}(B) = \int_B \varphi^{\pm} \, d\mathcal{P} \tag{7.14}$$

If we set

$$\begin{aligned} E^{\mathcal{A}'} X &= E^{\mathcal{A}'} X^+ - E^{\mathcal{A}'} X^- \\ &= \varphi^+ - \varphi^- \end{aligned} \tag{7.15}$$

then $E^{\mathcal{A}'} X$ will have the defining properties, and uniqueness follows from the uniqueness of φ^+ and φ^-.

If EX^+ and EX^- are both finite, that is, if $E|X|$ is finite, then $E^{\mathcal{A}'} X$ can always be taken to be finite valued so that it is a random variable as we have defined it. If not, $E^{\mathcal{A}'} X$ may have to assume values of $\pm \infty$ and is a random variable only in an extended sense. Of course, if we had defined random variables to be extended-valued functions to begin with, this difficulty would be avoided. This is done by many authors. However, this approach also has its disadvantages. For example, extended-valued functions cannot always be added, because the sum may involve $\infty - \infty$. We shall continue to define random variables as real-valued functions. When the need arises, we shall make free use of extended-valued measurable functions. While they are not random variables in our sense, the difference is seldom important.

Roughly speaking, a conditional expectation has (almost surely) all the properties of an expectation. We make this precise by the following proposition.

7. INDEPENDENCE AND CONDITIONAL EXPECTATION

Proposition 7.3. If $X = c$ a.s. then $E^{\alpha'}X = c$ a.s., and if $X \geq Y$ a.s., then $E^{\alpha'}X \geq E^{\alpha'}Y$ a.s. Furthermore $E^{\alpha'}$ is a linear operation, that is,

$$E^{\alpha'}(\alpha X + \beta Y) = \alpha E^{\alpha'}X + \beta E^{\alpha'}Y \qquad \text{a.s.} \tag{7.16}$$

Proof: Everything follows directly from the definition of $E^{\alpha'}$.

Conditional expectation also has the convergence property of expectation. The results corresponding to Propositions 5.2 to 5.4 and a result concerning convergence in $\nu.m.$ are stated in the following proposition, but proof will be omitted.

Proposition 7.4 (Convergence Properties).

(a) If $X_n \xrightarrow[n \to \infty]{\nu.m.} X$ and $\nu \geq 1$, then

$$E^{\alpha'}X_n \xrightarrow[n \to \infty]{\nu.m.} E^{\alpha'}X \tag{7.17}$$

(b) If $\{X_n\}$ is a monotone sequence of integrable random variables converging to X, then $\{E^{\alpha'}X_n\}$ is almost surely monotone and

$$E^{\alpha'}X_n \xrightarrow[n \to \infty]{\text{a.s.}} E^{\alpha'}X \tag{7.18}$$

(c) Let $\{X_n\}$ be a sequence of integrable random variables. Suppose that there exists an integrable random variable X such that $X_n \geq X$ for all n, then

$$\liminf E^{\alpha'}X_n \geq E^{\alpha'}\liminf X_n \qquad \text{a.s.} \tag{7.19}$$

(d) Suppose that $X_n \xrightarrow[n \to \infty]{\text{a.s.}} X$ and $|X_n| \leq Y$, for some integrable Y. Then

$$E^{\alpha'}X_n \xrightarrow[n \to \infty]{\text{a.s.}} E^{\alpha'}X \tag{7.20}$$

Conditional expectation is really a smoothing operation. Roughly speaking, $E^{\alpha'}X$ is obtained by averaging X on the atoms of α'. Proposition 7.5 below summarizes some of the properties related to the fact that $E^{\alpha'}X$ is a smoothed version of X.

Proposition 7.5 (Smoothing Properties).

(a) If B is an atom of α' and $\mathcal{P}(B) > 0$, then the value of $E^{\alpha'}X$ on B is given by

$$(E^{\alpha'}X)_B = \frac{1}{\mathcal{P}(B)} \int_B X \, d\mathcal{P} \tag{7.21}$$

(b) If every event in $\mathcal{Q}'$ is independent of every event of the form $\{\omega: X(\omega) < x\}$, then

$$E^{\mathcal{Q}'}X = EX \quad \text{a.s.} \tag{7.22}$$

In particular, if we denote $\mathcal{Q}_{\emptyset} = \{\Omega, \emptyset\}$, then $E^{\mathcal{Q}_{\emptyset}}X = EX$ a.s.

(c) If Y is $\mathcal{Q}'$ measurable, then

$$E^{\mathcal{Q}'}YX = YE^{\mathcal{Q}'}X \quad \text{a.s.} \tag{7.23}$$

In particular, $E^{\mathcal{Q}'}Y = Y$ almost surely, and for any random variable X, $E^{\mathcal{Q}}X = X$ a.s.

(d) If $\mathcal{Q}_1 \supset \mathcal{Q}_2$, then

$$E^{\mathcal{Q}_2}E^{\mathcal{Q}_1}X = E^{\mathcal{Q}_2}X \quad \text{a.s.} \tag{7.24}$$

Proof:

(a) If B is an atom of $\mathcal{Q}'$, then $E^{\mathcal{Q}'}X$ must be constant on B, because $E^{\mathcal{Q}'}X$ is $\mathcal{Q}'$ measurable. By definition,

$$EI_B E^{\mathcal{Q}'}X = EI_B X$$

Since $E^{\mathcal{Q}'}X$ is constant on B, we also have

$$EI_B E^{\mathcal{Q}'}X = (EI_B)(E^{\mathcal{Q}'}X)$$

Hence, $(E^{\mathcal{Q}'}X)_B = (1/EI_B)EI_B X$, which proves (a).

(b) Let $B \in \mathcal{Q}'$. Then I_B and X are independent. Therefore,

$$EI_B E^{\mathcal{Q}'}X = EI_B X = (EX)(EI_B)$$

On the other hand, EX is $\mathcal{Q}'$ measurable and

$$E(I_B EX) = (EX)(EI_B)$$

Hence, $E^{\mathcal{Q}'}X = EX$, a.s. by virtue of the uniqueness of $E^{\mathcal{Q}'}X$.

(c) If B is an event in $\mathcal{Q}'$, and $Y = I_B$, then for every $A \in \mathcal{Q}'$,

$$EI_A E^{\mathcal{Q}'}YX = EI_A I_B X = EI_{A \cap B} X$$
$$EI_A Y E^{\mathcal{Q}'}X = EI_A I_B E^{\mathcal{Q}'}X$$
$$= EI_{A \cap B} X$$

so that (7.23) follows. By linearity (7.23) must also be true when Y is a linear combination of indicator functions (that is, Y is a simple function). For the rest, we note that by virtue of linearity, we only need to prove (7.23) for the case when both X and Y are nonnegative. Since Y is nonnegative, we can find an increasing sequence of nonnegative simple functions $\{Y_n\}$ converging to Y. Because (7.23) is true for simple functions,

$$Y_n E^{\mathcal{Q}'}X = E^{\mathcal{Q}'}XY_n$$

7. INDEPENDENCE AND CONDITIONAL EXPECTATION

From Proposition 5.2 we have for every $A \in \mathcal{A}'$,

$$EI_A Y_n E^{\mathcal{A}'} X \xrightarrow[n\to\infty]{} EI_A Y E^{\mathcal{A}'} X$$

$$EI_A E^{\mathcal{A}'} X Y_n = EI_A X Y_n \xrightarrow[n\to\infty]{} EI_A X Y$$

$$= EI_A E^{\mathcal{A}'} X Y$$

whence (7.23) follows.

(d) Suppose that $\mathcal{A}_1 \supset \mathcal{A}_2$ and $A \in \mathcal{A}_2$. Then I_A is both $\mathcal{A}_2$ measurable and $\mathcal{A}_1$ measurable. Now for every $A \in \mathcal{A}_2$,

$$EI_A E^{\mathcal{A}_2} E^{\mathcal{A}_1} X = EI_A E^{\mathcal{A}_1} X$$
$$= EI_A X$$
$$= EI_A E^{\mathcal{A}_2} X$$

which proves (7.24). ∎

We define two σ algebras $\mathcal{A}_1$ and $\mathcal{A}_2$ to be independent if, whenever $B_1 \in \mathcal{A}_1$ and $B_2 \in \mathcal{A}_2$,

$$EI_{B_1} I_{B_2} = EI_{B_1} EI_{B_2} \tag{7.25}$$

Conditional independence can be defined in a similar way as follows: Two σ algebras $\mathcal{A}_1$ and $\mathcal{A}_2$ are said to be **conditionally independent** given $\mathcal{A}'$ if, whenever $B_1 \in \mathcal{A}_1$ and $B_2 \in \mathcal{A}_2$, we have

$$E^{\mathcal{A}'} I_{B_1} I_{B_2} = E^{\mathcal{A}'} I_{B_1} E^{\mathcal{A}'} I_{B_2} \tag{7.26}$$

Conditional independence plays a vital role in the theory of Markov processes.

Suppose that X and Y are two random variables, and $E|Y| < \infty$. Let $\mathcal{A}_X$ denote the smallest σ algebra with respect to which X is measurable. Now, we want to show that $E^{\mathcal{A}_X} Y$ can always be expressed as a Borel function of X, that is, there exists a real-valued Borel function $f(x)$, $x \in R$, such that

$$(E^{\mathcal{A}_X} Y)(\omega) = f(X(\omega)) \quad \text{a.s.} \tag{7.27}$$

The notation $E^{\mathcal{A}_X} Y$ is unnecessarily cumbersome, and one usually writes $E(Y|X)$ instead.

To show (7.27), we begin by recalling the probability measure $\mathbf{P}_X$ on the σ algebra of Borel sets $\mathcal{R}$ defined by

$$\mathbf{P}_X(B) = \mathcal{P}(\{\omega : X(\omega) \in B\}) \tag{7.28}$$

Let $I_B(x)$, $x \in R$, and $B \in \mathcal{R}$, be the indicator function

$$I_B(x) = \begin{cases} 1 & \text{if } x \in B \\ 0 & \text{if } x \notin B \end{cases} \tag{7.29}$$

It is clear that $EI_B(X)Y = 0$ whenever $\mathcal{P}(X \in B) = \mathbf{P}_X(B) = 0$. Therefore, by writing $Y = Y^+ - Y^-$ as the difference of two nonnegative functions and applying the Radon-Nikodym theorem, we find that there exists a Borel function f such that

$$EI_B(X)Y = \int_B f(x)\,\mathbf{P}_X(dx) \qquad B \in \mathcal{R}$$
$$= \int_{\{\omega:\, X(\omega) \in B\}} f(X(\omega))\,\mathcal{P}(d\omega) \qquad (7.30)$$

By the definition of $E^{\alpha_X}Y$, we also have

$$EI_B(X)Y = \int_{\{\omega:\, X(\omega) \in B\}} (E^{\alpha_X}Y)(\omega)\,\mathcal{P}(d\omega) \qquad (7.31)$$

Because α_X is generated by X, every event A in α_X is of the form $A = \{\omega : X(\omega) \in B\}$ for some $B \in \mathcal{R}$ (see Exercise 16). A comparison of (7.30) and (7.31) now yields (7.27).

Exactly the same procedure can now be used to show that if α_X is the smallest σ algebra with respect to which $\mathbf{X} = (X_1, X_2, \ldots, X_n)$ are all measurable, then

$$(E^{\alpha_X}Y)(\omega) = f(\mathbf{X}(\omega)) \qquad \text{a.s.} \qquad (7.32)$$

where Y satisfies $E|Y| < \infty$, and $f: R^n \to R$ is a Borel function. Again, we shall often write $E(Y|\mathbf{X})$ instead of $E^{\alpha_X}Y$.

Earlier in this section we defined the conditional distribution function $P(y|x_1, \ldots, x_n)$ [see (7.6) and (7.7)] and interpreted it as

$$P(y|x_1, \ldots, x_N) = \mathcal{P}(Y < y | X_\nu = x_\nu, \nu = 1, \ldots, n) \qquad (7.33)$$

However, $P(y|x_1, \ldots, x_n)$ was defined only if either of two conditions is satisfied. We can now remove this restriction. Let $I_{\{Y<y\}}$ denote the indicator function

$$I_{\{Y<y\}}(\omega) = \begin{cases} 1 & \text{if } Y(\omega) < y \\ 0 & \text{otherwise} \end{cases} \qquad (7.34)$$

According to (7.32) there exists a Borel function, say $g_y(\mathbf{x})$, $\mathbf{x} \in R^n$, such that

$$E(I_{\{Y<y\}}|\mathbf{X}) = g_y(\mathbf{X}) \qquad (7.35)$$

We now define

$$P(y|x_1, \ldots, x_n) = g_y(x_1, \ldots, x_n) \qquad (7.36)$$

By exactly the same procedure, we can define the joint distribution function

$$P(y_1, \ldots, y_m | x_1, \ldots, x_n)$$
$$= \mathcal{P}(Y_j < y_j, j = 1, \ldots, m | X_k = x_k, k = 1, \ldots, n) \qquad (7.37)$$

It is not hard to verify that (7.36) reduces to (7.6) and (7.7) under the corresponding conditions.

We close this chapter with an important result which is used repeatedly in Chapters 6 and 7. Let $\mathcal{H}$ be a vector space of bounded real-valued functions on a set Ω. $\mathcal{H}$ is said to be closed under uniform (resp. bounded monotone) convergence if whenever a sequence of f_n in $\mathcal{H}$ converges uniformly to a function f (resp. whenever there is a pointwise monotone increasing sequence of functions f_n such that the functions are uniformly bounded above by a constant and f is the limit) it holds that f is in $\mathcal{H}$.

Proposition 7.6 (Monotone Class Theorem). Let $\mathcal{H}$ be a vector space of bounded real-valued functions defined on Ω which contains the constants and which is closed under both uniform and bounded monotone convergence. Let $\mathcal{C}$ be a subset of $\mathcal{H}$ which is closed under multiplication. Then $\mathcal{H}$ contains all bounded functions measurable with respect to the σ-algebra generated by functions in $\mathcal{C}$. (See [Meyer and Dellacherie, 1978] for a proof.)

EXERCISES

1. Let $\Omega = [0, \infty)$.

    (a) Let C_1 be the class of all intervals of the form $[0,a)$. Show that C_1 is not an algebra.

    (b) Let C_2 be the class of all unions of a finite number of intervals of the form $[a,b)$. Show that C_2 is an algebra, but not a σ algebra.

    (c) Show that C_2 is the smallest algebra which contains C_1.

    (d) Show that the σ algebra generated by C_2 contains all intervals in $[0, \infty)$, closed or open at either end.

2. Suppose that $\Omega = [0, \infty)$, and C is the class of all intervals of the form $[0,a)$, $0 < a \leq \infty$. Let $P(x)$, $0 \leq x < \infty$, be a left-continuous nondecreasing function such that $P(0) = 0$ and $\lim_{x \to \infty} P(x) = 1$. We define $\mathcal{P}$ on C by

$$\mathcal{P}([0,a)) = P(a)$$

    (a) Show that $\mathcal{A}(C)$ includes all intervals in $[0, \infty)$.

    (b) How must $\mathcal{P}((a,b))$, $\mathcal{P}([a,b))$, and $\mathcal{P}([a,b])$ be defined in terms of P in order for $\mathcal{P}$ to be σ additive?

3. Let $\mathbf{X} = (X_1, \ldots, X_n)$ be random variables defined on a probability space $(\Omega, \mathcal{A}, \mathcal{P})$. We can uniquely define a probability measure $\mathbf{P}_X$ on $(R^n, \mathfrak{R}^n)$ by

$$\mathbf{P}_X(\{\mathbf{x}: \alpha_\nu \leq x_\nu < \beta_\nu, \nu = 1, \ldots, n\}) = \mathcal{P}(\{\omega: \alpha_\nu \leq X_\nu(\omega) < \beta_\nu, \nu = 1, \ldots, n\})$$

Let $f: R^n \to R$ be a nonnegative Borel function.

(a) Show that for arbitrary a and b

$$\mathbf{P}_X(\{\mathbf{x}: a \leq f(\mathbf{x}) < b\}) = \mathcal{P}(\{\omega: a \leq f(\mathbf{X}(\omega)) < b\})$$

(b) Both $\int_{R^n} f(\mathbf{x}) \mathbf{P}_X(d\mathbf{x})$ and $\int_\Omega f(\mathbf{X}(\omega)) \mathcal{P}(d\omega)$ are well defined by the procedure given in Sec. 1.5. Show that they are equal.

We note that $\int_{R^n} f(\mathbf{x}) \mathbf{P}_X(d\mathbf{x})$ so defined is called a Lebesgue-Stieltjes integral and is sometimes written as $\int_{R^n} f(\mathbf{x}) \, dP(\mathbf{x})$. By writing $f = f^+ - f^-$, we can extend the definition to functions which are not necessarily nonnegative.

4. Let $\{A_n, n = 1, 2, \ldots\}$ be a sequence of sets. We can show that a point ω belongs to $\bigcap_{n=1}^\infty \bigcup_{k \geq n} A_k$ if and only if it belongs to an infinite number of A_n's as follows.

Suppose that ω belongs to an infinite number of A_n's, then for every n, $\omega \in \bigcup_{k \geq n}^\infty A_k$. Therefore, $\omega \in \bigcap_{n=1}^\infty \bigcup_{k \geq n} A_k$. On the other hand, if ω belongs to only a finite number of A_n's, then there is some n_0 such that $\omega \notin \bigcup_{k \geq n_0} A_k$. Since $\bigcup_{k \geq n_0} A_k$ contains $\bigcap_{n=1}^\infty \bigcup_{k \geq n} A_k$, this proves that ω cannot belong to $\bigcap_{n=1}^\infty \bigcup_{k \geq n} A_k$ if ω belongs to only a finite number of A_n's. Show that $\omega \in \bigcup_{n=1}^\infty \bigcap_{k \geq n} A_k$ if and only if ω belongs to all but a finite number of A_n's.

5. Suppose that $X_1, \ldots, X_n$ are n random variables with a joint density function p_X. Let $Y_1, \ldots, Y_n$ be defined by $Y_i = f_i(X_1, \ldots, X_n)$. Suppose that $\mathbf{f}$ has a differentiable inverse $\mathbf{g}$ so that $X_i = g_i(Y_1, \ldots, Y_n)$. Show that the joint density function of $Y_1, \ldots, Y_n$ is given by

$$p_Y(y_1, \ldots, y_n) = p_X(g_1(\mathbf{y}), \ldots, g_n(\mathbf{y})) |J(\mathbf{y})|$$

where $|J|$ denotes the absolute value of the determinant of $J(\mathbf{y}) = \partial g_i / \partial y_j$. Suggestion: Consider the incremental "rectangle" in R^n with sides $dy_1, dy_2, \ldots, dy_n$ and located at the point $\mathbf{y}$. Under the transformation $\mathbf{g}$ this rectangle is mapped approximately into a "parallelepiped" located at $\mathbf{g}(\mathbf{y})$ with sides $J(\mathbf{y}) \, d\mathbf{y}$, the volume of which is $|J(\mathbf{y})| \, dy_1, dy_2, \ldots, dy_n$ [see, for example, Birkhoff and MacLane, 1953, pp. 307–310]. The desired result now follows from the interpretation of probability density function as probability per unit volume [see also, Thomasian, 1969, pp. 362–363].

EXERCISES

6. Suppose that X_1, X_2, X_3 are random variables with a joint density

$$p_X(x_1,x_2,x_3) = \frac{1}{(2\pi)^{3/2}} \exp\left[-\tfrac{1}{2}(x_1^2 + x_2^2 + x_3^2)\right]$$

Find the density function for $Y = \sqrt{X_1^2 + X_2^2 + X_3^2}$.
Hint: Introduce random variable Θ and Φ so that

$X_1 = Y \cos \Theta$
$X_2 = Y \sin \Theta \cos \Phi$
$X_3 = Y \sin \Theta \sin \Phi$

7. Suppose that $X_1, \ldots, X_n$ have a joint density function

$$p_X(x_1, \ldots, x_n) = \frac{1}{(2\pi)^{n/2}} \exp\left[-\tfrac{1}{2}(x_1^2 + \cdots + x_n^2)\right]$$

Let $Y_k = \sum_{j=1}^{k} X_j$, $k = 1, \ldots, n$. Find the joint density p_Y for $Y_1, \ldots, Y_n$.

8. Prove that

$$E\frac{|X|}{1+|X|} - \frac{\epsilon}{1+\epsilon} \leq \mathcal{P}(|X| \geq \epsilon) \leq \frac{1+\epsilon}{\epsilon} E\frac{|X|}{1+|X|}$$

9. Show that $X_n \xrightarrow[n\to\infty]{\text{in p.}} X$ if and only if

$$\lim_{n\to\infty} E\left(\frac{|X_n - X|}{1 + |X_n - X|}\right) = 0$$

10. Suppose that $\{X_n\}$ is a sequence of random variables such that

$$\mathcal{P}(X_n - X_m < x) = \frac{1}{\pi \sigma_{mn}} \int_{-\infty}^{x} \frac{1}{1 + \xi^2/\sigma_{mn}^2} d\xi$$

with $\sigma_{mn}^2 = |m - n|/mn$. Does $\{X_n\}$ converge in probability?
Hint: Compute $E[|X_n - X_m|/(1 + |X_n - X_m|)]$.

11. Starting from the Schwarz inequality $|EXY|^2 \leq EX^2 EY^2$, prove that $X_n \xrightarrow[n\to\infty]{\text{q.m.}} X$ implies that

$$EX_n \xrightarrow[n\to\infty]{} EX$$

and

$$EX_n^2 \xrightarrow[n\to\infty]{} EX^2$$

Hint: For the second part verify that

$$(\sqrt{EX_n^2} - \sqrt{EX^2})^2 \leq E(X_n - X)^2$$

12. If $\{X_n\}$ converges in q.m. to X and each X_n has a density function $p_n(x) = (1/\sqrt{2\pi\sigma_n^2}) \exp[-\frac{1}{2}(x - \mu_n)^2/\sigma_n^2]$, prove that X has a density

$$p(x) = (1/\sqrt{2\pi\sigma^2}) \exp[-\frac{1}{2}(x - \mu)^2/\sigma^2]$$

provided that $E(X - EX)^2 > 0$.

13. Prove that $X_1, \ldots, X_n$ are mutually independent if and only if

$$\mathcal{P}(X_\nu < x_\nu, \nu = 1, \ldots, n) = \prod_{\nu=1}^{n} \mathcal{P}(X_\nu < x_\nu)$$

14. Let X_1 and X_2 be two random variables such that $EX_1 = 0$ and $EX_2 = 0$.

(a) Suppose that we can find a linear combination $Y = X_1 + \alpha X_2$ which is independent of X_2. Show that $E(X_1|X_2) = -\alpha X_2$.

(b) If X_1 and X_2 have a joint density function

$$p(x_1, x_2) = \frac{1}{2\pi\sqrt{1-\rho^2}} \exp\left[-\frac{1}{2(1-\rho^2)}(x_1^2 - 2\rho x_1 x_2 + x_2^2)\right]$$

find $E(X_1|X_2)$.

15. Suppose that X_1 and X_2 have a joint density given by

$$p(x_1, x_2) = \frac{1}{2\pi} \exp[-\frac{1}{2}(x_1^2 + x_2^2)]$$

Let $Y = \sqrt{X_1^2 + X_2^2}$. Find $E(X_1|Y)$.
Hint: Introduce random variable Φ so that $X_1 = Y \cos \Phi$ and $X_2 = Y \sin \Phi$.

16. Let $\mathcal{A}_X$ denote the smallest σ algebra with respect to which the random variables $\mathbf{X} = (X_1, X_2, \ldots, X_n)$ are all measurable. Let $X^{-1}(B)$ denote $\{\omega : \mathbf{X}(\omega) \in B\}$. Show that $\mathcal{A}_X = \{X^{-1}(B), B \in \mathcal{R}^n\}$. In other words, $\mathcal{A}_X$ is the collection of all inverse images of Borel sets.

2
Stochastic Processes

1. DEFINITION AND PRELIMINARY CONSIDERATIONS

A **stochastic process** $\{X_t, t \in T\}$ is a family of random variables, indexed by a real parameter t and defined on a common probability space $(\Omega, \mathcal{A}, \mathcal{P})$. Unless otherwise specified, the parameter set T will always be taken to be an interval. By definition, for each t, X_t is an $\mathcal{A}$-measurable function. For each ω, $\{X_t(\omega), t \in T\}$ is a function defined on T and is called a **sample function** of the process.

If $T_n = \{t_1, \ldots, t_n\}$ is a finite set from T, we denote by P_{T_n} the joint distribution function of $\{X_{t_1}, \ldots, X_{t_m}\}$. The collection $\{P_{T_n}\}$ as T_n ranges over all finite sets in T is called the family of **finite-dimensional distributions** of the process $\{X_t, t \in T\}$. Loosely speaking, problems which can be answered directly in terms of the finite-dimensional distributions involve no mathematical difficulties of a measure-theoretical nature. The more elementary applications of stochastic processes are problems of this type. Let $\{X_t, t \in T\}$ be a stochastic process defined on $(\Omega, \mathcal{A}, \mathcal{P})$. Let $\mathcal{B}_X$ and $\mathcal{A}_X$ be, respectively, the smallest algebra and the smallest σ algebra with respect to which X_t is measurable for every $t \in T$. The difference between $\mathcal{B}_X$ and $\mathcal{A}_X$ is that $\mathcal{B}_X$ is only closed under all finite set operations while $\mathcal{A}_X$ is closed under all countable set operations. Now, if all finite-dimensional distributions are known, then the probability of every event in $\mathcal{B}_X$ is uniquely and directly determined, because every set in $\mathcal{B}_X$ involves only a finite number of X_t's. In other words, $\mathcal{P}$ is uniquely determined on $\mathcal{B}_X$. Now, since $\mathcal{P}$ is a probability measure, it must be σ additive. Hence, by the extension theorem (Proposition 1.1.1), we can extend $\mathcal{P}$ from $\mathcal{B}_X$ to the smallest σ algebra containing $\mathcal{B}_X$, and that is just $\mathcal{A}_X$. Therefore, the restriction of $\mathcal{P}$ on $\mathcal{A}_X$ is uniquely determined by the set of all finite-dimensional distributions.

We illustrate the previous remarks by an example. Suppose that $\{X_t,\ 0 \leq t \leq 1\}$ is a stochastic process. The event

$$\left\{\omega: X_t(\omega) \geq 0,\ t = \frac{1}{k},\ k = 1, 2, \ldots, n\right\} \tag{1.1}$$

is a set in $\mathcal{B}_X$. The events

$$\left\{\omega: X_t(\omega) \geq 0,\ t = \frac{1}{k},\ k = 1, 2, \ldots\right\} \tag{1.2}$$

and

$$\left\{\omega: X_t(\omega) \geq 0,\ t = \frac{l}{k},\ l = 0, 1, \ldots, k;\ k = 1, 2, \ldots\right\} \tag{1.3}$$

are sets in $\mathcal{A}_X$. However, the set

$$\{\omega: X_t(\omega) \geq 0,\ 0 \leq t \leq 1\} = \bigcap_{t \in [0,1]} \{\omega: X_t(\omega) \geq 0\} \tag{1.4}$$

may not even be in $\mathcal{A}$, that is, it may not be an event, because it involves uncountable set operations on events. Whether it is an event or not depends on the precise nature of the probability space and the process. We shall consider these questions in greater detail in Sec. 2.

In practice, one seldom begins with a given probability space and a given family of random variables defined on it. Instead one often starts with a proposed collection of finite-dimensional distributions $\{P_{T_n}$, all finite T_n in $T\}$, which is usually obtained by a combination of observations and hypotheses. The question then arises as to whether we can always find a stochastic process having these distributions. We shall answer this question as clearly as we can, because it is a source of some confusion.

First, the collection of finite-dimensional distributions must be **compatible** in the following sense: If T_n and T_m are two ordered finite

1. DEFINITION AND PRELIMINARY CONSIDERATIONS

sets from T such that T_n contains T_m, then P_{T_m} must be equal to P_{T_n} with the appropriate variables set to ∞. For example,

$$P_{t_1}(x_1) = P_{t_1,t_2}(x_1, \infty) \tag{1.5}$$

Given a compatible family of finite-dimensional distributions $\{P_{T_n}$, all finite T_n in $T\}$, we can always find a probability space $(\Omega, \mathcal{A}, \mathcal{P})$ and a family of random variables $\{X_t, t \in T\}$ having the given finite-dimensional distributions. The proof is by construction. Let $\Omega = R^T = \{$the set of all real-valued functions defined on $T\}$. Let $X_t(\omega) = $ the value of ω at t. Let $\mathcal{B}_X$ and $\mathcal{A}_X$ be, respectively, the smallest Boolean algebra and the smallest σ algebra with respect to which every X_t is measurable. Now, every set in $\mathcal{B}_X$ is of the form

$$\{\omega: (X_{t_1}(\omega), X_{t_2}(\omega), \ldots, X_{t_n}(\omega)) \in B\} \tag{1.6}$$

where B is an n-dimensional Borel set. Given a compatible family of finite-dimensional distributions $\{P_{T_n}\}$, we set

$$\mathcal{P}(\{\omega: (X_{t_1}(\omega), X_{t_2}(\omega), \ldots, X_{t_n}(\omega)) \in B\})$$
$$= \int_B dP_{t_1,t_2,\ldots,t_n}(x_1, \ldots, x_n) \tag{1.7}$$

This defines an elementary probability measure $\mathcal{P}$ on $(\Omega, \mathcal{B}_X)$. Now, it can be shown that $\mathcal{P}$, so defined, is not only finitely additive, but also σ additive. This means that $\mathcal{P}$ is a probability measure and can be uniquely extended to $\mathcal{A}_X$. To show that $\mathcal{P}$ is σ additive, and not merely finitely additive, is fairly difficult [see Neveu, 1965, pp. 82–83], and the proof will be omitted. To summarize, we take $\Omega = R^T$,

$$X_t(\omega) = \text{value of } \omega \text{ at } t \tag{1.8}$$

and $\mathcal{A}_X$ to be the minimal σ algebra generated by $\{X_t, t \in T\}$. The probability measure $\mathcal{P}$ is defined by (1.7). So defined, the process $\{X_t, t \in T\}$ has the prescribed finite-dimensional distributions. We note that $\{X_t(\omega), \omega \in R^T, t \in T\}$ as defined by (1.8) is called the **coordinate function,** because $X_t(\omega)$ is the tth coordinate of ω. Sets of the form (1.6) are called **cylinder sets.**

In the construction that we have just given, the basic space Ω was taken to be R^T, and the σ algebra was taken to be $\mathcal{A}_X$, the minimal σ algebra containing all cylinder sets. In a sense, R^T is too big and $\mathcal{A}_X$ is rather small. For example, sets of the form

$$\{\omega: a \leq X_t(\omega) \leq b \text{ for all } t \in T\}$$

are not in $\mathcal{A}_X$, if T is uncountable. Sometimes, it may be convenient to work with a different Ω. Suppose that we are given a compatible family of finite-dimensional distributions $\{P_{T_n}$, all finite T_n in $T\}$, and that we are also given a basic space Ω together with a family of functions $\{X_t(\omega), \omega \in \Omega, t \in T\}$. Let $\mathcal{A}_X$ be defined as before. The question is, "can we always find a probability measure $\mathcal{P}$ defined on $(\Omega, \mathcal{A}_X)$ so that $\{X_t, t \in T\}$ has the prescribed finite-dimensional distribution?" The answer is an immediate "no." In order for this to be possible, $\{P_{T_n}\}$ must satisfy other conditions, in addition to the compatibility condition. First of all, it may happen that for two different Borel sets A and B,

$$\{\omega: (X_{t_1}(\omega), X_{t_2}(\omega), \ldots, X_{t_n}(\omega)) \in A\}$$
$$= \{\omega: (X_{t_1}(\omega), X_{t_2}(\omega), \ldots, X_{t_n}(\omega)) \in B\} \quad (1.9)$$

Since the same set in $\mathcal{A}_X$ can have only one probability value, we must require

$$\int_A dP_{t_1,\ldots,t_n} = \int_B dP_{t_1,\ldots,t_n} \quad (1.10)$$

whenever A and B satisfy (1.9). This consistency condition is sufficient to ensure that (1.7) defines an elementary probability $\mathcal{P}$ on $(\Omega, \mathcal{B}_X)$. However, $\mathcal{P}$ need not be σ additive (or equivalently, monotone-sequential continuous at $\emptyset$). If $\mathcal{P}$ is not σ additive, it means that we cannot define a probability measure on $(\Omega, \mathcal{A}_X)$ so that $\{X_t, t \in T\}$ has the prescribed finite-dimensional distributions. We illustrate this discussion by an example. Let $T = [0,1]$, and take $\Omega = C[0,1] = \{$all continuous functions on $[0,1]\}$. Take $X_t(\omega)$ to be the coordinate function as before. Suppose that $\{P_{T_n}\}$ is given by

$$P_{t_1,t_2,\ldots,t_n}(x_1, \ldots, x_n) = \prod_{\nu=1}^n \int_{-\infty}^{x_\nu} \frac{1}{\sqrt{2\pi}} \exp\left(-\tfrac{1}{2}z^2\right) dz \quad (1.11)$$

In order to be consistent with (1.11), we must have

$$\mathcal{P}(X_t > \epsilon, X_s < -\epsilon) = \left[\int_\epsilon^\infty \frac{1}{\sqrt{2\pi}} \exp\left(-\tfrac{1}{2}z^2\right) dz\right]^2 \quad (1.12)$$

Because $\Omega = C[0,1]$, the sets

$$A_n = \{\omega: X_t(\omega) > \epsilon, X_{t+1/n}(\omega) < -\epsilon\}$$

must converge to $\emptyset$ for every $\epsilon > 0$. However,

$$\mathcal{P}(X_t > \epsilon, X_{t+1/n} < -\epsilon) = \left[\int_\epsilon^\infty \frac{1}{\sqrt{2\pi}} \exp\left(-\tfrac{1}{2}z^2\right) dz\right]^2 \underset{n \to \infty}{\not\to} 0$$

Hence, $\mathcal{P}$ defined by (1.7) and (1.11) cannot be sequentially continuous at $\emptyset$ and cannot be extended to $\mathcal{A}_X$. Intuitively, the reason is clear. Since $\Omega = C[0,1]$, any probability measure on $(\Omega, \mathcal{A}_X)$ gives $\mathcal{P}(\Omega) = 1$, which means that with probability 1, every sample function is continuous.

2. SEPARABILITY AND MEASURABILITY

Clearly, not every compatible family of finite-dimensional distributions is consistent with this fact. In particular, (1.11) implies that $\{X_t, t \in [0,1]\}$ is a family of independent and identically distributed random variables, and the independence between X_t and X_s, no matter how close t and s are, is incompatible with continuous sample functions. Later, we shall develop conditions on the finite-dimensional distributions which guarantee continuous sample functions.

2. SEPARABILITY AND MEASURABILITY

The definition of a stochastic process X_t requires it to be an α-measurable ω function for each t, but places no condition on it as a function of t. When T is an interval, questions of an analytical nature concerning sample functions of the process usually involve an uncountable number of events and random variables and are not always answerable without additional assumptions. For example, the set $\{\omega: X_t(\omega) \geq 0$ for all $t \in T\} = \bigcap_{t \in T} \{\omega: X_t(\omega) \geq 0\}$ may not be an event since it involves an *uncountable* intersection of events. Similarly, $Y = \sup_{t \in T} X_t$ may not be a random variable, because sets of the form

$$\{\omega: Y(\omega) \leq y\} = \bigcap_{t \in T} \{\omega: X_t(\omega) \leq y\} \tag{2.1}$$

may not be events. The same comment applies to $\inf_{t \in T} X_t$ and to

$$\limsup_{u \to t} X_u = \lim_{n \to \infty} \sup_{|u-t|<1/n} X_u \tag{2.2}$$

Hence, even when $\lim_{u \to t} X_u$ exists, it may not be a random variable. Further, an integral of the form

$$Z(\omega) = \int_a^b X_t(\omega)\, dt$$

may not be well defined for almost all ω, and may not be a random variable even when it is well defined. These difficulties motivated the introduction of the concepts of a separable process and a measurable process [Doob, 1953, pp. 50-71].

Definition. A process $\{X_t, t \in T\}$ is said to be **separable** if there exist a countable set $S \subset T$ and a fixed null event Λ such that for any closed set $K \subset [-\infty, \infty]$ and any open interval I, the two sets

$$\{\omega: X_t(\omega) \in K, t \in I \cap T\} \qquad \{\omega: X_t(\omega) \in K, t \in I \cap S\}$$

differ by a subset of Λ.

The countable set S is called the **separant** or **separating set**. If $\{X_t, t \in T\}$ is separable, then every set of the form $\{\omega: X_t(\omega) \in K, t \in I \cap T\}$ differs from an event by at most a null set and can be rendered an event by completing the probability space on which the process is defined. If the process $\{X_t, t \in T\}$ is separable and $\omega \notin \Lambda$, then for any open interval I, $X_t(\omega) \leq a$, $t \in I \cap S$ implies $X_t(\omega) \leq a$ for all $t \in I \cap T$. Thus,

$$\sup_{t \in I \cap S} X_t(\omega) \geq \sup_{t \in I \cap T} X_t(\omega)$$

Since $T \supset S$, the opposite inequality holds also. Hence for any open interval I

$$\sup_{t \in I \cap S} X_t(\omega) = \sup_{t \in I \cap T} X_t(\omega) \qquad (2.3)$$

for all $\omega \notin \Lambda$. Because S is countable, $\sup_{t \in I \cap S} X_t$ is a random variable, thus $\sup_{t \in I \cap T} X_t$ is equal almost everywhere to a random variable. If the probability space is complete, then $\sup_{t \in I \cap T} X_t$ is itself a random variable. Similar results hold for $\inf_{t \in I \cap T} X_t$ and, hence, also for

$$\limsup_{s \to t} X_s = \lim_{n \to \infty} \sup_{|s-t|<1/n} X_s$$

and

$$\liminf_{s \to t} X_s = \lim_{n \to \infty} \inf_{|s-t|<1/n} X_s$$

Thus, $\lim_{s \to t} X_s$, when it exists, is equal almost everywhere to a random variable and can be made a random variable by completing the underlying probability space. Henceforth, the underlying probability space will always be assumed to be complete.

Given a probability space $(\Omega, \mathcal{Q}, \mathcal{P})$ and a process $\{X_t, t \in T\}$ defined on it, then $\{X_t, t \in T\}$ is either separable or not, and there is nothing one can do about it. In practice, the situation is never this rigid. Instead, one is usually free to choose the way in which $\{X_t, t \in T\}$ is to be defined, as long as some specified finite-dimensional distributions are satisfied. Thus, the following result is an exceedingly important one.

Proposition 2.1. For every stochastic process $\{X_t, t \in T\}$ there exists a process $\{\tilde{X}_t, t \in T\}$, defined on the same probability space, such that

(a) $\{\tilde{X}_t, t \in T\}$ is separable
(b) $\mathcal{P}(X_t = \tilde{X}_t) = 1$ for each $t \in T$

2. SEPARABILITY AND MEASURABILITY

Remarks:

(a) Although the set $\{\omega: X_t(\omega) \neq \tilde{X}_t(\omega)\}$ is a null event for each t, the set

$$\{\omega: X_t(\omega) \neq \tilde{X}_t(\omega) \text{ for at least one } t \text{ in } T\}$$
$$= \bigcup_{t \in T} \{\omega: X_t \neq \tilde{X}_t(\omega)\} \qquad (2.4)$$

need not be an event and need not have zero probability even if it is an event. If it is a null event, then $\{X_t, t \in T\}$ is itself a separable process.

(b) Obviously, $\{X_t, t \in T\}$ and $\{\tilde{X}_t, t \in T\}$ have the same finite-dimensional distributions.

(c) It may be necessary for $\tilde{X}_t$ to assume values $\pm \infty$.

A proof of Proposition 2.1 will be omitted. Suffice it to note that the standard proof is by construction, and the separating set S that results is, in general, quite complicated. The situation improves if a continuity condition is satisfied by the finite-dimensional distributions. A process $\{X_t, t \in T\}$ is said to be **continuous in probability at t** if

$$\mathcal{P}(|X_s - X_t| \geq \epsilon) \xrightarrow[s \to t]{} 0 \qquad (2.5)$$

for every $\epsilon > 0$. If $\{X_t, t \in T\}$ is continuous in probability at every point in T, then we shall say simply $\{X_t, t \in T\}$ is continuous in probability. We note that continuity in probability is verifiable in terms of two-dimensional distributions.

Proposition 2.2. Let $\{X_t, t \in T\}$ be a separable process which is continuous in probability. Then every countable set dense in T is a separating set.

Thus, whenever a process $\{X_t, t \in T\}$ is both separable and continuous in probability, the probability of an event involving an uncountable number of X_t can be computed by choosing a sequence of partitions of T in the usual manner. For example, suppose $\{X_t, 0 \leq t \leq 1\}$ is both separable and continuous in probability, and we wish to compute $\mathcal{P}(X_t \geq 0, 0 \leq t \leq 1)$. First, we note that we can take S to be the set of all dyadic rationals, that is,

$$S = \left\{ \frac{k}{2^n}, 0 \leq k \leq 2^n, n = 0, 1, \ldots \right\}$$

Hence,

$$\mathcal{P}(X_t \geq 0, 0 \leq t \leq 1) = \mathcal{P}(X_{k/2^n} \geq 0, 0 \leq k \leq 2^n, n = 0, 1, \ldots)$$
$$= \mathcal{P}\left(\bigcap_{n=0}^{\infty} \{\omega: X_{k/2^n}(\omega) \geq 0, 0 \leq k \leq 2^n\} \right)$$

Because $A_n = \{\omega: X_{k/2^n}(\omega) \geq 0, 0 \leq k \leq 2^n\}$ is a decreasing sequence in n, and because every probability measure is sequentially continuous, we have

$$\mathcal{P}(X_t \geq 0, 0 \leq t \leq 1) = \lim_{n \to \infty} \mathcal{P}(\{\omega: X_{k/2^n}(\omega) \geq 0, 0 \leq k \leq 2^n\}) \quad (2.6)$$

As an example of a nonseparable process, consider the following. Let $\Omega = [0,1]$, $\mathcal{A}$ be the σ algebra of Lebesgue measurable sets, and let $\mathcal{P}$ be the Lebesgue measure. Consider a process $\{X_t, t \in [0,1]\}$ defined on $(\Omega, \mathcal{A}, \mathcal{P})$ by

$$X_t(\omega) = \begin{cases} 1 & \text{if } \omega = t \\ 0 & \text{otherwise} \end{cases} \quad (2.7)$$

The process $\{X_t, t \in [0,1]\}$ is nonseparable because for any set $S \subset [0,1]$,

$$\{\omega: X_t(\omega) = 0 \text{ for all } t \in S\} = [0,1] - S$$

Therefore,

$$\mathcal{P}(\{\omega: X_t(\omega) = 0 \text{ for all } t \text{ in } [0,1]\}) = 0$$

while if S is any countable set, then

$$\mathcal{P}(\{\omega: X_t(\omega) = 0 \text{ for all } t \text{ in } S\}) = 1$$

Now, let $\{\tilde{X}_t, t \in [0,1]\}$ be defined by

$$\tilde{X}_t(\omega) = 0 \quad \text{for all } t \text{ and } \omega \quad (2.8)$$

The process $\{\tilde{X}_t, t \in [0,1]\}$ is clearly separable. Indeed, for every closed set K,

$$\{\omega: \tilde{X}_t(\omega) \in K \text{ for all } t \in [0,1]\} = \{\omega: X_0(\omega) \in K\}$$
$$= \begin{cases} [0,1] & \text{if } K \ni 0 \\ \emptyset & \text{if } K \not\ni 0 \end{cases}$$

For each t,

$$\{\omega: X_t(\omega) = \tilde{X}_t(\omega)\} = [0,1] - \{t\}$$

which is an event with probability 1.

It is often desirable to be able to define integrals of the form $\int_a^b X_t(\omega)\,dt$. If the integral is to be interpreted as a Lebesgue integral of sample functions, Lebesgue integrability of almost all sample functions is clearly a necessity. Even if almost all sample functions of $\{X_t, t \in T\}$ are Lebesgue integrable, the resulting integral $\int_a^b X_t(\omega)\,dt$ still may not be a random variable. What is needed is that $X_t(\omega)$ defines a (t,ω) function measurable with respect to $\mathcal{L} \otimes \mathcal{A}$, where $\mathcal{L}$ denotes the σ algebra

2. SEPARABILITY AND MEASURABILITY

of Lebesgue measurable sets in T.† We recall that $\mathcal{L}$ is the smallest σ algebra containing all intervals and completed with respect to the measure which assigns lengths to intervals.

Definition. A process $\{X_t, t \in T\}$ with a Lebesgue measurable parameter set T is said to be a **measurable process** if $X_t(\omega)$ is a (t,ω) function measurable with respect to $\mathcal{L} \otimes \mathcal{A}$, where $\mathcal{L}$ is the σ algebra of Lebesgue measurable sets in T, and $\mathcal{A}$ is the σ algebra of events in the defining probability space, i.e., for every $x \in (-\infty, \infty)$,

$$\{(t,\omega): X_t(\omega) \leq x\} \in \mathcal{L} \otimes \mathcal{A}$$

We note that whereas separability imposes no restriction on the finite-dimensional distributions, in general, measurability does. For example, it can be shown that if T is an interval, and if $\{X_t, t \in T\}$ represents a collection of independent and identically distributed random variables, then it cannot be measurable unless the distribution of X_t is concentrated at a point (that is, X_t is a.s. a constant). The following proposition gives a sufficient condition on the finite-dimensional distribution for a measurable process to exist and also summarizes some preceding results [Doob, 1953, pp. 61–62].

Proposition 2.3. Let $\{X_t, t \in T\}$ be a process continuous in probability, and let T be an interval. Then, there exists a process $\{\tilde{X}_t, t \in T\}$ defined on the same probability space such that
  (a) $\mathcal{P}(X_t = \tilde{X}_t) = 1$ for each $t \in T$
  (b) $\{\tilde{X}_t, t \in T\}$ is separable
  (c) $\{\tilde{X}_t, t \in T\}$ is measurable
  (d) Any countable set dense in T is a separating set for $\{\tilde{X}_t, t \in T\}$

We call the process $\{\tilde{X}_t, t \in T\}$ in Proposition 2.3 a separable and measurable modification of $\{X_t, t \in T\}$. Whenever a process $\{X_t, t \in T\}$ is continuous in probability, we can always assume that it has already been replaced by a separable and measurable modification. Doing so is almost never incompatible with any important assumption.

Finally, we note that if A is a Lebesgue measurable set of the real line, and if

$$\int_A E|X_t|\, dt < \infty$$

then almost all sample functions are Lebesgue integrable on A, and $\int_A X_t\, dt$ defines a random variable. This follows directly from Fubini's theorem.

† Let $\mathcal{A}_1$ and $\mathcal{A}_2$ be, respectively, σ algebras of subsets of Ω_1 and Ω_2. $\mathcal{A}_1 \otimes \mathcal{A}_2$ denotes the smallest σ algebra which includes all sets of the form $A_1 \times A_2$, $A_1 \in \mathcal{A}_1$, $A_2 \in \mathcal{A}_2$.

3. GAUSSIAN PROCESSES AND BROWNIAN MOTION

Many random phenomena in physical problems are well approximated by stochastic processes that are called Gaussian processes. This is fortunate, because Gaussian processes have distributions which enjoy great analytical simplicity. Brownian motion, or Wiener process as it is sometimes called, is a specific kind of Gaussian process and plays a vital role in the modern theory of stochastic processes. Generalizing on some of the properties of Brownian motion has led to the theory of diffusion processes and sample-continuous martingales. Brownian motion is also the key to a proper understanding of "white noise," a widely used model for noise phenomena. Through this intervening role of Brownian motion, systems with white-noise disturbances can be studied using results of diffusion theory and martingale theory. This theme will be developed in considerable detail in Chaps. 4 and 6.

Let Z be a random variable such that $EZ^2 < \infty$. Let $\mu = EZ$ and $\sigma^2 = E(Z - \mu)^2$. The random variable Z is said to be Gaussian either if $\sigma^2 = 0$, in which case Z is equal to μ with probability 1, or if

$$\mathcal{P}(Z < a) = \int_{-\infty}^{a} \frac{1}{\sqrt{2\pi\sigma^2}} \exp\left[-\frac{1}{2}\frac{(z-\mu)^2}{\sigma^2}\right] dz \tag{3.1}$$

In other words, Z has a density function

$$p_Z(z) = \frac{1}{\sqrt{2\pi\sigma^2}} \exp\left[-\frac{1}{2}\frac{(z-\mu)^2}{\sigma^2}\right] \tag{3.2}$$

whenever $\sigma^2 > 0$. We can compute the characteristic function and find

$$F(u) = Ee^{iuZ} = \int_{-\infty}^{\infty} \frac{1}{\sqrt{2\pi\sigma^2}} \exp\left[-\frac{1}{2\sigma^2}(z-\mu)^2\right] e^{iuz} dz$$
$$= e^{iu\mu - \frac{1}{2}u^2\sigma^2} \tag{3.3}$$

Since the distribution function is uniquely determined by the characteristic function, a necessary and sufficient condition for Z to be Gaussian is that Z should satisfy

$$Ee^{iuZ} = e^{iuEZ - \frac{1}{2}E(Z-EZ)^2} \tag{3.4}$$

We note that this condition is valid even for the case $E(Z - EZ)^2 = 0$.

A stochastic process $\{X_t, t \in T\}$ is said to be a **Gaussian process** if every finite linear combination of the form

$$Z = \sum_{i=1}^{N} \alpha_i X_{t_i} \tag{3.5}$$

is a Gaussian random variable. For $\{X_t, t \in T\}$ to be a Gaussian process, it is clearly necessary that for each t, X_t be a Gaussian random variable.

3. GAUSSIAN PROCESSES AND BROWNIAN MOTION

But this is not enough. A necessary and sufficient condition is given by the following.

Proposition 3.1. A process $\{X_t, t \in T\}$ is Gaussian if and only if
(a) $EX_t^2 < \infty$ for each $t \in T$
(b) For every finite collection $(t_1, \ldots, t_N) \subset T$

$$E \exp\left(i \sum_{k=1}^{N} u_k X_{t_k}\right) = \exp\left[i \sum_{k=1}^{N} u_k \mu(t_k) - \frac{1}{2} \sum_{k,l=1}^{N} u_k u_l R(t_k, t_l)\right] \tag{3.6}$$

where

$$\mu(t) = EX_t \tag{3.7}$$

and

$$R(t,s) = E[X_t - \mu(t)][X_s - \mu(s)] \tag{3.8}$$

Proof: Suppose that $\{X_t, t \in T\}$ is a Gaussian process. Then by definition, $Z = \sum_{k=1}^{N} u_k X_{t_k}$ is Gaussian. Therefore,

$$Ee^{iZ} = e^{iEZ} e^{-\frac{1}{2}E(Z-EZ)^2}$$

By direct computation we find

$$EZ = \sum_{k,l=1}^{N} u_k \mu(t_k)$$

and

$$E(Z - EZ)^2 = \sum_{k,l=1}^{N} u_k u_l R(t_k, t_l)$$

Conversely, suppose conditions (a) and (b) are satisfied. Let $Z = \sum_{k=1}^{N} \alpha_k X_{t_k}$ be an arbitrary finite linear combination. Then, using (3.6) we find

$$Ee^{iuZ} = E \exp\left(i \sum_{k=1}^{N} u\alpha_k X_{t_k}\right)$$
$$= \exp\left[iu \sum_{k=1}^{N} \alpha_k \mu(t_k) - \frac{1}{2}u^2 \sum_{k,l} \alpha_k \alpha_l R(t_k, t_l)\right]$$
$$= e^{iuEZ} e^{-\frac{1}{2}u^2 E(Z-EZ)^2}$$

Therefore, Z is Gaussian. Since this is true for every linear combination, $\{X_t, t \in T\}$ is a Gaussian process by definition. ∎

Remark: Suppose that $\{X_n\}$ is a sequence of Gaussian random variables converging to a random variable X. Then (see Chap. 1, Exercise 10)

$$\mu_n = EX_n \xrightarrow[n \to \infty]{} \mu = EX$$

and

$$\sigma_n^2 = E(X_n - EX_n)^2 \xrightarrow[n \to \infty]{} \sigma^2 = E(X - EX)^2$$

Since X_n has a density function $p_n(\cdot)$ given by

$$p_n(x) = \frac{1}{\sqrt{2\pi\sigma_n^2}} \exp\left[-\frac{1}{2\sigma_n^2}(x - \mu_n)^2\right]$$

and the distribution functions of X_n converge to that of X, the density function $p(\cdot)$ of X must be given by

$$p(x) = \frac{1}{\sqrt{2\pi\sigma^2}} \exp\left[-\frac{1}{2\sigma^2}(x - \mu)^2\right]$$

In other words, the limit of a q.m. convergent sequence of Gaussian random variables is always a Gaussian random variable. Therefore, if $\{X_t, t \in T\}$ is a Gaussian process, then not only is every sum of the form

$$\sum_{k=1}^{N} \alpha_k X_{t_k}$$

a Gaussian random variable, but so is the limit of every q.m. convergent sequence of such sums. This makes precise the often stated result that a random variable obtained by a linear operation on a Gaussian process is always Gaussian. Linear operation is taken to mean the q.m. limit of a sequence of finite linear combinations.

The function $\mu(t)$ is called the **mean function** of $\{X_t, t \in T\}$ or simply the **mean**. The function $R(t,s)$ defined by (3.8) is called the **covariance function**. Covariance functions will be considered in some detail in Chap. 3. One important property of covariance functions is that for any finite collection $(t_1, t_2, \ldots, t_n) \subset T$, the matrix **R** formed by setting $R_{ij} = R(t_i, t_j)$ is nonnegative definite.[1] This is simply because for any complex constants $a_1, a_2, \ldots, a_n$, we have

$$\sum_{i,j=1}^{n} a_i \bar{a}_j R_{ij} = \sum_{i,j}^{n} E\{a_i[X_{t_i} - \mu(t_i)]\}\{\bar{a}_j[X_{t_j} - \mu(t_j)]\}$$

$$= E\left|\sum_{i=1}^{n} a_i[X_{t_i} - \mu(t_i)]\right|^2 \geq 0 \qquad (3.9)$$

---

[1] The term positive semidefinite is more conventional than nonnegative definite. We have adopted the latter for a closer correspondence with the terminology usually associated with covariance functions.

3. GAUSSIAN PROCESSES AND BROWNIAN MOTION

It is apparent from (3.6) that the characteristic function F for every finite collection $X_{t_1}, \ldots, X_{t_N}$ is completely determined by the mean $\mu(t)$ and the covariance function $R(t,s)$, viz.,

$$F(u_1, \ldots, u_N) = E \exp\left(i \sum_{k=1}^{N} u_k X_{t_k}\right)$$

$$= \exp\left[i \sum_{k=1}^{N} u_k \mu(t_k) - \frac{1}{2} \sum_{k,l=1}^{N} u_k u_l R(t_k, t_l)\right] \quad (3.10)$$

If the matrix $\mathbf{R} = [R(t_k, t_l)]$ is positive definite and not merely nonnegative definite, then the inversion formula for Fourier integral in R^n can be used to obtain the probability density function for $X_{t_1}, \ldots, X_{t_N}$. Specifically, we have

$$p(x_1, t_1; x_2, t_2; \ldots; x_N, t_N)$$

$$= \frac{1}{(2\pi)^N} \int_{-\infty}^{\infty} \cdots \int_{-\infty}^{\infty} F(u_1, \ldots, u_N) \exp\left(-i \sum_{k=1}^{N} u_k x_k\right) du_1$$

$$\cdots du_N = \frac{1}{(2\pi)^{N/2} |\mathbf{R}|^{\frac{1}{2}}} e^{-\frac{1}{2}(\mathbf{x}-\mathbf{\mu})' \mathbf{R}^{-1}(\mathbf{x}-\mathbf{\mu})} \quad (3.11)$$

where $\mathbf{R}^{-1}$ denotes the inverse of $\mathbf{R}$, $\mathbf{x}$ denotes the column vector with components $x_1, \ldots, x_N$, $\mathbf{\mu}$ denotes the column vector with components $\mu(t_1), \ldots, \mu(t_N)$, $|\mathbf{R}|$ denotes the determinant of $\mathbf{R}$, and prime denotes transpose. When the matrix $\mathbf{R}$ is not positive definite, then (3.11) fails, and $X_{t_1}, \ldots, X_{t_N}$ do not have a joint density function. However, the joint distribution function can still be obtained from the characteristic function given by (3.6). In particular, if A is a rectangle such that the distribution function is continuous on its boundary, then we have

$$\mathcal{P}((X_{t_1}, \ldots, X_{t_N}) \epsilon A)$$
$$= \int_{R^N} F(u_1, \ldots, u_N) \psi_A(u_1, \ldots, u_N) du_1 \cdots du_N \quad (3.12)$$

where

$$\psi_A(u_1, \ldots, u_N) = \int_A \exp\left(-i \sum_{k=1}^{N} u_k x_k\right) dx_1 \cdots dx_k$$

Finally, the distribution function $\mathcal{P}(X_{t_i} < x_i, i = 1, \ldots, N)$ can be obtained by using (3.12) and taking limits.

These considerations show that all finite-dimensional distributions of a Gaussian process are completely determined once we specify its mean $\mu(t)$ and covariance function $R(t,s)$. This is indeed the simplest way of specifying the finite-dimensional distributions of a Gaussian process.

As an example, consider the **Brownian motion** process $\{X_t, t \geq 0\}$ defined by

$\{X_t, t \geq 0\}$ is a Gaussian process (3.13a)
$EX_t = 0 \quad EX_t X_s = \min(t,s)$ (3.13b)

If $0 < t_1 < t_2 < \cdots < t_N$, the matrix $\mathbf{R} = [\min(t_k, t_l)]$ is positive definite. Thus, the density function can be written down immediately by the use of (3.11). After a little rearrangement, we find

$$p(x_1, t_1; x_2, t_2; \ldots ; x_N, t_N)$$
$$= \prod_{\nu=1}^{N} \frac{1}{\sqrt{2\pi(t_\nu - t_{\nu-1})}} \exp\left[-\frac{1}{2} \frac{(x_\nu - x_{\nu-1})^2}{t_\nu - t_{\nu-1}}\right] \quad (3.14)$$

where $t_0 = 0 = x_0$. Equation (3.14) shows that $\{X_{t_1}, X_{t_2} - X_{t_1}, \ldots, X_{t_N} - X_{t_{N-1}}\}$ must be a collection of independent random variables for every increasing $(t_1, \ldots, t_N)$. Any process having this property is called a **process with independent increments.** Thus, a Brownian motion process is a process with independent increments.

A Brownian motion is also a Markov process, which is defined as follows.

Definition. A process $\{X_t, t \in T\}$ is said to be a **Markov process** if for any increasing collection $t_1, t_2, \ldots, t_n$ in T,

$$\mathcal{P}(X_{t_n} \leq x_n | X_{t_\nu} = x_\nu, \nu = 1, \ldots, n-1)$$
$$= \mathcal{P}(X_{t_n} \leq x_n | X_{t_{n-1}} = x_{n-1}) \quad (3.15)$$

In other words, given the past $(X_{t_1}, \ldots, X_{t_{n-2}})$ and the present $(X_{t_{n-1}})$, the future (X_{t_n}) depends only on the present. For a Brownian motion, (3.15) is easily verified as follows:

$$\mathcal{P}(X_{t_n} \leq x_n | X_{t_\nu} = x_\nu, \nu = 1, \ldots, n-1)$$
$$= \int_{-\infty}^{x_n} p(\xi, t_n; x_1, t_1; \ldots ; x_{n-1}, t_{n-1}) \, d\xi$$
$$= \int_{-\infty}^{x_n} \frac{p(x_1, t_1; \ldots ; \xi, t_n)}{p(x_1, t_1; \ldots ; x_{n-1}, t_{n-1})} \, d\xi$$
$$= \int_{-\infty}^{x_n} \frac{1}{\sqrt{2\pi(t_n - t_{n-1})}} \exp\left[-\frac{1}{2} \frac{(\xi - x_{n-1})^2}{t_n - t_{n-1}}\right] d\xi$$
$$= \mathcal{P}(X_{t_n} \leq x_n | X_{t_{n-1}} = x_{n-1})$$

Indeed, a little reflection would show that any process with independent increments $\{X_t, t \in T\}$ is also a Markov process.

A Brownian motion $\{X_t, t \geq 0\}$ also has the property (see Exercises 8 and 9)

$$E(X_t | X_\tau, 0 \leq \tau \leq s) = X_s \quad \text{a.s.}$$

3. GAUSSIAN PROCESSES AND BROWNIAN MOTION

for any $t \geq s$. A process having this property is called a martingale. We shall define a martingale a little more generally as follows.

Definition. Let $\{X_t, t \in T\}$ be a stochastic process, and let $\{\mathcal{A}_t, t \in T\}$ be an increasing family of σ algebras such that for each t, X_t is $\mathcal{A}_t$ measurable. $\{X_t, \mathcal{A}_t, t \in T\}$ is said to be a **martingale** if $t > s$ implies

$$E^{\mathcal{A}_s} X_t = X_s \quad \text{a.s.} \tag{3.16}$$

The process is said to be a **submartingale (supermartingale)** if the equality (3.16) is replaced by $\geq$ (respectively, ≤ 0).

For any Brownian motion $\{X_t, t \geq 0\}$, if we take $\mathcal{A}_{xt}$ to be the smallest σ algebra with respect to which $\{X_s, s \leq t\}$ are all measurable, then $\{X_t, \mathcal{A}_{xt}, t \geq 0\}$ is a martingale. Suppose that $\{X_t, t \geq 0\}$ is a Brownian motion, and $\{\mathcal{A}_t, t \geq 0\}$ is an increasing family of σ algebras such that for each t, X_t is $\mathcal{A}_t$ measurable and for each t, $\{X_s - X_t, s \geq t\}$ is independent of $\mathcal{A}_t$. We shall emphasize this relationship between X_t and $\mathcal{A}_t$ by saying that $\{X_t, \mathcal{A}_t, t \geq 0\}$ is a Brownian motion. If $\{X_t, \mathcal{A}_t, t \geq 0\}$ is a Brownian motion, then it is a martingale. To prove this we merely note that for $t > s$,

$$E^{\mathcal{A}_s} X_t = E^{\mathcal{A}_s}[X_s + (X_t - X_s)]$$
$$= X_s + E(X_t - X_s) = X_s \quad \text{a.s.}$$

We should note that instead of taking the parameter set to be $[0, \infty)$, we can define a Brownian motion on $(-\infty, \infty)$ by replacing (3.13) with

$$EX_t = 0 \quad EX_t X_s = \tfrac{1}{2}(|t| + |s| - |t - s|) \tag{3.17}$$

What results is a pair of independent Brownian motions $\{X_t, t \geq 0\}$ and $\{X_{-t}, t \geq 0\}$ pieced together at $t = 0$ (see Exercise 7).

If $\{X_t, t \geq 0\}$ is a Brownian motion, then $E(X_t - X_s)^2 = |t - s|$. Therefore, by virtue of the Chebyshev inequality, we have

$$\mathcal{P}(|X_{t+h} - X_t| \geq \epsilon) \leq \frac{|h|}{\epsilon^2} \xrightarrow[h \to 0]{} 0$$

for every $\epsilon > 0$. Thus, a Brownian motion is continuous in probability. It follows from Proposition 2.3 that every Brownian motion has a separable and measurable modification. A separable Brownian motion has some important sample function properties. The first of those that we shall consider is an inequality that it shares with all separable second-order martingales.

Proposition 3.2. Let $\{X_t, a \leq t \leq b\}$ be a separable martingale such that $EX_t^2 < \infty$ for every t in $[a,b]$. Then, for every positive ϵ,

$$\mathcal{P}(\sup_{a \leq t \leq b} |X_t| \geq \epsilon) \leq \frac{EX_b^2}{\epsilon^2} \tag{3.18}$$

Proof: Since $\{X_t, t \in [a,b]\}$ is separable, there is a countable set S such that
$$\mathcal{P}(\sup_{a \leq t \leq b} |X_t| \geq \epsilon) = \mathcal{P}(\sup_{t \in S} |X_t| \geq \epsilon)$$

Now, since S is countable, we can write $S = \bigcup_n S_n$ where $\{S_n\}$ is an increasing sequence of finite collections of points in $[a,b]$. Because $\mathcal{P}$ is monotone-sequential continuous,

$$\mathcal{P}(\sup_{t \in S} |X_t| \geq \epsilon) = \lim_{n \to \infty} \mathcal{P}(\max_{t \in S_n} |X_t| \geq \epsilon)$$

For a fixed n, let $t_1, t_2, \ldots, t_N$ be the points in S_n with $t_1 < t_2 < \cdots < t_N$. Let $\nu(\omega)$ be the first i (if any) such that $|X_{t_i}(\omega)| \geq \epsilon$, and write

$$EX_b^2 = \sum_{k=1}^{N} \mathcal{P}(\nu = k) E(X_b^2 | \nu = k)$$
$$+ \mathcal{P}(\max_{t \in S_n} |X_t| < \epsilon) E(X_b^2 | \max_{t \in S_n} |X_t| < \epsilon)$$
$$\geq \sum_{k=1}^{N} \mathcal{P}(\nu = k) E(X_b^2 | \nu = k)$$

Writing $X_b^2 = (X_b - X_{t_k} + X_{t_k})^2$, we find
$$E(X_b^2 | \nu = k) = E[(X_b - X_{t_k})^2 | \nu = k] + E(X_{t_k}^2 | \nu = k)$$
$$+ 2[X_{t_k}(X_b - X_{t_k}) | \nu = k]$$

From the definition of ν, we have $E(X_{t_k}^2 | \nu = k) \geq \epsilon^2$. Because the event $\nu = k$ depends only on $X_{t_1}, X_{t_2}, \ldots, X_{t_k}$, we have

$$E[X_{t_k}(X_b - X_{t_k}) | \nu = k] = E\{X_{t_k} E[(X_b - X_{t_k}) | X_{t_1}, \ldots, X_{t_k}] | \nu = k\}$$
$$= 0$$

due to the martingale property. It follows that $E(X_b^2 | \nu = k) \geq \epsilon^2$ and

$$EX_b^2 \geq \epsilon^2 \sum_{k=1}^{N} \mathcal{P}(\nu = k)$$
$$= \epsilon^2 \mathcal{P}(\max_{t \in S_n} |X_t| \geq \epsilon)$$
$$\xrightarrow[n \to \infty]{} \epsilon^2 \mathcal{P}(\sup_{a \leq t \leq b} |X_t| \geq \epsilon)$$

proving (3.18). ∎

Remarks:

(a) (3.18) holds for separable complex-valued martingales with $E|X_b|^2$ replacing EX_b^2. The proof is almost identical.

3. GAUSSIAN PROCESSES AND BROWNIAN MOTION

(b) If $\{X_t, a \leq t \leq b\}$ is a separable process with independent increments such that $\mathcal{P}(X_t - X_s > 0) = \mathcal{P}(X_t - X_s < 0)$, $a \leq t$, $s \leq b$. Then we can prove

$$\mathcal{P}(\sup_{a \leq t \leq b} |X_t| \geq \epsilon) \leq 2\mathcal{P}(|X_b| \geq \epsilon)$$

which is a formula in the same spirit as (3.18) and can be proved in a similar way.

Proposition 3.3. With probability 1, every sample function of a separable Brownian motion is uniformly continuous on every finite interval.

The proof of this proposition will be deferred until the next section where it can be easily proved using some sufficient conditions for sample continuity. The sample-continuity property of a separable Brownian motion can be interpreted in a different way as follows: Let C denote the space of all continuous real-valued functions on $[0, \infty)$. Let $X_t(\omega)$, $\omega \in C$, be the value of ω at t. Let $\mathcal{Q}$ denote the smallest σ algebra of subsets of C such that every X_t is measurable. Then it can be shown that there exists a unique probability measure $\mathcal{P}$ on $(C, \mathcal{Q})$ such that $\{X_t, 0 \leq t < \infty\}$ is a Brownian motion. So constructed, $\{X_t, t \geq 0\}$ is necessarily separable. Sample continuity in this context says no more than $\mathcal{P}(C) = 1$. The measure $\mathcal{P}$ on $(C, \mathcal{Q})$ will be called the **Wiener measure.** Although the sample functions of a separable Brownian motion are almost surely continuous, they are highly irregular. For example, with probability 1, the sample functions of a separable Brownian motion are nowhere differentiable and are of unbounded variation on every interval. Roughly speaking, $(X_{t+\delta} - X_t)$ is of order $0(\sqrt{\delta})$, which is incompatible with either differentiability or bounded variation. This behavior of the Brownian motion is made precise by the following proposition.

Proposition 3.4. Let $T = [a,b]$ be a closed subinterval of $[0, \infty)$. Let $T_n = [a = t_0^{(n)} < t_1^{(n)} < \cdots < t_{N(n)}^{(n)} = b]$, $n = 1, 2, \ldots$, be a sequence of partitions of T such that

$$\Delta_n = \max_{1 \leq \nu \leq N(n)} (t_\nu^{(n)} - t_{\nu-1}^{(n)}) \xrightarrow[n \to \infty]{} 0$$

Let $\{X_t, t \geq 0\}$ be a Brownian motion, then

$$\sum_{\nu=1}^{N(n)} (X_{t_\nu^{(n)}} - X_{t_{\nu-1}^{(n)}})^2 \xrightarrow[n \to \infty]{\text{q.m.}} b - a$$

If $\sum_{n=1}^{\infty} \Delta_n < \infty$ then the convergence is also almost sure.

Proof: To prove quadratic-mean convergence, we write

$$S_n = \sum_{\nu=1}^{N(n)} (X_{t_\nu^{(n)}} - X_{t_{\nu-1}^{(n)}})^2 - (b-a)$$

$$= \sum_{\nu=1}^{N(n)} [(X_{t_\nu^{(n)}} - X_{t_{\nu-1}^{(n)}})^2 - (t_\nu^{(n)} - t_{\nu-1}^{(n)})]$$

Since S_n is a sum of independent random variables, each with zero mean, we have

$$ES_n = 0$$

$$ES_n^2 = \sum_{\nu=1}^{N(n)} E[(X_{t_\nu^{(n)}} - X_{t_{\nu-1}^{(n)}})^2 - (t_\nu^{(n)} - t_{\nu-1}^{(n)})]^2$$

$$= 2 \sum_{\nu=1}^{N(n)} (t_\nu^{(n)} - t_{\nu-1}^{(n)})^2 \leq 2\Delta_n (b-a)$$

Since $\Delta_n \xrightarrow[n \to \infty]{} 0$, this proves $S_n \xrightarrow[n \to \infty]{\text{q.m.}} 0$.

For the second part of the proposition, using the Chebyshev inequality (1.5.5), we find

$$\mathcal{P}(|S_n| > \epsilon) \leq \frac{ES_n^2}{\epsilon^2} \leq 2(b-a)\frac{\Delta_n}{\epsilon^2}$$

Since by assumption, $\sum_n \Delta_n < \infty$, we have for every $\epsilon > 0$,

$$\sum_n \mathcal{P}(|S_n| > \epsilon) < \infty$$

By the Borel-Cantelli lemma, $|S_n| > \epsilon$ for at most a finite number of times with probability 1, proving the theorem. ∎

Remark: The condition $\sum_n \Delta_n < \infty$ for a.s. convergence can be replaced by the condition that T_n be nested, that is, $T_{n+1} \supset T_n$ for each n [Doob, 1953, pp. 395–396].

As we said earlier, an intuitive interpretation of Proposition 3.4 is that $dX_t \sim \sqrt{dt}$. This has some important consequences. For example, we would expect the following:

$$df(X_t) = f(X_{t+dt}) - f(X_t)$$
$$= f'(X_t)\,dX_t + \tfrac{1}{2}f''(X_t)\,dt$$

This relationship will be made precise subsequently by means of the stochastic integral.

4. CONTINUITY

For a stochastic process $\{X_t, t \in T\}$ defined on an interval T, we distinguish the following types of continuity. The process is said to be

1. **Continuous in probability** at t if for any $\epsilon > 0$,

$$\mathcal{P}(|X_{t+h} - X_t| > \epsilon) \xrightarrow[h \to 0]{} 0 \tag{4.1}$$

2. **Continuous in νth mean** at t (in quadratic mean if $\nu = 2$), if

$$E|X_{t+h} - X_t|^\nu \xrightarrow[h \to 0]{} 0 \tag{4.2}$$

3. **Almost-surely continuous** at t, if

$$\mathcal{P}(\{\omega: \lim_{h \to 0} |X_{t+h}(\omega) - X_t(\omega)| = 0\}) = 1 \tag{4.3}$$

4. Almost-surely **sample continuous,** if $\{X_t, t \in T\}$ is separable and

$$\mathcal{P}\left(\bigcup_{t \in T} \{\omega: \lim_{h \to 0} |X_{t+h}(\omega) - X_t(\omega)| \neq 0\}\right) = 0 \tag{4.4}$$

If a process $\{X_t, t \in T\}$ is continuous in probability (in νth mean) at every t of T, then it is said to be continuous in probability (in νth mean). We have already encountered continuity in probability in connection with questions of separability. Continuity in quadratic mean ($\nu = 2$) will play an important role in our discussion of second-order properties in Chap. 3. In this section we are primarily interested in a.s. continuity and a.s. sample continuity.

First of all, we note that almost-sure continuity at every t of T is not equivalent to a.s. sample continuity. This is simply because $A_t = \{\omega: \lim_{h \to 0} |X_{t+h}(\omega) - X_t(\omega)| \neq 0\}$, being a null event for every t, does not imply that the uncountable union $\bigcup_{t \in T} A_t$ is a null event. The defining condition (4.4) for sample continuity is not one that can be easily verified in terms of finite-dimensional distributions. Our primary objective here is to replace it by simpler sufficient conditions. To begin with, if a process is a.s. sample continuous, then it is necessarily a.s. continuous at every t in T, which in turn implies that the process is continuous in probability. Thus, every sample-continuous process is both separable (by definition) and continuous in probability. From Proposition 2.2 we know that for every such process, every countable set S dense in T is a separating set. Hence, with probability 1

$$\sup_{\substack{t,s \in S \\ |t-s| < h}} |X_t - X_s| = \sup_{\substack{t,s \in T \\ |t-s| < h}} |X_t - X_s| \tag{4.5}$$

Assume T to be a closed interval. With no loss of generality we can take T to be [0,1]. Then S can be taken to be

$$S = \left\{\frac{k}{2^n}, \ k = 0, 1, \ldots, 2^n - 1; \ n = 0, 1, 2, \ldots\right\} \quad (4.6)$$

Proposition 4.1. Let S be given by (4.6) and define

$$Z_\nu(\omega) = \sup_{0 \leq k \leq 2^\nu} |X_{(k+1)/2^\nu}(\omega) - X_{k/2^\nu}(\omega)| \quad (4.7)$$

Then,

$$\sup_{\substack{t,s \in S \\ |t-s|<2^{-n}}} |X_t - X_s| \leq 2 \sum_{\nu=n+1}^{\infty} Z_\nu \quad (4.8)$$

Proof: If $|t - s| < 2^{-n}$, then we can find k such that $0 < k < 2^n$ and $|t - k/2^n| < 2^{-n}$, $|s - k/2^n| < 2^{-n}$. If $t \in S$ and $|t - k/2^n| < 2^{-n}$, t must be of the form

$$t = \frac{k}{2^n} \pm \sum_{\nu=1}^{m} t_\nu 2^{-(n+\nu)} \quad (t_\nu = 0,1)$$

Thus,

$$|X_t - X_{k/2^n}| \leq \sum_{\nu=n+1}^{n+m} Z_\nu \leq \sum_{\nu=n+1}^{\infty} Z_\nu$$

Since, $|X_t - X_s| \leq |X_t - X_{k/2^n}| + |X_s - X_{k/2^n}|$, we have

$$\sup_{\substack{t,s \in S \\ |t-s|<2^{-n}}} |X_t - X_s| \leq 2 \sum_{\nu=n+1}^{\infty} Z_\nu$$

which proves the proposition. ∎

Proposition 4.1 shows that if

$$\lim_{n \to \infty} \sum_{\nu=n}^{\infty} Z_\nu = 0 \quad (4.9)$$

with probability 1, then

$$\lim_{n \to \infty} \sup_{\substack{t,s \in S \\ |t-s|<2^{-n}}} |X_t - X_s| = 0 \quad (4.10)$$

with probability 1, which proves that every sample function is uniformly continuous on T with probability 1, since in (4.10) S can be replaced by T.

4. CONTINUITY

An extremely useful condition which implies (4.9) is the Kolmogorov condition given in the following proposition.

Proposition 4.2. Let $\{X_t, t \in T\}$ be a separable process and let T be a finite interval. If there exist strictly positive constants α, β, C such that

$$E|X_{t+h} - X_t|^\alpha \leq Ch^{1+\beta} \tag{4.11}$$

then

$$\sup_{\substack{t,s \in T \\ |t-s| < h}} |X_t - X_s| \xrightarrow[h \to 0]{\text{a.s.}} 0 \tag{4.12}$$

so that almost every sample function is uniformly continuous on T.

Remark: The great advantage of the Kolmogorov condition is simply that it only involves two-dimensional distributions and can usually be verified by direct computation.

Proof: From the Markov inequality (1.5.5), we have

$$\mathcal{P}(|X_{t+h} - X_t| \geq \epsilon) \leq \frac{E|X_{t+h} - X_t|^\alpha}{\epsilon^\alpha} \qquad \epsilon > 0$$

Therefore,

$$\mathcal{P}(|X_{t+h} - X_t| \geq h^\gamma) \leq Ch^{1+\beta-\alpha\gamma}$$

Let $0 \leq \gamma < \beta/\alpha$, and set $\delta = \beta - \alpha\gamma > 0$. Then

$$\mathcal{P}(|X_{t+h} - X_t| \geq h^\gamma) \leq Ch^{1+\delta}$$

Hence,

$$\mathcal{P}\left(Z_\nu \geq \left(\frac{1}{2^\nu}\right)^\gamma\right) = \mathcal{P}\left(\sup_{0 \leq k \leq 2^\nu - 1} |X_{k+1/2^\nu} - X_{k/2^\nu}| \geq \left(\frac{1}{2^\nu}\right)^\gamma\right)$$

$$\leq \sum_{k=0}^{2^\nu - 1} \mathcal{P}\left(|X_{k+1/2^\nu} - X_{k/2^\nu}| \geq \left(\frac{1}{2^\nu}\right)^\gamma\right)$$

$$\leq C 2^\nu \left(\frac{1}{2^\nu}\right)^{1+\delta}$$

$$= C 2^{-\delta\nu}$$

Since $\sum_{\nu=0}^{\infty} 2^{-\delta\nu} < \infty$, we have

$$\sum_{\nu=0}^{\infty} \mathcal{P}\left(Z_\nu \geq \left(\frac{1}{2^\nu}\right)^\gamma\right) < \infty$$

By the Borel-Cantelli lemma, $Z_\nu \geq 1/2^{\nu\gamma}$ for at most a finite number of ν's with probability 1. That is, there exists $N(\omega)$ almost-surely finite such that

$$Z_\nu(\omega) < \frac{1}{2^{\nu\gamma}} \quad \text{for all } \nu \geq N(\omega)$$

and $\lim_{n \to \infty} \sum_{\nu=n+1}^{\infty} Z_\nu(\omega) = 0$ with probability 1. From Proposition 4.1,

$$\sup_{\substack{t,s \in S \\ |t-s| < 2^{-n}}} |X_t - X_s| \xrightarrow[n \to \infty]{\text{a.s.}} 0$$

Since S is a separating set in T, this proves Proposition 4.2. ∎

The Kolmogorov condition (4.11) is sufficient to prove the sample continuity of Brownian motion, viz., Proposition 3.3. We recall that for a Brownian motion

$$\mathcal{P}(X_t - X_s < x) = \int_{-\infty}^{x} \frac{1}{\sqrt{2\pi|t-s|}} \exp\left(-\frac{u^2}{2|t-s|}\right) du$$

Therefore,

$$E|X_t - X_s|^4 = \int_{-\infty}^{\infty} \frac{u^4}{\sqrt{2\pi|t-s|}} \exp\left(-\frac{1}{2}\frac{u^2}{|t-s|}\right) du$$
$$= 3(t-s)^2$$

and

$$E|X_{t+h} - X_t|^4 = 3h^2$$

which verifies the Kolmogorov condition with $C = 3$, $\alpha = 4$, $\beta = 1$. Therefore, with probability 1, every sample function of a separable Brownian motion is uniformly continuous on every finite interval.

Let T be a finite and closed interval which we take to be [0,1] with no loss of generality. Let C be the space of all continuous functions on [0,1]. If $\omega \in C$, we define

$$X_t(\omega) = \text{value of } \omega \text{ at } t \quad \begin{array}{l} \omega \in C \\ t \in [0,1] \end{array}$$

The space C is a Banach space (complete normed linear space) with a norm given by

$$\|\omega\| = \max_{0 \leq t \leq 1} |X_t(\omega)|$$

Let $\mathcal{B}$ be the minimal Boolean algebra with respect to which every X_t is measurable. $\mathcal{B}$ is not a σ algebra. Now, every consistent family of finite-dimensional distributions for $\{X_t, t \in [0,1]\}$ defines an elementary probability measure $\mathcal{P}'$ on $(C,\mathcal{B})$. However, $\mathcal{P}'$ is, in general, only finitely additive, but not σ additive. If $\mathcal{P}'$ is σ additive (equivalently, if $\mathcal{P}'$ is sequentially continuous), then it can be extended to a probability measure $\mathcal{P}$ on $(C,\mathcal{A})$ where $\mathcal{A}$ is the minimal σ algebra with respect to which every X_t is measurable. What results then is a stochastic process $\{X_t, t \in [0,1]\}$ defined on $(C,\mathcal{A},\mathcal{P})$. Since $\mathcal{P}(C) = 1$, every sample function of X_t is obviously continuous. We can interpret the Kolmogorov condition to mean that every consistent family of finite-dimensional distributions which satisfies (4.11) defines an elementary probability measure $\mathcal{P}'$ on $(C,\mathcal{B})$ which is σ *additive* [Prokhorov, 1956]. It is possible to show this directly, but we won't do it here. In the case of Brownian motion, the resulting $\mathcal{P}$ is known as the **Wiener measure**. We should note that every process $\{X_t, t \in T\}$ defined on $(C,\mathcal{A},\mathcal{P})$ in this way is necessarily separable and measurable.

If a process is not sample continuous, then it is clearly of interest to know the nature of its discontinuities. We shall say that a function $f(t), 0 \le t \le 1$ has only **discontinuities of the first kind** if (1) f is bounded, (2) at every $t \in (0,1)$ limit from the left $\lim_{s\uparrow t} f(s) = f^-(t)$ and limit from the right $\lim_{s\downarrow t} f(s) = f^+(t)$ exist. Clearly, $[0,1]$ can easily be replaced by $[a,b]$ in everything that we say. We give without proof a condition, similar to the Kolmogorov condition, which guarantees that a process $\{X_t, t \in [0,1]\}$ has sample functions which have only discontinuities of the first kind with probability 1 [Cramér, 1966].

Proposition 4.3. Let $\{X_t, 0 \le t \le 1\}$ be a separable process. If there exist strictly positive constants α, β, C such that

$$\sup_{t \le s \le t+h} E(|X_{t+h} - X_s|^\alpha |X_s - X_t|^\alpha) \le Ch^{1+\beta} \tag{4.13}$$

then with probability 1, every sample function of $\{X_t, 0 \le t \le 1\}$ has only discontinuities of the first kind.

Remark: Condition (4.13) is weaker than (4.12), as it should be. To show this, we need only to apply the Schwarz inequality and find

$$E(|X_{t+h} - X_s|^\alpha |X_s - X_t|^\alpha) \le \sqrt{E|X_{t+h} - X_s|^{2\alpha} E|X_s - X_t|^{2\alpha}}$$

Hence, if condition (4.12) is satisfied with α, β and C, then (4.13) is satisfied with $\alpha/2$, β and C.

As an example of the application of (4.13), consider the random telegraph process $\{X_t, -\infty < t < \infty\}$ defined as follows:

(a) $\quad \mathcal{P}(X_t = 1) = \mathcal{P}(X_t = -1) = \frac{1}{2}$

(b) $\{X_t, -\infty < t < \infty\}$ is a process with independent increments, that is,

$$(X_{t_n} - X_{t_{n-1}}), (X_{t_{n-1}} - X_{t_{n-2}}), \ldots, (X_{t_2} - X_{t_1}), X_{t_1}$$

are independent whenever $t_n > t_{n-1} > \cdots > t_1$

(c) $\mathcal{P}(X_t - X_s = 0) = \frac{1}{2}(1 + e^{-|t-s|})$
$\mathcal{P}(X_t - X_s = 2) = \mathcal{P}(X_t - X_s = -2) = \frac{1}{4}(1 - e^{-|t-s|})$ (4.14)

It is clear that with probability 1, $\{X_t, -\infty < t < \infty\}$ takes only values ± 1. It has independent increments,

$$E(|X_{t+h} - X_s| \, |X_s - X_t|) = E|X_{t+h} - X_s| E|X_s - X_t|$$
$$= (1 - e^{-|t+h-s|})(1 - e^{-|s-t|})$$
$$\leq h^2/4 \quad \text{for all } s \text{ in } [t, t+h]$$

Hence, (4.13) is satisfied and with probability 1, every sample function has only discontinuities of the first kind.

As a final topic, we note that the proof of Proposition 4.2 contains an estimate of the modulus of continuity which we now make explicit. As in Proposition 4.2, we assume $\{X_t, 0 \leq t \leq 1\}$ to be separable and for some positive constants C, α, β

$$E|X_{t+h} - X_t|^\alpha \leq Ch^{1+\beta}$$

We then found that there existed an a.s. finite random variable $N(\omega)$ so that

$$Z_\nu(\omega) = \sup_{0 \leq k \leq 2^\nu} |X_{(k+1)/2^\nu}(\omega) - X_{k/2^\nu}(\omega)|$$
$$< 1/2^{\nu\gamma} \quad \text{for all } \nu \geq N(\omega) \quad (4.15)$$

where γ is any constant in $[0, \beta/\alpha)$. If we take $S = \{k/2^n, k = 0, \ldots, 2^n - 1, n = 0, 1, \ldots\}$, then from Proposition 4.1

$$\sup_{\substack{t,s \in S \\ |t-s| < 2^{-n}}} |X_t(\omega) - X_s(\omega)| \leq 2 \sum_{\nu=n+1}^{\infty} Z_\nu(\omega)$$

so that for $n \geq N(\omega)$

$$\sup_{\substack{t,s \in S \\ |t-s| < 2^{-n}}} |X_t(\omega) - X_s(\omega)| \leq 2 \sum_{\nu=n+1}^{\infty} \frac{1}{2^{\nu\gamma}} = 2 \left(\frac{1}{2^\gamma}\right)^{n+1}$$

for any γ in $[0, \beta/\alpha)$. For any $h < 2^{-N(\omega)}$,

$$\sup_{\substack{t,s \in S \\ |t-s| \leq h}} |X_t(\omega) - X_s(\omega)| \leq 2h^\gamma \quad 0 \leq \gamma < \frac{\beta}{\alpha} \quad (4.16)$$

5. MARKOV PROCESSES

Recalling that S is a separating set in [0,1], we have proven that

$$\lim_{h \downarrow 0} h^{-\beta/\alpha+\epsilon} \sup_{\substack{t,s \in [0,1] \\ |t-s| \leq h}} |X_t - X_s| = 0 \tag{4.17}$$

with probability 1 for any $\epsilon > 0$. For a Brownian motion process, the largest β/α that we can take is $\frac{1}{2}$. Therefore, for a separable Brownian motion process,

$$\lim_{h \downarrow 0} \frac{\sup_{\substack{t,s \in [0,1] \\ |t-s| \leq h}} |X_t - X_s|}{h^{\frac{1}{2}-\epsilon}} = 0 \text{ with probability 1} \tag{4.18}$$

for any $\epsilon > 0$.

5. MARKOV PROCESSES

In this section we shall give a preliminary discussion of Markov processes. In later chapters a more detailed treatment of sample-continuous Markov processes will be given in connection with the diffusion equation and stochastic differential equations. We recall the definition of a Markov process given in Sec. 3.

Definition. A process $\{X_t, t \in T\}$ is said to be a **Markov process** if for any increasing collection $t_1, t_2, \ldots, t_n$ in T

$$\mathcal{P}(X_{t_n} < x_n | X_{t_\nu} = x_\nu, \nu = 1, \ldots, n-1)$$
$$= \mathcal{P}(X_{t_n} < x_n | X_{t_{n-1}} = x_{n-1}) \tag{5.1}$$

If the finite-dimensional distributions of a Markov process $\{X_t, t \in T\}$ can be expressed in terms of density functions, then whenever $t_1 < t_2 < \cdots < t_n$ we can write

$$p(x_1,t_1; x_2,t_2; \ldots; x_n,t_n)$$
$$= p(x_n,t_n | x_1,t_1; \ldots; x_{n-1},t_{n-1}) p(x_1,t_1; \ldots; x_{n-1},t_{n-1})$$
$$= p(x_n,t_n | x_{n-1},t_{n-1}) p(x_1,t_1; \ldots; x_{n-1},t_{n-1}) \tag{5.2}$$

Therefore, by using (5.2) repeatedly, we find for $t_1 < t_2 < \cdots < t_n$,

$$p(x_1,t_1; x_2,t_2; \ldots; x_n,t_n) = p(x_1,t_1) \prod_{\nu=2}^{n} p(x_\nu,t_\nu | x_{\nu-1},t_{\nu-1}) \tag{5.3}$$

which means that all finite-dimensional distributions are completely determined by the two-dimensional distributions. This fact is true with all Markov processes and does not depend on the existence of density functions.

To obtain an expression relating the n-dimensional distribution function to the two-dimensional distribution, we first adopt the notation

$$P(x_1,t_1; x_2,t_2; \ldots ; x_n,t_n) = \mathcal{P}(X_{t_\nu} < x_\nu; \nu = 1, \ldots, n) \quad (5.4)$$
$$P(x_\nu,t_\nu|x_{\nu-1},t_{\nu-1}) = \mathcal{P}(X_{t_\nu} < x_\nu|X_{t_{\nu-1}} = x_{\nu-1}) \quad (5.5)$$

The conditional distribution $P(x,t|\xi,s)$, $t > s$, is often called the **transition function**. Suppose $t_1 < t_2 < t_3$, then we can write $P(x_1,t_1; x_2,t_2; x_3,t_3)$ as

$$P(x_1,t_1; x_2,t_2; x_3,t_3) = \int_{-\infty}^{x_1} \int_{-\infty}^{x_2} \mathcal{P}(X_{t_3} < x_3; X_{t_2} \in d\xi_2; X_{t_1} \in d\xi_1)$$

where $d\xi_i$ stands for $[\xi_i, \xi_i + d\xi_i)$. Using the Markov property, we get

$$P(x_1,t_1; x_2,t_2; x_3,t_3) = \int_{-\infty}^{x_1} \int_{-\infty}^{x_2} P(x_3,t_3|\xi_2,t_2) \, dP(\xi_1,t_1; \xi_2,t_2)$$
$$= \int_{-\infty}^{x_1} \int_{-\infty}^{x_2} P(x_3,t_3|\xi_2,t_2) \, dP(\xi_2,t_2|\xi_1,t_1) \, dP(\xi_1,t_1)$$

More generally, we have for $t_1 < t_2 < \cdots < t_n$

$$P(x_1,t_1; x_2,t_2; \ldots ; x_n,t_n)$$
$$= \int_{-\infty}^{x_1} \cdots \int_{-\infty}^{x_{n-1}} P(x_n,t_n|\xi_{n-1},t_{n-1}) \, dP(\xi_{n-1},t_{n-1}|\xi_{n-2},t_{n-2})$$
$$\cdots dP(\xi_2,t_2|\xi_1,t_1) \, dP(\xi_1,t_1) \quad (5.6)$$

which expresses the n-dimensional distribution of a Markov process in terms of the one-dimensional distribution $P(x,t)$ and the conditional distribution $P(x,t|\xi,s)$, $t > s$, both of which are obtainable from $P(\xi,s,x,t)$.

However, we should note that we cannot construct the finite-dimensional distributions of a Markov process out of just any two-dimensional distribution. There are two consistency conditions that must be satisfied. The first one is obvious,

$$P(x,t) = \mathcal{P}(X_t < x, -\infty < X_s < \infty)$$
$$= \int_{-\infty}^{\infty} P(x,t|\xi,s) \, dP(\xi,s) \quad (5.7)$$

This condition relates $P(x,t)$ to $P(x,t|\xi,s)$. The second condition is obtained by noting that if $t_0 < s < t$, then

$$P(x,t|x_0,t_0) = \mathcal{P}(X_t < x|X_{t_0} = x_0)$$
$$= \mathcal{P}(X_t < x, -\infty < X_s < \infty|X_{t_0} = x_0)$$
$$= \int_{-\infty}^{\infty} \mathcal{P}(X_t < x|X_s = \xi, X_{t_0} = x_0)\mathcal{P}(X_s \in d\xi|X_{t_0} = x_0)$$
$$= \int_{-\infty}^{\infty} P(x,t|\xi,s) \, dP(\xi,s|t_0,x_0)$$

This yields a condition that must be satisfied by the transition function $P(x,t|\xi,s)$, $t > s$, namely,

$$P(x,t|x_0,t_0) = \int_{-\infty}^{\infty} P(x,t|\xi,s) \, dP(\xi,s|t_0,x_0) \qquad t_0 < s < t \quad (5.8)$$

5. MARKOV PROCESSES

Equation (5.8) is known as the **Chapman-Kolmogorov equation**. Given distribution functions $P(x,t)$ and $P(x,t|\xi,s)$ satisfying (5.7) and (5.8), the finite-dimensional distribution functions constructed by using (5.6) will indeed satisfy the Markov property. Therefore, (5.7) and (5.8) are both necessary and sufficient for $P(x,t)$ and $P(x,t|\xi,s)$ to be the one-dimensional distribution and the transition function of a Markov process.

As an example, suppose that we want to define a Markov process $\{X_t, -\infty < t < \infty\}$ with the following properties:

1. X_t takes only values $+1$ or -1
2. $\mathcal{P}(X_t = 1) = \frac{1}{2} = \mathcal{P}(X_t = -1)$
3. For $t \geq s$,

$$\mathcal{P}(X_t = X_s) = q(t-s)$$
$$\mathcal{P}(X_t = -X_s) = 1 - q(t-s)$$

where $q(t)$ is continuous and $q(0) = 1$

It is easy to verify that (5.7) is automatically satisfied. However, (5.8) imposes a constraint on q, namely,

$$q(t - t_0) = q(t-s)q(s-t_0) + [1 - q(t-s)][1 - q(s-t_0)] \quad (5.9)$$

A second equation can also be obtained, but it turns out to be redundant. Equation (5.9) looks a little simpler if we make a few changes of variables and write

$$q(t+s) = q(t)q(s) + [1 - q(t)][1 - q(s)]$$

By setting $q(t) = \frac{1}{2}[1 + f(t)]$, we find

$$f(t+s) = f(t)f(s) \quad (5.10)$$

The only continuous solution of (5.10) not identically equal to 0 is

$$f(t) = e^{-\lambda t}$$

and λ must be nonnegative, because $f(t) \leq 1$. This means that $q(t)$ must be of the form

$$q(t) = \frac{1}{2}(1 + e^{-\lambda t}) \quad (5.11)$$

The resulting process is precisely the **random telegraph process** that we introduced in Sec. 4 (4.14), except for the trivial scale factor λ.

As a second example of using (5.8), we shall derive a set of conditions which must be satisfied by the covariance function of a Gaussian-Markov process. We can always assume that $\{X_t, t \in T\}$ has zero mean, because X_t is Markov if and only if $X_t - \mu(t)$ is Markov. If $\{X_t, t \in T\}$ is Gaussian, zero mean, and if it has covariance function $R(t,s)$, then

$$E(X_t|X_s = \xi) = \frac{R(t,s)}{R(s,s)}\xi \qquad (5.12)$$

To prove (5.12), we merely note that X_s and $X_t - [R(t,s)]/[R(s,s)]X_s$ are Gaussian and uncorrelated, hence are independent. Thus,

$$\begin{aligned} E(X_t|X_s = \xi) &= \frac{R(t,s)}{R(s,s)}\xi + E\left[X_t - \frac{R(t,s)}{R(s,s)}X_s\right] \\ &= \frac{R(t,s)}{R(s,s)}\xi \end{aligned}$$

Using (5.12) and (5.8) together, we get

$$\begin{aligned} E(X_t|X_{t_0} = x_0) &= \int_{-\infty}^{\infty} x\, dP(x,t|x_0,t_0) \\ &= \int_{-\infty}^{\infty} \frac{R(t,s)}{R(s,s)}\xi\, dP(\xi,s|x_0,t_0) \\ &= \frac{R(t,s)}{R(s,s)}\frac{R(s,t_0)}{R(t_0,t_0)} x_0 \\ &= \frac{R(t,t_0)}{R(t_0,t_0)} x_0 \qquad t > s > t_0 \end{aligned}$$

Therefore, for $\{X_t, t \in T\}$ to be Gaussian and Markov we must have

$$R(t,t_0) = \frac{R(t,s)R(s,t_0)}{R(s,s)} \qquad t > s > t_0 \qquad (5.13)$$

Suppose that T is an interval and $R(t,t)$ is strictly positive for t in the interior of T. Then we can show that any solution of (5.13) must have the form

$$R(t,s) = f(\max(t,s))g(\min(t,s)) \qquad t, s \in T \qquad (5.14)$$

It turns out that (5.14) is also a sufficient condition for a Gaussian process to be Markov. The simplest way to show this is as follows. First, we note that a Brownian motion is Gaussian, Markov, and the $R(t,s) = \min(t,s)$ (see Sec. 3). Set

$$\tau(t) = \frac{g(t)}{f(t)} \qquad (5.15)$$

Because $R(t,s) \leq \sqrt{R(t,t)R(s,s)}$, $\tau(t)$ must be monotone nondecreasing in t. Now, if X_t is a Gaussian process with zero mean and covariance function given by (5.14), then it can be represented as

$$X_t = f(t)Y_{\tau(t)} \qquad (5.16)$$

5. MARKOV PROCESSES

where Y_t is a Brownian motion. Since $f(t)$ is a deterministic function, and $\tau(t)$ is nondecreasing, the fact that Y_t is Markov implies that X_t is Markov.

As was pointed out in Sec. 1.7, rather than to deal with conditional distributions, it is far simpler to deal with conditional expectations. The setting is however more abstract. We shall now reexamine the Markov property in terms of conditional expectations. First, we adopt some notations. Let $\{X_t, t \in T\}$ be a process defined on a fixed probability space $(\Omega, \mathcal{A}, \mathcal{P})$. Let $\mathcal{A}_t$ denote the smallest σ algebra with respect to which every X_s, $s \leq t$, is measurable. Let $\mathcal{A}_t^+$ denote the smallest σ algebra with respect to which X_s is measurable for every $s \geq t$. Suppose that $\{X_t, t \in T\}$ is a Markov process, then the following relations hold: All equalities are understood to hold up to sets of zero probability.

1. The future and the past are conditionally independent given the present. That is, if Z is $\mathcal{A}_t^+$ measurable and Y is $\mathcal{A}_t$ measurable, then

$$E(ZY|X_t) = E(Z|X_t)E(Y|X_t) \tag{5.17}$$

2. The future, given the past and the present, is equal to the future given only the present. That is, if Z is $\mathcal{A}_t^+$ measurable then

$$E^{\mathcal{A}_t}Z = E(Z|X_t) \tag{5.18}$$

Either 1 or 2 can be taken as the defining condition for Markov process.

Roughly speaking, the counterparts of (5.17) and (5.18) in terms of density functions are, respectively, given as follows:

$$p(x,t; x_0,t_0|\xi,s) = p(x,t|\xi,s)p(x_0,t_0|\xi,s) \quad t > s > t_0$$
$$p(x,t|x_0,t_0; \xi,s) = p(x,t|\xi,s) \quad t > s > t_0$$

In this form, the equivalence between conditions 1 and 2 is intuitively clear. To get (5.17) from (5.18) is easy. We merely note that if Z is $\mathcal{A}_t^+$ measurable and Y is $\mathcal{A}_t$ measurable, then

$$E(ZY|X_t) = E(E^{\mathcal{A}_t}ZY|X_t) = E(YE^{\mathcal{A}_t}Z|X_t)$$
$$= E[YE(Z|X_t)|X_t] = E(Z|X_t)E(Y|X_t)$$

To get (5.18) from (5.17), take an arbitrary set $B \in \mathcal{A}_t$ and compute

$$E[I_B E(Z|X_t)] = E\{E[I_B E(Z|X_t)|X_t]\}$$
$$= E[E(Z|X_t)E(I_B|X_t)] = E[E(I_B Z|X_t)] = E(I_B Z)$$

Since $EI_B E(Z|X_t) = EI_B Z$ for every set B in $\mathcal{A}_t$ and $E(Z|X_t)$ is $\mathcal{A}_t$ measurable, (5.18) follows by the definition of conditional expectation.

There is still another consequence of the Markov property that is useful to express. Suppose that Z is $\mathcal{A}_t^+$ measurable and $t_0 < t$. Then we have

$$E(Z|X_{t_0}) = E[E(Z|X_t)|X_{t_0}] \tag{5.19}$$

To prove (5.19), we merely have to use (5.18) and write

$$E(Z|X_{t_0}) = E(E^{\alpha_t}Z|X_{t_0}) = E[E(Z|X_t)|X_{t_0}]$$

Equation (5.19) is really the Chapman-Kolmogorov equation (5.8) in disguise. If we set $B = \{\omega: X_t < x\}$, then (5.8) can be rewritten as

$$E\{I_B|X_{t_0} = x_0\} = E\{E(I_B|X_s)|X_{t_0} = x_0\} \qquad t_0 < s < t \qquad (5.20)$$

which follows immediately from (5.19). A comparison between (5.8) and (5.19) reveals the notational simplicity gained by using conditional expectation.

6. STATIONARITY AND ERGODICITY

Definition. A process $\{X_t, -\infty < t < \infty\}$ is said to be a **stationary process** if for any $(t_1, \ldots, t_n)$ the joint distribution of $\{X_{t_1+t_0}, X_{t_2+t_0}, \ldots, X_{t_n+t_0}\}$ does not depend on t_0.

Thus, if $\{X_t, t \in T\}$ is stationary, the n-dimensional distribution function $P(x_1,t_1; x_2,t_2; \ldots; x_n,t_n)$ is equal to $P(x_1,0; x_2,t_2 - t_1; \ldots; x_n,t_n - t_1)$, and hence, it depends only on the time differences $t_2 - t_1, \ldots, t_n - t_1$.

Definition. A process $\{X_t, -\infty < t < \infty\}$ is said to be a **wide-sense stationary process** if
 (a) $EX_t^2 < \infty$
 (b) $EX_t = \mu$ is a constant
 (c) $E(X_t - \mu)(X_s - \mu)$ depends only on $t - s$

Because a stationary process may have $EX_t^2 = \infty$, a stationary process is also wide-sense stationary if and *only if* $EX_t^2 < \infty$. Wide-sense stationarity will be discussed in Chap. 3 in connection with second-order processes. In this section, we restrict our consideration to stationary processes, which are sometimes referred to as strict-sense stationary processes to emphasize the difference from wide-sense stationarity. The random telegraph process is an example of a stationary process. The Brownian motion is not stationary. A Gaussian process with zero mean and a covariance function given by

$$R(t,s) = e^{-|t-s|}$$

is both stationary and Markov. It is known as the Ornstein-Uhlenbeck process.

Let $\{X_t, -\infty < t < \infty\}$ be a stationary process defined on a completed probability space $(\Omega, \mathcal{A}, \mathcal{P})$. We suppose that $\mathcal{A}$ is the completion of the smallest σ algebra with respect to which every X_t is measurable.

6. STATIONARITY AND ERGODICITY

We also suppose that $\{X_t, -\infty < t < \infty\}$ is a separable and measurable process. Let S denote the space of all $\mathcal{A}$-measurable random variables. We agree that two random variables which are equal almost surely count as the same element of S. Then S is a linear space closed under almost-sure convergence. Now, we define a family $\{T_t, -\infty < t < \infty\}$ of linear mappings of S onto S as follows:

$$T_t X_s = X_{t+s} \qquad -\infty < t, s < \infty \qquad (6.1a)$$

If $Z(\omega) = f(X_{t_1}(\omega), \ldots, X_{t_n}(\omega))$, then

$$T_t Z = f(T_t X_{t_1}, \ldots, T_t X_{t_n}) \qquad (6.1b)$$

If $\{Z_n\}$ is a sequence in S converging almost surely to Z, then

$$T_t Z = \lim_{n \to \infty} \text{a.s.} \ T_t Z_n \qquad (6.1c)$$

Since every Z in S can be approximated by (a.s. limit) a sequence of Borel functions of a finite number of X_t's, (6.1a) to (6.1c) adequately define $\{T_t, -\infty < t < \infty\}$.

Since $\{X_t, -\infty < t < \infty\}$ is stationary, $\{T_t X_{t_\nu}, \nu = 1, \ldots, n\}$ and $\{X_{t_\nu}, \nu = 1, \ldots, n\}$ have the same distribution. Therefore, $T_t Z$ and Z have identical distributions, whenever Z is of the form $f(X_{t_1}, \ldots, X_{t_n})$. By virtue of the fact that convergence a.s. implies convergence in distribution, $T_t Z$ and Z are always identically distributed. It is also easy to see that $\{T_t, -\infty < t < \infty\}$ is a translation group, that is, $T_{t+s} = T_t T_s$, $T_0 = I$, $T_{-t} = T_t^{-1}$. Summarizing these properties of T_t, we have the following:

T_t is a linear mapping of S onto S (6.2a)
T_t preserves a.s. convergence (6.2b)
$T_{t+s} = T_t T_s$, $T_0 = I$, $T_{-t} = T_t^{-1}$ (6.2c)
$T_t Z$ and Z are identically distributed, $-\infty < t < \infty, Z \in S$ (6.2d)

Now, suppose a random variable Z in S is such that for each t, $T_t Z = Z$ almost surely. That is, for each t, $T_t Z$ and Z differ only on a set Λ_t such that $\mathcal{P}(\Lambda_t) = 0$. Then, Z is said to be an **invariant random variable** of the process $\{X_t, -\infty < t < \infty\}$.

Definition. A stationary process $\{X_t, -\infty < t < \infty\}$ is said to be **ergodic** if every invariant random variable of the process is almost surely equal to a constant.

We note that a random variable almost surely equal to a constant is always invariant. The great interest in ergodic processes, from the point of view of applications, is largely due to the following theorem.

Proposition 6.1. Let $\{X_t, -\infty < t < \infty\}$ be a separable and measurable ergodic process. Let f be any Borel function such that $E|f(X_0)| < \infty$. Then

$$\lim_{T \to \infty} \frac{1}{2T} \int_{-T}^{T} f(X_t) \, dt = Ef(X_0) \qquad \text{almost surely} \tag{6.3}$$

Remark: We note that because of stationarity, $Ef(X_t) = Ef(X_0)$ for every t. We interpret Proposition 6.1 as saying that for an ergodic process, time average equals ensemble average.

The proof of Proposition 6.1 will be omitted. We consider, instead, an example illustrating the foregoing discussion. Suppose

$$X_t(\omega) = A(\omega) \cos [2\pi t + \theta(\omega)] \qquad -\infty < t < \infty \tag{6.4}$$

where A and θ are independent random variables, and θ is uniformly distributed on $[0,2\pi)$. It is easy to show (e.g., by computing the characteristic functions) that $\{X_t, -\infty < t < \infty\}$ is stationary. Now, for this simple example, every Z in S is some Borel function of the pair (A,θ), and

$$T_t f(A,\theta) = f(A, \text{mod}_{2\pi} (\theta + 2\pi t)) \tag{6.5}$$

Every Z in S which depends only on A and not on θ is an invariant random variable. Thus $\{X_t, -\infty < t < \infty\}$ is ergodic if and only if A is almost surely a constant. This result can also be inferred from Proposition 6.1. For an arbitrary Borel function f such that $E|f(X_0)| < \infty$, we have

$$\lim_{T \to \infty} \frac{1}{2T} \int_{-T}^{T} f(X_t(\omega)) \, dt = \lim_{N \to \infty} \frac{1}{2N} \int_{-N}^{N} f(A(\omega) \cos [2\pi t + \theta(\omega)]) \, dt$$

$$= \lim_{N \to \infty} \frac{1}{2N} \int_{-N}^{N} f(A(\omega) \cos 2\pi t) \, dt$$

$$= \int_{0}^{1} f(A(\omega) \cos 2\pi t) \, dt \tag{6.6}$$

where we made repeated use of the fact that $f(A \cos (2\pi t + \theta))$ is periodic in t with period 1. On the other hand

$$Ef(X_0) = Ef(A \cos \theta)$$

$$= \int_{-\infty}^{\infty} \left[\int_{0}^{2\pi} \frac{1}{2\pi} f(a \cos \varphi) \, d\varphi \right] dP_A(a)$$

$$= E \int_{0}^{1} f(A \cos 2\pi t) \, dt \tag{6.7}$$

6. STATIONARITY AND ERGODICITY

If we denote $\int_0^1 f(A \cos 2\pi t)\, dt$ by $\hat{f}(A)$, then the time average is $\hat{f}(A(\omega))$. These two are equal if and only if $A(\omega)$ is almost surely equal to a constant.

In general, it is not easy to give a simple condition which would ensure ergodicity. For Gaussian processes, however, we have the following sufficient condition for ergodicity:[1] Assume that $\{X_t, -\infty < t < \infty\}$ is Gaussian and stationary, with mean μ and covariance function

$$R(\tau) = E(X_{t+\tau} - \mu)(X_t - \mu)$$

Then $\{X_t, -\infty < t < \infty\}$ is ergodic if

$$\int_{-\infty}^{\infty} |R(\tau)|\, d\tau < \infty \tag{6.8}$$

The question of possible equality between time average and ensemble average can be asked in a different and more direct way. Suppose for a fixed stationary process $\{X_t, -\infty < t < \infty\}$ and a fixed function f, we seek a sufficient condition for

$$\frac{1}{2T} \int_{-T}^{T} f(X_t)\, dt \xrightarrow[T \to \infty]{\text{q.m.}} Ef(X_0) \tag{6.9a}$$

that is,

$$\lim_{T \to \infty} E\left[\frac{1}{2T} \int_{-T}^{T} f(X_t)\, dt - Ef(X_0)\right]^2 = 0 \tag{6.9b}$$

We note that $Ef(X_t) = Ef(X_0)$ for every t. Therefore,

$$E\left[\frac{1}{2T} \int_{-T}^{T} f(X_t)\, dt - Ef(X_0)\right]^2$$

$$= \frac{1}{(2T)^2} \iint_{-T}^{T} E\{[f(X_t) - Ef(X_t)][f(X_s) - Ef(X_s)]\}\, dt\, ds$$

$$= \frac{1}{(2T)^2} \iint_{-T}^{T} R_f(t - s)\, dt\, ds \tag{6.10}$$

where we have denoted the covariance function of $f(X_t)$ by R_f. By a change in the variables of integration $\tau = t - s$, $\sigma = t + s$, the last integral in (6.10) can be rewritten as

$$\frac{1}{(2T)^2} \frac{1}{2} \int_{-2T}^{2T} \left[\int_{-2T+|\tau|}^{2T-|\tau|} d\sigma\right] R_f(\tau)\, d\tau$$

$$= \frac{1}{2T} \int_{-2T}^{2T} R_f(\tau)\left(1 - \frac{|\tau|}{2T}\right) d\tau \leq \frac{1}{2T} \int_{-\infty}^{\infty} |R_f(\tau)|\, d\tau$$

---
[1] A necessary and sufficient condition is for the spectral-distribution function, defined by (3.5.21) and (3.5.22), to be continuous [Grenander, 1950, pp. 257–260].

Therefore, a sufficient condition for (6.9b) is given by

$$\int_{-\infty}^{\infty} |R_f(\tau)| \, d\tau < \infty \tag{6.11}$$

Unlike ergodicity, condition (6.11) has to be verified for each f separately.

EXERCISES

1. Suppose that $\{X_t, 0 \leq t \leq 1\}$ is a family of independent random variables, i.e., every finite subcollection $X_{t_1}, \ldots, X_{t_n}$ is mutually independent. Show that $\{X_t, 0 \leq t \leq 1\}$ cannot be continuous in probability unless there is a continuous function $f(\cdot)$ such that for each t, $X_t(\omega) = f(t)$ for almost all ω.

2. Suppose that X is a Gaussian random variable with $EX = 0$ and $EX^2 = 1$. Let $X_t = t + X$, $-\infty < t < \infty$.

    (a) Let T_n be any countable set in $(-\infty, \infty)$, say $T_n = \{t_1, t_2, \ldots\}$. Show that $\mathcal{P}(X_t = 0 \text{ for at least one } t \text{ in } T_n) = 0$
    (b) Let T be an interval, say $[0,1]$. Show that
    $\mathcal{P}(X_t = 0 \text{ for at least one } t \text{ in } T) = \mathcal{P}\left(\bigcup_{t \in T} \{\omega : X_t(\omega) = 0\}\right) > 0$

    Note that even though $\{\omega : X_t(\omega) = 0 \text{ for at least one } t \text{ in } T\}$ is an uncountable union, it is an event in this particular case.

3. For an extremely explicit example of stochastic process consider the following. Let $\Omega = [0,1]$, let α be the σ algebra of Borel sets in $[0,1]$, and let $\mathcal{P}$ be the Lebesgue measure so that $\mathcal{P}$ (interval) = length. Define

    $X_t(\omega) = t\omega \quad 0 \leq t \leq 1$

    Find the one-dimensional distribution function P_t for this process. Repeat for the two-dimensional distribution function $P_{t,s}$.

4. Find the mean function $\mu(t) = EX_t$ and the covariance function $R(t,s) = E[X_t - \mu(t)][X_s - \mu(s)]$ for the process defined in Exercise 3.

5. Verify that the process defined in Exercise 3 is both separable and measurable.

6. Let Z and θ be independent random variables such that Z has a density function p_Z given by

    $$p_Z(z) = \begin{cases} 0 & z < 0 \\ ze^{-\frac{1}{2}z^2} & z \geq 0 \end{cases}$$

    and θ is uniformly distributed in the interval $[0, 2\pi)$. Define

    $X_t = Z \cos(2\pi t + \theta)$

    Show that $\{X_t, -\infty < t < \infty\}$ is a Gaussian process.

7. Let $\{X_t, -\infty < t < \infty\}$ be a Gaussian process with $EX_t = 0$ and

    $EX_s X_t = \frac{1}{2}(|t| + |s| - |t - s|)$

Now define $Y_t = X_{-t}$ for $t \leq 0$. Show that $\{Y_t, 0 \leq t < \infty\}$ and $\{X_t, 0 \leq t < \infty\}$ are two independent processes.

8. Let $\{X_t, t \in T\}$ be a stochastic process such that $E|X_t| < \infty$ for every $t \in T$, and let $\mathfrak{A}_{xt}$ denote the smallest σ algebra with respect to which X_s is measurable for every $s \leq t$. Suppose that

$$E(X_{t_n}|X_{t_1}, X_{t_2}, \ldots, X_{t_{n-1}}) = X_{t_{n-1}} \quad \text{a.s.}$$

whenever $t_1 < t_2 < \cdots < t_n$. Prove that whenever $t \geq s$

$$E^{\mathfrak{A}_{xs}} X_t = X_s \quad \text{a.s.}$$

Note: $E^{\mathfrak{A}_{xs}} X_t$ can also be written in a more suggestive way as $E(X_t|X_\tau, \tau \leq s)$.

9. Let $\{X_t, t \geq 0\}$ be a Brownian motion. Use the result of Exercise 8 to prove that for $t \geq s$,

$$E(X_t|X_\tau, \tau \leq s) = X_s \quad \text{a.s.}$$

10. Let $\{X_t, t \in [0,1]\}$ be a separable Gaussian process with zero mean. If

$$E(X_{t+h} - X_t)^2 = h^\gamma \quad \gamma > 0$$

show that $\{X_t \in [0,1]\}$ must be sample continuous no matter how small γ may be.

11. Suppose that $\{X_t, -\infty < t < \infty\}$ is a Gaussian process with zero mean and $EX_t X_s = e^{-|t-s|}$. Express X_t in the form

$$X_t = f(t) W_{g(t)/f(t)}$$

where $\{W_s, 0 \leq s < \infty\}$ is a standard Brownian motion.

12. Let $\{X_t, -\infty < t < \infty\}$ be a stationary Markov process which assumes only a finite number of values $x_1, x_2, \ldots, x_n$. Let $\mathbf{p}(\tau)$ be an $n \times n$ matrix with elements

$$p_{ij}(\tau) = \mathcal{P}(X_{t+\tau} = x_i | X_t = x_j)$$

(a) Suppose that $\lim_{\tau \downarrow 0} (1/\tau)[\mathbf{p}(\tau) - \mathbf{I}] = \mathbf{A}$ exists and is finite. Show that $\mathbf{p}(\tau)$ must have the form

$$\mathbf{p}(\tau) = e^{\tau \mathbf{A}}$$

(b) Let $\mathbf{q}$ be an n vector with

$$q_i = \text{prob}(X_t = x_i)$$

and let $\mathbf{l}$ denote the n vector $\begin{bmatrix} 1 \\ 1 \\ \cdot \\ \cdot \\ \cdot \\ 1 \end{bmatrix}$. Show that

$$\mathbf{p}(\tau)\mathbf{q} = \mathbf{q}$$

and

$\mathbf{p}'(\tau)\mathbf{I} = \mathbf{I}$ ($\mathbf{p}'$ = transpose of $\mathbf{p}$)

Suppose that $n = 2$ and $\mathbf{q} = \begin{bmatrix} \frac{1}{2} \\ \frac{1}{2} \end{bmatrix}$. Show that $\mathbf{p}(\tau)$ must be of the form

$$\mathbf{p}(\tau) = \begin{bmatrix} \dfrac{1 + e^{-\lambda\tau}}{2} & \dfrac{1 - e^{-\lambda\tau}}{2} \\ \dfrac{1 - e^{-\lambda\tau}}{2} & \dfrac{1 + e^{-\lambda\tau}}{2} \end{bmatrix} \qquad \lambda \geq 0$$

13. Let T be an interval and $R(t,s)$, $t, s \in T$, be a continuous covariance function such that $R(t,t) > 0$ for every t in the interior of T. Suppose that R satisfies

$$R(t,t_0) = \frac{R(t,s)R(s,t_0)}{R(s,s)} \qquad t_0 < s < t$$

(a) Let $\rho(t,s) = R(t,s)/\sqrt{R(t,t)R(s,s)}$ and show that ρ satisfies

$$\rho(t,t_0) = \rho(t,s)\rho(s,t_0) \qquad \begin{array}{l} t_0 < s < t \\ t_0, s, t \text{ in int } (T) \end{array}$$

(b) Show that $\rho(t,s) > 0$ for all t and s in the interior of T.

(c) Let a be a fixed point in int (T) and define

$$\alpha(t) = \begin{cases} \rho(t,a) & t \leq a \\ \dfrac{1}{\rho(t,a)} & t \geq a \end{cases}$$

Show that

$$\rho(t,s) = \frac{\alpha(\min (t,s))}{\alpha(\max (t,s))} \qquad t, s \in \text{int } (T)$$

Note: This *proves* (5.14).

14. Let $\{X_t, -\infty < t < \infty\}$ be a q.m. continuous and stationary Gaussian-Markov process. Find its covariance function.

15. Suppose that $\{X_t, -\infty < t < \infty\}$ is given by the form

$$X_t(\omega) = A(\omega) \cos [2\pi t + \theta(\omega)]$$

where A and θ are independent. A is nonnegative and θ is uniformly distributed on $[0,2\pi)$.

(a) Show that $\{X_t, -\infty < t < \infty\}$ is stationary.
(b) Show that $EX_t = 0$ for all t provided that $EA < \infty$. Does

$$M_T = \frac{1}{2T} \int_{-T}^{T} X_t \, dt \text{ converge in probability to } EX_t \text{ as } T \to \infty?$$

(c) Show that $\{X_t, -\infty < t < \infty\}$ is a Gaussian process if and only if A has a Rayleigh distribution, that is, A has a density p_A given by

$$p_A(r) = \frac{r}{\sigma^2} \exp\left(-\frac{1}{2}\frac{r^2}{\sigma^2}\right) \quad r \geq 0$$

Is $\{X_t, -\infty < t < \infty\}$ Markov then?

16. Suppose that $\{X_t, t \geq 0\}$ and $\{Y_t, t \geq 0\}$ are two independent standard Brownian motions. Let $Z_t = \sqrt{X_t^2 + Y_t^2}$.

(a) Suppose that X_t, Y_t are observed at $t = 1, 2, 3$ and the data are summarized as follows:

t	1	2	3
x	.3	1	-2
y	$-.1$	$-.2$	1

Find $E\{Z_4^2|\text{observed data}\}$ and show that

$$\sqrt{5} \leq E\{Z_4|\text{observed data}\} \leq \sqrt{7}$$

(b) Is $\{Z_t, t \geq 0\}$ a Markov process?

3
Second-Order Processes

1. INTRODUCTION

In this chapter it will actually be easier to deal with complex-valued random variables and stochastic processes. A complex-valued random variable X is a complex-valued function on Ω such that its real and imaginary parts are real random variables.

Definition. A **second-order random variable** X is one which satisfies $E|X|^2 < \infty$. A **second-order stochastic process** $\{X_t, t \in T\}$ is a one-parameter family of second-order random variables.

If $\{X_t, t \in T\}$ is a second-order process, then we can define the **mean, correlation function,** and **covariance function,** respectively, as follows:

$$EX_t = \mu(t) \tag{1.1}$$
$$EX_t \bar{X}_s = \mathcal{R}(t,s) \tag{1.2}$$
$$E[X_t - \mu(t)]\overline{[X_s - \mu(s)]} = R(t,s) \tag{1.3}$$

1. INTRODUCTION

where the overbar denotes complex conjugate. As far as second-order properties are concerned, we are primarily interested in "linear operations" on processes. Since the output of a linear operation on $\{X_t, t \in T\}$ is equal to the sum of the outputs of the same operation on $X_t - \mu(t)$ and $\mu(t)$ separately, we can always assume zero mean with little loss in generality. In this chapter we shall always assume that the mean is zero. If the mean is zero, the correlation function and the covariance function are the same thing. We use the term covariance function exclusively. Roughly speaking, second-order properties of a process are those properties that can be deduced knowing only its covariance function.

A covariance function satisfies a number of important properties, some of which are listed below.

1. *Nonnegative definite*
   A complex-valued function $f(t,s)$ defined on a square $T \times T$ is said to be **nonnegative definite,** if for any finite collection $t_1, t_2, \ldots, t_n$ in T and any complex constants $\alpha_1, \alpha_2, \ldots, \alpha_n$

$$\sum_{j=1}^{n} \sum_{k=1}^{n} \alpha_j \bar{\alpha}_k f(t_j, t_k) \geq 0 \tag{1.4}$$

   Every covariance function is nonnegative definite, because

$$\sum_{j=1}^{n} \sum_{k=1}^{n} \alpha_j \bar{\alpha}_k R(t_j, t_k) = E \left| \sum_{j=1}^{n} \alpha_j X_{t_j} \right|^2 \geq 0$$

   Conversely, if $R(t,s)$ is a nonnegative definite function on $T \times T$, then we can always find a second-order process $\{X_t, t \in T\}$ whose covariance function is $R(t,s)$. In fact, we can always find a pair of real processes $\{Y_t, Z_t, t \in T\}$ such that $Y_t + iZ_t$ has covariance function $R(t,s)$, and Y_t, Z_t are jointly Gaussian, i.e., any real linear combination $\alpha Y_t + \beta Z_t$ is again a Gaussian process. Summarizing, *a function $R(t,s)$ on $T \times T$ is a covariance function if and only if it is nonnegative definite.*

2. *Hermitian symmetry*
   A covariance function $R(t,s)$ always satisfies

$$R(t,s) = \bar{R}(s,t) \tag{1.5}$$

   because $E X_t \bar{X}_s = \overline{E \bar{X}_t X_s}$.

3. *Schwarz inequality*
   If Y and Z are a pair of second-order random variables, then

$$|E Y \bar{Z}| \leq \sqrt{E|Y|^2 E|Z|^2} \tag{1.6}$$

   which is a special case of the Schwarz inequality. To prove (1.6) we need only note that

$$E|Y|^2 E|Z|^2 - E|Y\bar{Z}|^2 = E|Z|^2 E|Y - \frac{EY\bar{Z}}{E|Z|^2} Z|^2 \geq 0 \qquad (1.7)$$

Applying (1.6) to a second-order process, we get immediately

$$|R(t,s)| \leq \sqrt{R(t,t)R(s,s)} \qquad (1.8)$$

whenever $R(t,s)$ is a covariance function.

4. *Closure properties*

(a) *Multiplication.* If $R_1(t,s)$ and $R_2(t,s)$ are two covariance functions, then we can always find two independent Gaussian processes Y_t and Z_t with covariance functions $R_1(t,s)$ and $R_2(t,s)$, respectively. The covariance function of $X_t = Y_t Z_t$ is given by $R_1(t,s)R_2(t,s)$. Therefore, the product of two covariance functions is again a covariance function.

(b) *Addition.* The sum $R_1(t,s) + R_2(t,s)$ of two covariance functions is again a covariance function. The argument for this assertion is almost identical to that of (a).

(c) *Positive sums.* A positive constant C is always a covariance function, simply because we can always take $X_t = X$ to be a Gaussian random variable with zero mean and $EX^2 = C$. It follows that if $R_1(t,s)$, $R_2(t,s)$, ..., $R_n(t,s)$ are covariance functions on $T \times T$ and $C_1, C_2, \ldots, C_n$ are positive constants, then

$$R(t,s) = \sum_{\nu=1}^{n} C_\nu R_\nu(t,s) \qquad (1.9)$$

is again a covariance function on $T \times T$.

(d) *Pointwise limit.* If $\{R_\nu(t,s), \nu = 1, \ldots\}$ is a sequence of covariance functions on $T \times T$ converging pointwise to $R(t,s)$ at every point of $T \times T$, then $R(t,s)$ is a covariance function. We only need to verify the definition of a nonnegative definite function,

$$\sum_{j=1}^{n} \sum_{k=1}^{n} \alpha_j \bar{\alpha}_k R(t_j, t_k) = \lim_{\nu \to \infty} \sum_{j=1}^{n} \sum_{k=1}^{n} \alpha_j \bar{\alpha}_k R_\nu(t_j, t_k) \geq 0$$

(e) *Bilinear forms.* For any function $\sigma(t)$, $\sigma(t)\bar{\sigma}(s)$ is a covariance function, simply because we can always set $X_t = \sigma(t) X$ where X is normal $N(0,1)$. It follows that any finite bilinear sum

$$\sum_{\nu=1}^{n} \bar{\sigma}_\nu(t) \sigma_\nu(s) \qquad (1.10)$$

is a covariance function, and so is any convergent infinite series

$\sum_{\nu=1}^{\infty} \sigma_\nu(t)\bar{\sigma}_\nu(s)$. More generally, any pointwise limit of a sequence of bilinear sums of the form (1.10) is a covariance function. This includes not only infinite sum such as $\sum_{\nu=1}^{\infty} \sigma_\nu(t)\bar{\sigma}_\nu(s)$, but also integrals of the form $\int_a^b \sigma(t,\lambda)\bar{\sigma}(s,\lambda)\,d\lambda$. It will be seen later that most covariance functions can be represented in the form of bilinear sums and/or integrals, and these representations play an extremely useful role in the application of stochastic processes.

2. SECOND-ORDER CONTINUITY

Definition. A second-order process $\{X_t, t \in T\}$ is said to be **continuous in quadratic mean** (q.m. continuous) at t if

$$E|X_{t+h} - X_t|^2 \xrightarrow[h \to 0]{} 0$$

If a process is q.m. continuous at every t of T, we shall say that it is a q.m. continuous process.

Q.m. continuity of a second-order process must, of course, be closely related to questions of continuity of the covariance function of the process. The following proposition summarizes succinctly the relationship between q.m. continuity and continuity of the covariance function.

Proposition 2.1. Let $\{X_t, t \in T\}$ be a second-order process on an interval T, and let $R(t,s) = EX_t\bar{X}_s$ denote its covariance function.
(a) $\{X_t, t \in T\}$ is q.m. continuous at t if and only if $R(\cdot,\cdot)$ is continuous at the diagonal point (t,t).
(b) If $\{X_t, t \in T\}$ is q.m. continuous at every $t \in T$, then $R(\cdot,\cdot)$ is continuous at every point of the square $T \times T$.
(c) If a nonnegative definite function on $T \times T$ is continuous at every diagonal point, then it is continuous everywhere on $T \times T$.

Proof:
(a) If R is continuous at (t,t), then

$$\begin{aligned}E|X_{t+h} - X_t|^2 &= R(t+h, t+h) - R(t, t+h) \\ &\qquad - R(t+h, t) + R(t,t) \\ &= [R(t+h, t+h) - R(t,t)] - [R(t, t+h) - R(t,t)] \\ &\qquad - [R(t+h, t) - R(t,t)] \xrightarrow[h \to 0]{} 0\end{aligned}$$

Conversely, if $\{X_t, t \in T\}$ is q.m. continuous at t, then

$$\begin{aligned}R(t+h, t+h') - R(t,t) &= EX_{t+h}\bar{X}_{t+h'} - EX_t\bar{X}_t \\ &= E(X_{t+h} - X_t)\bar{X}_{t+h'} + EX_t\overline{(X_{t+h'} - X_t)}\end{aligned}$$

From the Schwarz inequality (1.6), we have

$$|R(t + h, t + h') - R(t,t)| \leq [E|X_{t+h} - X_t|^2 E|\bar{X}_{t+h'}|^2]^{\frac{1}{2}}$$
$$+ [E|X_t|^2 E|X_{t+h'} - X_t|^2]^{\frac{1}{2}} \xrightarrow[h,h'\to 0]{} 0$$

(b) If $\{X_t, t \in T\}$ is q.m. continuous at every $t \in T$, then

$$|R(t + h, s + h') - R(t,s)| = |EX_{t+h}\bar{X}_{s+h'} - EX_t\bar{X}_s|$$
$$= |E(X_{t+h} - X_t)\bar{X}_{s+h'} + EX_t(\overline{X_{s+h'} - X_s})|$$
$$\leq \sqrt{E|X_{t+h} - X_t|^2 E|X_{s+h'}|^2} + \sqrt{E|X_t|^2 E|X_{s+h'} - X_s|^2} \xrightarrow[h,h'\to 0]{} 0$$

(c) Since every nonnegative definite function on $T \times T$ is the covariance function of some second-order process on T, part (c) follows immediately from (a) and (b). ∎

3. LINEAR OPERATIONS AND SECOND-ORDER CALCULUS

Let $\{X_t, t \in T\}$ be a second-order process. A random variable Y is said to be derived from a **linear operation** on $\{X_t, t \in T\}$ if either

(a) $\quad Y(\omega) = \sum_{\nu=1}^{N} \alpha_\nu X_{t_\nu}(\omega)$ \hfill (3.1)

or

(b) $\quad Y$ is the q.m. limit of a sequence of such finite linear combinations

We denote the collection of all such random variables derived from a given process $\{X_t, t \in T\}$ by $\mathcal{H}_X$. Two elements Y and Y' are not distinguished whenever $E|Y - Y'|^2 = 0$.[1] Equality between second-order random variables will always be understood to be up to zero q.m. differences. The space $\mathcal{H}_X$ becomes an inner product space if we define the inner product $\langle Y, Z \rangle$ by

$$\langle Y, Z \rangle = EY\bar{Z} \tag{3.2}$$

The inner product automatically defines a norm

$$\|Y\| = \sqrt{\langle Y, Y \rangle} \tag{3.3}$$

and a metric

$$d(Y,Z) = \|Y - Z\| \tag{3.4}$$

A sequence $\{Y_n\}$ in $\mathcal{H}_X$ is said to be a **Cauchy sequence** if

$$\|Y_m - Y_n\| \xrightarrow[m,n\to\infty]{} 0$$

---

[1] Strictly speaking, $\mathcal{H}_X$ is a space of equivalence class, as is generally the case of L_2 spaces.

3. LINEAR OPERATIONS AND SECOND-ORDER CALCULUS

Since $\|Y_m - Y_n\|^2 = E|Y_m - Y_n|^2$, a Cauchy sequence in $\mathcal{H}_X$ is a mutually q.m. convergent sequence, which must converge in q.m. That is, for every Cauchy sequence $\{Y_n\}$ in $\mathcal{H}_X$, there exists a Y such that $\|Y_n - Y\| \xrightarrow[n\to\infty]{} 0$. It is easy to show that Y is in $\mathcal{H}_X$. This means that $\mathcal{H}_X$ is an inner product space which, as a metric space, is *complete*. By definition, $\mathcal{H}_X$ is a **Hilbert space**.

Second-order properties of a process $\{X_t, t \in T\}$ are those properties that are derivable from its covariance function. Every property of $\mathcal{H}_X$ is a second-order property, and conversely. It is easy to see, therefore, that there is inherently a close relationship between linear operations (defined by membership in $\mathcal{H}_X$) and second-order properties. Often, questions on second-order properties are most lucidly answered in the framework of the Hilbert space theory. A good example of this is the series representation of X_t to be treated in the next section.

Quadratic-mean derivative of a process $\{X_t, t \in T\}$ at a point t is defined by

$$\dot{X}_t = \lim_{h \to 0} \text{in q.m.} \frac{1}{h}(X_{t+h} - X_t) \tag{3.5}$$

Because of the equivalence between mutual q.m. convergence and q.m. convergence, $\dot{X}_t$ exists if and only if

$$\lim_{h,h' \to 0} E \left| \frac{1}{h}(X_{t+h} - X_t) - \frac{1}{h'}(X_{t+h'} - X_t) \right|^2 = 0 \tag{3.6}$$

which holds if the mixed partial derivative of the covariance function, that is, $\partial^2 R(t_1, t_2)/\partial t_1 \partial t_2$, exists in a neighborhood of (t, t) and is continuous at (t, t). We note that

$$E \dot{X}_t \bar{\dot{X}}_s = \frac{\partial^2 R(t,s)}{\partial t \, \partial s} \tag{3.7}$$

Since if $\dot{X}_t$ exists, it is equal to

$$\dot{X}_t = \lim_{n \to \infty} \text{in q.m.} \, n \, (X_{t+1/n} - X_t) \tag{3.8}$$

$\dot{X}_t \in \mathcal{H}_X$ by definition.

Quadratic-mean integrals arise even more frequently than q.m. derivatives. Let $\{X_t, t \in T\}$ be a second-order process, and let $f(t)$ be a complex-valued function defined on the interval T. We define the q.m. integral $\int_T f(t) X_t \, dt$ as an element in $\mathcal{H}_X$ as follows. Let $\{T_n\}$ be a sequence of partitions of T such that as $n \to \infty$ T_n becomes dense in T. To be specific, let T be a finite interval $[a,b]$ and

$$T_n = \{a = t_0^{(n)} < t_1^{(n)} < \cdots < t_n^{(n)} = b\} \tag{3.9}$$

$$\max_{1 \leq \nu \leq n} (t_\nu^{(n)} - t_{\nu-1}^{(n)}) \xrightarrow[n \to \infty]{} 0 \tag{3.10}$$

We define the integral $\int_a^b f(t) X_t \, dt$ as

$$\int_a^b f(t) X_t \, dt = \lim_{n \to \infty} \text{in q.m.} \sum_{\nu=0}^{n-1} f(t'_{\nu n}) X_{t'_{\nu n}} (t_{\nu-1}^{(n)} - t_\nu^{(n)}) \tag{3.11}$$

where $t'_{\nu n}$ are any sequence of points satisfying $t_\nu^{(n)} \leq t'_{\nu n} < t_{\nu+1}^{(n)}$. The integral $\int_a^b f(t) X_t \, dt$ is well defined by this procedure, provided that the q.m. limit exists and is independent of the choice of $\{T_n\}$ and, for each $\{T_n\}$, is independent of the choice of $\{t'_{\nu n}\}$. In that case, we say that the integral $\int_a^b f(t) X_t \, dt$ exists.

Proposition 3.1. The q.m. integral $\int_a^b f(t) X_t \, dt$ exists if and only if $\int_a^b \int_c^b f(t) \bar{f}(s) R(t,s) \, dt \, ds$ exists as a Riemann integral.

Remark:

(a) We note that if X_t is q.m. continuous, that is, if $R(t,s)$ is continuous, then $f(t)$ being piecewise continuous on $[a,b]$ is sufficient to ensure the existence of $\int_a^b f(t) X_t \, dt$.

(b) The q.m. integral can easily be generalized to include infinite intervals by taking q.m. limit as one or both of the endpoints goes to infinity.

(c) A common way in which q.m. integrals arise is when a process $\{Y_t, t \in T'\}$ is generated from a second process $\{X_t, t \in T\}$ according to the formula

$$Y_t = \int_T h(t,s) X_s \, ds \qquad t \in T' \tag{3.12}$$

Equation (3.12) often admits the interpretation of a linear system with input X_t, output Y_t, and impulse response $h(t,s)$.

(d) From the point of view of applications one might well prefer $\int_T f(t) X_t(\omega) \, dt$ to be defined as the Lebesgue integral of a sample function $\{X_t(\omega), t \in T\}$. The existence of such a Lebesgue integral is ensured if the process is a measurable process and if

$$\int_T |f(t)| E|X_t| \, dt < \infty$$

(cf. discussion at the end of Sec. 2.2).

4. ORTHOGONAL EXPANSIONS

A family $\mathfrak{F}$ of elements of $\mathcal{H}_X$ is said to be an **orthonormal** (O-N) family if any two distinct elements Y and Z of $\mathfrak{F}$ satisfy

$$\|Y\| = 1 = \|Z\|$$
$$\langle Y,Z \rangle = EY\bar{Z} = 0 \qquad (4.1)$$

The second of these conditions is called orthogonality. An O-N family $\mathfrak{F}$ is said to be **complete** in $\mathcal{H}_X$ if there exists no element of $\mathcal{H}_X$, except the zero element, which is orthogonal to every element of $\mathfrak{F}$.

Suppose that $\{X_t, t \in T\}$ is a q.m. continuous process, where T is an interval, finite or infinite. Let T' be the set of all rational points in T. For every $t \in T$, there exists a sequence $\{t_n\}$ in T' such that $t_n \xrightarrow[n \to \infty]{} t$. Since $\{X_t, t \in T\}$ is q.m. continuous, we have for every t,

$$X_t = \lim_{n \to \infty} \text{in q.m. } X_{t_n}$$

It follows that every element in $\mathcal{H}_X$ is a linear combination of $\{X_t, t \in T'\}$ or the q.m. limit of a sequence of such linear combinations. In short, the countable family $\{X_t, t \in T'\}$ is dense in $\mathcal{H}_X$. It follows that every O-N family in $\mathcal{H}_X$ is at most countable [see, e.g., Taylor, 1961, Sec. 3.2].

For a q.m. continuous process $\{X_t, t \in T\}$, let $\{Z_n, n = 1, 2, \ldots\}$ be an O-N family in $\mathcal{H}_X$. If $Y \in \mathcal{H}_X$, then

$$E\left| Y - \sum_{n=1}^{N} \langle Y,Z_n \rangle Z_n \right|^2 = E|Y|^2 - \sum_{n=1}^{N} |\langle Y,Z_n \rangle|^2 \geq 0$$

Therefore,

$$\infty > E|Y|^2 \geq \sum_{n=1}^{\infty} |\langle Y,Z_n \rangle|^2$$

so that $\sum_{n=1}^{\infty} \langle Y,Z_n \rangle Z_n$ is well defined and

$$Y - \sum_{n=1}^{\infty} \langle Y,Z_n \rangle Z_n$$

is orthogonal to every Z_n. It follows that if $\{Z_n\}$ is complete in $\mathcal{H}_X$, then every Y in $\mathcal{H}_X$ has the representation

$$Y = \sum_{n=1}^{\infty} \langle Y,Z_n \rangle Z_n \qquad (4.2)$$

Suppose that $\{Z_n, n = 1, 2, \ldots\}$ is a given complete O-N family in $\mathcal{H}_X$ and we set

$$\sigma_n(t) = \langle X_t, Z_n \rangle \qquad (4.3)$$

Then, from (4.2) we have

$$X_t = \sum_{n=1}^{\infty} \sigma_n(t) Z_n \qquad t \in T \tag{4.4}$$

The functions $\sigma_n(t)$, $t \in T$, are continuous, because $\{X_t, t \in T\}$ is q.m. continuous. Further, the set of functions $\{\sigma_n, n = 1, 2, \ldots\}$ is also **linearly independent,** i.e., for every N

$$\sum_{n=1}^{N} \alpha_n \sigma_n(t) = 0 \qquad \text{for all } t \in T$$

implies $\alpha_n = 0$, $n = 1, \ldots, N$. The linear independence of $\{\sigma_n, n = 1, 2, \ldots\}$ is due to the fact that $\sum_{n=1}^{N} \alpha_n \sigma_n(t) = 0$ for all $t \in T$ implies that $\sum_{n=1}^{N} \alpha_n Z_n$ is orthogonal to X_t for every $t \in T$, hence, also orthogonal to every Z_n which implies $\alpha_n = 0$ for every n. It follows from (4.4) that

$$E X_t \bar{X}_s = R(t,s) = \sum_{n=1}^{\infty} \sigma_n(t) \bar{\sigma}_n(s) \qquad t, s \in T \tag{4.5}$$

Conversely, suppose that $\{\sigma_n(t), t \in T, n = 1, 2, \ldots\}$ is a linearly independent family of continuous functions such that

$$R(t,s) = \sum_{n=1}^{\infty} \sigma_n(t) \bar{\sigma}_n(s) \qquad t, s \in T$$

Then, it follows from a very general representation theorem of Karhunen [Karhunen, 1947] that there exists a complete O-N family $\{Z_n, n = 1, 2, \ldots\}$ in $\mathcal{H}_X$ such that

$$X_t(\omega) = \sum_{n=1}^{\infty} \sigma_n(t) Z_n(\omega) \qquad t \in T$$

Thus, (4.4) and (4.5) imply each other. Representations of the form (4.4) are useful because they permit the continuum of random variables $\{X_t, t \in T\}$ to be represented by a countable number of orthonormal random variables $\{Z_n\}$. However, their use is, in general, limited by the fact that it is usually difficult to express the random variables Z_n explicitly in terms of $\{X_t, t \in T\}$. An exceptional case is when $\{\sigma_n\}$ are orthogonal, that is, $\int_T \sigma_m(t) \bar{\sigma}_n(t) \, dt = 0$ whenever $m \neq n$. This motivates the expansion widely known as the Karhunen-Loève expansion.

Consider a q.m. continuous process $\{X_t, a \leq t \leq b\}$, where the parameter set is explicitly assumed to be a closed and finite interval.

4. ORTHOGONAL EXPANSIONS

Suppose that there exists an expansion of the form

$$X_t(\omega) = \sum_{n=1}^{\infty} \sigma_n(t) Z_n(\omega) \qquad a \leq t \leq b \tag{4.6}$$

where $\{Z_n\}$ and $\{\sigma_n\}$ satisfy

$$E Z_m \bar{Z}_n = \delta_{mn} \tag{4.7}$$

$$\int_a^b \sigma_m(t) \bar{\sigma}_n(t) \, dt = \lambda_n \delta_{mn} \tag{4.8}$$

Now, from (4.5), the covariance function R must satisfy

$$R(t,s) = \sum_{n=1}^{\infty} \sigma_n(t) \bar{\sigma}_n(s) \tag{4.9}$$

for each (t,s) in $[a,b] \times [a,b]$. Now from the Schwarz inequality and the fact that R is continuous on $[a,b] \times [a,b]$ we have

$$\sup_{a \leq t,s \leq b} \left| \sum_{n=1}^{N} \sigma_n(t) \bar{\sigma}_n(s) \right| \leq \sup_{a \leq t \leq b} \sum_{n=1}^{N} |\sigma_n(t)|^2$$

$$\leq \sup_{a \leq t \leq b} R(t,t) < \infty \tag{4.10}$$

Therefore, the convergence in (4.9) is bounded. It follows that for every m

$$\int_a^b R(t,s) \sigma_m(s) \, ds = \lambda_m \sigma_m(t) \qquad a \leq t \leq b \tag{4.11}$$

What we have shown is that if an expansion (4.6), satisfying (4.7) and (4.8), exists, then $\{\sigma_n\}$ must be solutions to the integral equation (4.11). We shall see that under our assumptions, such an expansion always exists.

The above considerations suggest that we investigate integral equations of the form of (4.11). Fortunately, such equations are well known. We shall now summarize some of the important facts concerning it. First, we shall introduce a few definitions. Consider the integral equation

$$\int_a^b R(t,s) \varphi(s) \, ds = \lambda \varphi(t) \qquad a \leq t \leq b \tag{4.12}$$

where we have explicitly denoted the interval T by $[a,b]$. We assume that $[a,b]$ is finite and $R(t,s)$ is continuous on $[a,b] \times [a,b]$. A nonzero number λ, for which there exists a φ satisfying both (4.12) and the condition $\int_a^b |\varphi(t)|^2 \, dt < \infty$, is called an **eigenvalue** of the integral equation. The corresponding φ is called an **eigenfunction**.

1. Any eigenvalue of (4.12) must be real and positive. The fact that λ is real follows from the Hermitian symmetry $R(t,s) = \bar{R}(s,t)$. The fact that λ is positive follows from the nonnegative definiteness of R.

2. There is at least one eigenvalue for (4.12), if R is not identically zero. The largest eigenvalue λ_0 is given by

$$\lambda_0 = \max_{\|\varphi\|=1} \int\int_a^b R(t,s)\varphi(s)\bar{\varphi}(t)\,ds\,dt$$
$$\|\varphi\| = \left[\int_a^b |\varphi(t)|^2\,dt\right]^{\frac{1}{2}} \qquad (4.13)$$

This fact is not easily proved. It depends on both the nonnegative definiteness of R and its continuity [see, e.g., Taylor, 1961, pp. 334–336].

3. Let $\varphi_0(t)$ denote the normalized eigenfunction corresponding to λ_0, then $\varphi_0(t)$ is continuous on $[a,b]$. This is because we can write

$$\varphi_0(t) = \frac{1}{\lambda_0}\int_a^b R(t,s)\varphi_0(s)\,ds \qquad a \leq t \leq b \qquad (4.14)$$

and the continuity of $\varphi_0(t)$ follows from the continuity of $R(t,s)$ in t.

4. Let $R_1(t,s) = R(t,s) - \lambda_0\varphi_0(t)\bar{\varphi}_0(s)$. Then $R_1(t,s)$ is both continuous and nonnegative definite. The continuity of R_1 is obvious from 3. To show nonnegative definiteness, let

$$Y_t = X_t - \varphi_0(t)\int_a^b X_\tau\varphi_0(\tau)\,d\tau \qquad (4.15)$$

Then, we have

$$\begin{aligned}EY_t\bar{Y}_s &= EX_t\bar{X}_s - E\left[X_t\bar{\varphi}_0(s)\int_a^b \bar{X}_\sigma\varphi_0(\sigma)\,d\sigma\right]\\&\quad - E\left[\bar{X}_s\varphi_0(t)\int_a^b X_\tau\bar{\varphi}_0(\tau)\,d\tau\right]\\&\quad + E\left[\varphi_0(t)\bar{\varphi}_0(s)\int\int_a^b X_\tau X_\sigma\bar{\varphi}_0(\tau)\varphi_0(\sigma)\,d\tau\,d\sigma\right]\\&= R(t,s) - \lambda_0\varphi_0(t)\bar{\varphi}_0(s)\\&= R_1(t,s)\end{aligned} \qquad (4.16)$$

Hence, $R_1(t,s)$, being a covariance function, must be nonnegative definite.

5. We observe that $\int_a^b R_1(t,s)\varphi_0(s)\,ds = 0$. Therefore, if we repeat step 2 and obtain λ_1, φ_1, then

$$\int_a^b R_1(t,s)\varphi_1(s)\,ds = \lambda_1\varphi_1(t) \qquad a \leq t \leq b$$

It follows that $\int_a^b \varphi_1(t)\bar{\varphi}_0(t)\,dt = \frac{1}{\lambda}\int_a^b \varphi_1(s)\left[\overline{\int_a^b R_1(s,t)\varphi_0(t)\,dt}\right]ds = 0$

so that φ_1 is orthogonal to φ_0. In addition,

$$\begin{aligned}\int_a^b R(t,s)\varphi_1(s)\,ds &= \int_a^b R_1(t,s)\varphi_1(s)\,ds - \lambda_0\varphi_0(t)\int_a^b \varphi_1(s)\bar{\varphi}_0(s)\,ds\\&= \lambda_1\varphi_1(t) \qquad a \leq t \leq b\end{aligned} \qquad (4.17)$$

In other words, λ_1 and φ_1 are eigenvalue and eigenfunction of (4.12).

4. ORTHOGONAL EXPANSIONS

6. It is clear that the above procedure can be iterated, and we get a nonincreasing sequence of eigenvalues $\lambda_0, \lambda_1, \ldots$ and a corresponding sequence of eigenfunctions $\varphi_0, \varphi_1, \ldots$ which are orthonormal, that is,

$$\int_a^b \varphi_n(t)\bar\varphi_m(t)\, dt = \delta_{mn} \tag{4.18}$$

7. The sequence $\lambda_0, \lambda_1, \ldots,$ may terminate after a finite number of terms, in which case we have

$$R(t,s) = \sum_{n=0}^{N} \lambda_n \varphi_n(t) \bar\varphi_n(s) \tag{4.19}$$

If the sequence $\lambda_0, \lambda_1, \ldots$ does not terminate, then $\lambda_n \xrightarrow[n\to\infty]{} 0$ (see Exercise 2).

8. If the number of eigenvalues is infinite, then

$$\lim_{N\to\infty} \sup_{a\leq t,s \leq b} \left| R(t,s) - \sum_{n=0}^{N} \lambda_n \varphi_n(t)\bar\varphi_n(s) \right| = 0 \tag{4.20}$$

In other words,

$$\lim_{N\to\infty} \sum_{n=0}^{N} \lambda_n \varphi_n(t)\bar\varphi_n(s) = R(t,s) \quad \text{uniformly on } [a,b]^2 \tag{4.21}$$

This important result is known as the **Mercer's theorem** [Riesz and Sz.-Nagy, 1952, p. 245]. The fact that the convergence is uniform is a strong result and immediately implies convergence in mean square, that is,

$$\lim_{N\to\infty} \int\int_a^b \left| R(t,s) - \sum_{n=0}^{N} \lambda_n \varphi_n(t)\bar\varphi_n(s) \right|^2 dt\, ds = 0 \tag{4.22}$$

9. In general, the O-N family $\{\varphi_n\}$ is not complete in the space $L_2(a,b)$. The most that can be said is that given $f \in L_2(a,b)$ (that is,

$$\int_a^b |f(t)|^2\, dt < \infty)$$

we can write

$$f(t) = f_0(t) + \underset{N\to\infty}{\text{l.i.m.}} \sum_{n=0}^{N} \varphi_n(t) \int_a^b f(s)\bar\varphi_n(s)\, ds \tag{4.23}$$

where l.i.m. means limit in mean square and f_0 satisfies

$$\int_a^b R(t,s) f_0(s)\, ds = 0 \quad \text{for all } t \in [a,b] \tag{4.24}$$

It follows that $\{\varphi_n\}$ is complete in $L_2(a,b)$, if and only if (4.24) implies $\int_a^b |f_0(t)|^2\, dt = 0$.

We are now in a position to state precisely the theorem concerning the biorthogonal series for a second-order process. This expansion is often referred to as the Karhunen-Loève expansion [Loève, 1963, pp. 478–479].

Proposition 4.1. Let $\{X(\omega,t),\ t \in [a,b]\}$ be a q.m. continuous second-order process with covariance function $R(t,s)$.
(a) If $\{\varphi_n\}$ are the orthonormal eigenfunctions of

$$\int_a^b R(t,s)\varphi(s)\,ds = \lambda\varphi(t) \qquad a \leq t \leq b \tag{4.25}$$

and $\{\lambda_n\}$ the eigenvalues, then

$$X(\omega,t) = \lim_{N \to \infty} \text{in q.m.} \sum_{n=0}^{N} \sqrt{\lambda_n}\,\varphi_n(t)b_n(\omega)$$

$$\text{uniformly for } a \leq t \leq b \tag{4.26}$$

where $\{b_n\}$ satisfy

$$b_n(\omega) \stackrel{\text{q.m.}}{=} \left(\sqrt{\lambda_n}\right)^{-1} \int_a^b \bar{\varphi}_n(t) X(\omega,t)\,dt \tag{4.27}$$

and

$$E b_m \bar{b}_n = \delta_{mn} \tag{4.28}$$

(b) Conversely, if $X(\omega,t)$ has an expansion of the form (4.26) with $\int_a^b \varphi_m(t)\bar{\varphi}_n(t)\,dt = \delta_{mn} = E b_m \bar{b}_n$, then $\{\varphi_n\}$ and $\{\lambda_n\}$ must be eigenfunctions and eigenvalues of (4.25).

Proof:
(a) By direct computation we have

$$E\left| X_t - \sum_{n=0}^{N} \sqrt{\lambda_n}\,\varphi_n(t)b_n \right|^2 = R(t,t) - \sum_{n=0}^{N} \lambda_n |\varphi_n(t)|^2$$

which goes to zero as $N \to \infty$ uniformly in t by virtue of the Mercer's theorem.
(b) Suppose X_t has the stated expansion. Then, we have

$$R(t,s) = E X_t \bar{X}_s = \sum_{n=0}^{\infty} \lambda_n \varphi_n(t) \bar{\varphi}_n(s)$$

Hence,

$$\int_a^b R(t,s)\varphi_m(s)\,ds = \int_a^b \sum_{n=0}^{\infty} \lambda_n \varphi_n(t)\bar{\varphi}_n(s)\varphi_m(s)\,ds$$
$$= \lambda_n \varphi_m(t) \qquad a \leq t \leq b$$

The proof is complete. ∎

4. ORTHOGONAL EXPANSIONS

Consider now an example of how an integral equation of the form of (4.25) might be solved. Let $R(t,s) = \min(t,s)$ and consider

$$\int_0^T \min(t,s)\varphi(s)\,ds = \lambda\varphi(t) \qquad 0 \le t \le T \tag{4.29}$$

or

$$\int_0^t s\varphi(s)\,ds + t\int_t^T \varphi(s)\,ds = \lambda\varphi(t)$$

Differentiating once with respect to t yields

$$\int_t^T \varphi(s)\,ds = \lambda\dot\varphi(t) \qquad 0 < t < T \tag{4.30}$$

Differentiating again, we find

$$-\varphi(t) = \lambda\ddot\varphi(t) \qquad 0 < t < T \tag{4.31}$$

We also have the obvious boundary conditions $\varphi(0) = 0$ and $\dot\varphi(T) = 0$. Equation (4.31) with initial condition $\varphi(0) = 0$ yields

$$\varphi(t) = A \sin \frac{1}{\sqrt{\lambda}} t$$

Applying the condition $\dot\varphi(T) = 0$ obtained from (4.30), we find

$$\cos \frac{1}{\sqrt{\lambda}} T = 0$$

In other words the eigenvalues are given by

$$\lambda_n = \frac{T^2}{(n+\tfrac{1}{2})^2\pi^2} \qquad n = 0, 1, 2, \ldots \tag{4.32}$$

The normalized eigenfunctions are given by

$$\varphi_n(t) = \sqrt{\frac{2}{T}} \sin\left[(n+\tfrac{1}{2})\pi\left(\frac{t}{T}\right)\right] \tag{4.33}$$

It is rather interesting to note that because of Mercer's theorem, we have

$$\min(t,s) = \frac{2}{T} \sum_{n=0}^\infty \frac{T^2}{\pi^2(n+\tfrac{1}{2})^2} \sin\left[(n+\tfrac{1}{2})\pi\left(\frac{t}{T}\right)\right]$$
$$\sin\left[(n+\tfrac{1}{2})\pi\left(\frac{s}{T}\right)\right] \qquad \text{uniformly on } [0,T]^2 \tag{4.34}$$

which is by no means an obvious result.

Analytical solution to the integral equation (4.25) is, in general, difficult. For some special cases, the integral equation can be reduced to a differential equation and then solved in a way similar to the example above. In particular, if the covariance function $R(t,s)$ has the form

$$R(t,s) = \int_{-\infty}^{\infty} e^{i2\pi\lambda(t-s)}\phi(\lambda)\,d\lambda \tag{4.35}$$

where $\phi(\lambda)$ is a ratio of polynomials in λ, then the integral equation can be reduced to a differential equation with constant coefficient. The details are given in Davenport and Root [1958, pp. 241–242].

Before leaving the subject of orthogonal expansion, consider a second example. Suppose $EX_t\bar{X}_s = R(t,s) = \cos 2\pi(t-s)$. Then we can show that the Karhunen-Loève expansion has only two terms. In fact, any expansion in terms of an orthonormal basis in $\mathcal{H}_X$ has only two terms. Furthermore, in this instance, the integral equation (4.25) can be solved rather easily. To show all this, we first observe that

$$EX_0\bar{X}_{\frac{1}{4}} = 0$$
$$E|X_0|^2 = E|X_{\frac{1}{4}}|^2 = 1$$

Therefore $\{X_0, X_{\frac{1}{4}}\}$ is an O-N family in $\mathcal{H}_X$. Next, we observe that

$$E|X_t - X_0 \cos 2\pi t - X_{\frac{1}{4}} \sin 2\pi t|^2 = 0$$

Therefore, $\{X_0, X_{\frac{1}{4}}\}$ is also complete. This means that $\mathcal{H}_X$ is two dimensional. Hence, the Karhunen-Loève expansion has only two terms. The eigenvalues and eigenfunctions depend on the interval T. Suppose the interval is $[0,k]$, where $k = $ integer. Then, the eigenfunction can be taken to be $\{(1/\sqrt{k})e^{i2\pi t}, (1/\sqrt{k})e^{-i2\pi t}\}$ or $\{(2/\sqrt{k})\cos 2\pi t, (2/\sqrt{k})\sin 2\pi t\}$. If the interval $[0,T]$ is not integral, then the eigenfunctions can be taken to be suitable linear combinations of $\sin 2\pi t$ and $\cos 2\pi t$.

5. WIDE-SENSE STATIONARY PROCESSES

A second-order process $\{X_t, -\infty \leq t \leq \infty\}$ is said to be wide-sense stationary if its covariance function is a function of only the time difference, that is,

$$EX_t\bar{X}_s = R(t-s) \tag{5.1}$$

We recall once again that the mean EX_t is assumed to be zero throughout this chapter. Wide-sense stationarity means that

$$EX_{t+t_0}\bar{X}_{s+t_0} = EX_t\bar{X}_s \tag{5.2}$$

for any t_0. As before, let $\mathcal{H}_X$ denote the Hilbert space containing all linear combinations of $\{X_t, -\infty < t < \infty\}$. Suppose we define a linear operator U_t for t, mapping $\mathcal{H}_X$ onto $\mathcal{H}_X$, as follows:

$$U_t X_s = X_{t+s} \tag{5.3a}$$
$$\|Z_n - Z\| \xrightarrow[n\to\infty]{} 0 \quad \text{implies} \quad \|U_t Z_n - U_t Z\| \xrightarrow[n\to\infty]{} 0 \tag{5.3b}$$

5. WIDE-SENSE STATIONARY PROCESSES

It is clear from the definition of $\mathfrak{IC}_X$ (3.1) that $U_t Z$ is well defined for every Z in $\mathfrak{IC}_X$ by (5.3). Equation (5.2) can now be rewritten as

$$\langle U_{t_0} X_t, U_{t_0} X_s \rangle = \langle X_t, X_s \rangle \qquad \forall\, t_0, t, s$$

which is easily extended to

$$\langle U_{t_0} Z, U_{t_0} Y \rangle = \langle Z, Y \rangle \qquad \begin{array}{c} -\infty < t_0 < \infty \\ Y, Z \in \mathfrak{IC}_X \end{array} \tag{5.4}$$

Equation (5.4) means that for each t, U_t is a unitary operator. It is also easy to see that $U_t U_s = U_{t+s}$, $U_0 = I$, $U_t^{-1} = U_{-t}$, so that $\{U_t, -\infty < t < \infty\}$ is a **translation group**. Summarizing these results, we can state the following.

Every wide-sense stationary process $\{X_t, -\infty < t < \infty\}$ can be represented as

$$X_t = U_t X_0 \tag{5.5}$$

where $\{U_t, -\infty < t < \infty\}$ is a translation group of unitary operators mapping $\mathfrak{IC}_X$ onto $\mathfrak{IC}_X$. This result should be compared with the corresponding result for the strictly stationary process given in Sec. 2.6.

The invariance of covariance function with respect to time shifts suggests that harmonic analysis (i.e., representation in terms of complex sinusoids $e^{i 2\pi \nu t}$) should play a useful role in the theory of wide-sense stationary processes. To those who are familiar with application of Fourier integrals to the analysis of time-invariant linear systems, this is certainly no surprise. We begin our discussion with a brief review of Fourier integrals and their application to the analysis of linear time-invariant systems. Let L_p, C_0, and $\mathcal{S}$ denote, respectively, the following function spaces of Lebesgue measurable complex functions defined on $(-\infty, \infty)$:

L_p is the space of all functions satisfying[1]

$$\int_{-\infty}^{\infty} |f(t)|^p\, dt < \infty \qquad 1 \leq p < \infty \tag{5.6a}$$

C_0 is the space of all bounded-continuous functions such that

$$f(t) \xrightarrow[|t| \to \infty]{} 0 \tag{5.6b}$$

$\mathcal{S}$ is the space of all infinitely differentiable functions such that for any integers m and n,

$$|t|^m \left| \frac{d^n f(t)}{dt^n} \right| \xrightarrow[|t| \to \infty]{} 0 \tag{5.6c}$$

---

[1] Strictly speaking, L_p is a space of equivalence classes. Two functions f_1 and f_2 belong to the same equivalence class if and only if $\int_{-\infty}^{\infty} |f_1(t) - f_2(t)|^p\, dt = 0$.

The spaces L_p and C_0 are complete normed linear spaces (Banach spaces) with respective norms $\left[\int_{-\infty}^{\infty} |f(t)|^p \, dt\right]^{1/p}$ and $\sup_t |f(t)|$. The space $\mathcal{S}$ is dense in both L_p and C_0. That is, the completion of $\mathcal{S}$ with respect to the norm of L_p is L_p, and its completion with respect to the norm of C_0 is C_0. Therefore, for every f in L_p, we can find a sequence $\{f_n\}$ in $\mathcal{S}$ such that

$$\int_{-\infty}^{\infty} |f(t) - f_n(t)|^p \, dt \xrightarrow[n \to \infty]{} 0$$

and for every f in C_0, we can find $\{f_n\}$ in $\mathcal{S}$ such that

$$\sup_t |f_n(t) - f(t)| \xrightarrow[n \to \infty]{} 0$$

For $f \in L_1$, the **Fourier integral** (or Fourier transform)

$$\hat{f}(\nu) = \int_{-\infty}^{\infty} e^{-i2\pi \nu t} f(t) \, dt \tag{5.7}$$

is an absolutely convergent integral for each $\nu \in (-\infty, \infty)$ and defines a function $\hat{f}$ in C_0. If $f \in L_1$ and is of bounded variation in a neighborhood of t, then the **inversion formula** is given by

$$\frac{f(t - 0) + f(t + 0)}{2} = \lim_{N \to \infty} \int_{-N}^{N} e^{i2\pi \nu t} \hat{f}(\nu) \, d\nu \tag{5.8}$$

where $f(t - 0)$ and $f(t + 0)$ denote, respectively, the limit from the left and the limit from the right of f at t. The right-hand side of (5.8) is often written simply as $\int_{-\infty}^{\infty} e^{i2\pi \nu t} \hat{f}(\nu) \, d\nu$, although the integral is not absolutely convergent unless $\hat{f}$ is also in L_1.

Roughly speaking, if $f \in L_1$, L_2, or $\mathcal{S}$, we have a pair of equations relating a function f and its Fourier transform $\hat{f}$:

$$\begin{aligned} \hat{f}(\nu) &= \int_{-\infty}^{\infty} e^{-i2\pi \nu t} f(t) \, dt \\ f(t) &= \int_{-\infty}^{\infty} e^{i2\pi \nu t} \hat{f}(\nu) \, d\nu \end{aligned} \tag{5.9}$$

These two equations are nearly identical, the only difference being the terms $e^{\pm i2\pi \nu t}$ in the integrals. By convention, $\hat{f}$ is called the Fourier transform of f, and f is called the **inverse Fourier transform** of $\hat{f}$. Depending on f, one or both of the integrals in (5.9) may have to be defined as the limit in some sense of a sequence of finite integrals, and the equality may only hold in a restricted sense. For example, if $f \in L_1$ and is of bounded variation, then the first integral is absolutely convergent, but the second equation, strictly speaking, should be replaced by (5.8). If $f \in \mathcal{S}$, then $\hat{f}$ is also in $\mathcal{S}$. In this case both integrals are absolutely convergent, and equality holds for every ν and t in $(-\infty, \infty)$. If $f \in L_2$,

5. WIDE-SENSE STATIONARY PROCESSES

then the first equation really says that there exists $\hat{f} \in L_2$ such that

$$\int_{-\infty}^{\infty} \left| \hat{f}(\nu) - \int_{-T_1}^{T_2} f(t) e^{-i2\pi\nu t} \, dt \right| d\nu \xrightarrow[T_1, T_2 \to \infty]{} 0$$

The second equation is also interpreted in a similar way. When we use (5.9), it is understood that these equations are subject to such qualifications.

The convolution product between two functions f and h, when it exists, is defined by[1]

$$(f * h)(t) = \int_{-\infty}^{\infty} f(t - s) h(s) \, ds \tag{5.10}$$

We note that if f and h are in S, then $f * h$ is also in S. The convolution product is symmetric in f and h. This is easily seen by a change in the variable of integration on the right-hand side of (5.10). If we denote Fourier transformation by $\wedge$, then a most important property of convolution is given by the relationship

$$\widehat{f * h} = \hat{f}\hat{h} \tag{5.11}$$

Equations (5.10) and (5.11) are well-known representations of linear time-invariant filtering operations. Roughly speaking, a linear filter is a linear mapping of some input function space V_i into some output function space V_0. Suppose we define time shift T_α by

$$(T_\alpha f)(t) = f(t + \alpha) \tag{5.12}$$

A linear filter A is said to be time invariant if for every $\alpha \in (-\infty, \infty)$,

$$AT_\alpha f = T_\alpha A f$$

If the input space V_i is L_1, and a filter A is defined by $Af = h * f$, then A is linear and time invariant. The function h is known as the impulse response of the filter A. The Fourier transform $\hat{h}$ of the impulse response is known as the transfer function. More generally, suppose a filter A is defined by

$$(Af)(t) = \int_{-\infty}^{\infty} e^{-i2\pi\nu t} \hat{h}(\nu) f(\nu) \, d\nu \tag{5.13}$$

Then, A is again time invariant and linear. The function $\hat{h}$ is again called a transfer function. In general, $\hat{h}(\nu)$ may not be the Fourier integral of any impulse response. To put it in another way, the inverse Fourier transform of $\hat{h}$ may not exist except as a generalized function. We have been deliberately vague in specifying the input space V_i and the output space V_0, because they depend very much on the filter. For example,

---

[1] If $f, h \in L_1$, then $f * h \in L_1$. If $f, h \in L_2$, then $f * h$ is bounded. If $f \in L_1$ and h is bounded, then $f * h$ is again bounded.

$\hat{h}(\nu) = 2\pi i \nu$ represents the transfer function of a differentiator. Clearly, the input to a differentiator must be differentiable. In any event, our main interest here is not the filtering of known t functions, but stochastic processes.

Let $\{X_t, -\infty < t < \infty\}$ be a q.m. continuous and wide-sense stationary process with covariance function $R(\tau)$. Assume $R \in L_1$, then clearly R also belongs to $L_1 \cap C_0$ because of the q.m. continuity of X_t. Let S be the Fourier transform of R defined by

$$S(\nu) = \int_{-\infty}^{\infty} e^{-i2\pi\nu t} R(\tau) \, d\tau \qquad -\infty < \nu < \infty \qquad (5.14)$$

The function S is nonnegative because of the nonnegative definiteness of R (to be proved), and if $R \in L_1 \cap C_0$ as is assumed, then $S \in L_1 \cap C_0$, also. The inversion integral gives us

$$R(\tau) = \int_{-\infty}^{\infty} e^{i2\pi\nu t} S(\nu) \, d\nu \qquad (5.15)$$

The function S is called the **spectral-density function** for the process $\{X_t, -\infty < t < \infty\}$, and has the interpretation of being the density of average power distribution per unit frequency. To make this notion precise, we need to introduce the concept of linear time-invariant filtering of wide-sense stationary processes.

Imagine a time-invariant linear filter characterized by an impulse response function h, and suppose that the input is a sample function $X_t(\omega)$, $-\infty < t < \infty$, of a wide-sense stationary process. It is natural to view the output $Y_t(\omega)$, $-\infty < t < \infty$, as a sample function of another second-order process and write

$$Y_t(\omega) = \int_{-\infty}^{\infty} h(t - \tau) X_\tau(\omega) \, d\tau \qquad (5.16)$$

The difficulty is that if the convolution integral is to be viewed as an absolutely convergent Lebesgue integral involving a sample function of the X process, then we need to impose conditions on the X process which are not in the province of second-order properties. Mathematically, it is much neater to regard the integral in (5.16) as a q.m. integral as defined in Sec. 3. From Proposition 3.1 we have that for the integral in (5.16) to exist as a q.m. integral, it is necessary and sufficient that

$$\iint_{-\infty}^{\infty} h(t - \tau) \bar{h}(t - \sigma) R(\tau - \sigma) \, d\tau \, d\sigma$$

exists as a Riemann integral. If $R \in L_1 \cap C_0$ as was assumed, and h

5. WIDE-SENSE STATIONARY PROCESSES

is square integrable, then

$$\iint_{-\infty}^{\infty} h(t-\tau)h(t-\sigma)R(\tau-\sigma)\,d\tau\,d\sigma$$
$$= \int_{-\infty}^{\infty} |\hat{h}(\nu)|^2 S(\nu)\,d\nu < \infty \quad (5.17)$$

and the existence of (5.16) as a q.m. integral is ensured. We should not lose sight of the fact that to treat (5.16) as a q.m. integral is a mathematical convenience. In applications, since we never observe more than one sample function at a time, filtering operations should be interpreted as being performed on a sample function. Thus, in practice, we really want (5.16) to be interpreted both as a q.m. integral, which is the limit of a q.m. convergent sequence of random variables, and as a Lebesgue integral for almost all sample functions. In this chapter, we focus our attention only on the former interpretation (cf. Sec. 2.2).

With $\{Y_t, -\infty < t < \infty\}$ defined by (5.16) with h square integrable, then

$$EY_t \bar{Y}_s = \iint_{-\infty}^{\infty} h(t-\tau)\bar{h}(s-\tau)R(\tau-\sigma)\,d\tau\,d\sigma$$
$$= \int_{-\infty}^{\infty} e^{i2\pi\nu(t-s)} |\hat{h}(\nu)|^2 S(\nu)\,d\nu \quad (5.18)$$

which means that (1) $\{Y_t, -\infty < t < \infty\}$ is again wide-sense stationary, and (2) the spectral-density function of $\{Y_t, -\infty < t < \infty\}$ is given by

$$S_Y(\nu) = |\hat{h}(\nu)|^2 S(\nu) \quad (5.19)$$

If we take $\hat{h}(\nu)$ to be the indicator function for $[\nu_1, \nu_2]$, then from the usual interpretation of filtering, Y process is just those components of X process lying in the frequency range $\nu_1 \leq \nu \leq \nu_2$. Since

$$E|Y_t|^2 = \int_{-\infty}^{\infty} |\hat{h}(\nu)|^2 S(\nu)\,d\nu$$
$$= \int_{\nu_1}^{\nu_2} S(\nu)\,d\nu \quad (5.20)$$

it follows that $\int_{\nu_1}^{\nu_2} S(\nu)\,d\nu$ is just the average power of the X process in $[\nu_1, \nu_2]$. This justifies our earlier assertion that S is nonnegative and that it has the interpretation of average power per unit frequency.

Equation (5.15) is a representation of the covariance function as a positive linear combination of sinusoids. This idea can be generalized to situations where $R \notin L_1$ and the spectral-density function may not exist. Intuitively, the situation is quite simple. When $R \in L_1 \cap C_0$, the average power is smoothly distributed over the continuum of all frequencies $(-\infty, \infty)$, with no single frequency having a finite amount of power.

In the more general situation, there may be spectral lines, i.e., distinct frequencies with finite amount of power. Even more complicated situations involving continuous, but not absolutely continuous, distributions may arise. The general statement concerning the harmonic representation of a stationary covariance function is given by the **Bochner's theorem** stated below.

Proposition 5.1. A function $R(\tau)$, $-\infty < \tau < \infty$, is the covariance function of a q.m. continuous and wide-sense stationary process if and only if it is of the form

$$R(\tau) = \int_{-\infty}^{\infty} e^{i2\pi\nu\tau} F(d\nu) \tag{5.21}$$

where F is a finite Borel measure on the real line $(-\infty, \infty)$ and is called the **spectral measure.**

Remarks:

(a) We note that $[1/R(0)]F$ is a probability measure defined on the σ algebra of Borel sets of the real line.

(b) The integral in (5.21) is defined exactly as the expectation was defined in (Sec. 1.5).

(c) Alternatively, if we define a **spectral-distribution function** ϕ by

$$\phi(\nu) = F((-\infty, \nu)) \tag{5.22}$$

then ϕ is a bounded nondecreasing function, and the integral in (5.21) can be replaced by a Stieltjes integral

$$\int_{-\infty}^{\infty} e^{i2\pi\nu\tau} \, d\phi(\nu)$$

(d) Finally, we note that if ϕ is absolutely continuous (that is, F is an absolutely continuous measure with respect to Lebesgue measure), then there exists a nonnegative function S such that

$$\phi(\lambda) = \int_{-\infty}^{\lambda} S(\nu) \, d\nu \tag{5.23}$$

Naturally, $S(\nu)$ is called the **spectral-density function,** agreeing with and generalizing our earlier definition.

Proof: First, if F is a finite Borel measure, then

$$\int_{-\infty}^{\infty} e^{i2\pi\nu(t-s)} F(d\nu) \tag{5.24}$$

is nonnegative definite function, because it is the limit of a sequence of nonnegative definite functions of the form $\sum_{n=0}^{N} \sigma_n(t)\bar{\sigma}_n(s)$. It is continuous

5. WIDE-SENSE STATIONARY PROCESSES

at $t = s$, because

$$\left| \int_{-\infty}^{\infty} (e^{i2\pi\nu\tau} - 1) F(d\nu) \right| \xrightarrow[\tau \to 0]{} 0$$

Hence, it is continuous everywhere on $(-\infty, \infty) \times (-\infty, \infty)$. Therefore, (5.22) defines the covariance function of some q.m. continuous wide-sense stationary process. Thus, the first half of Proposition 5.1 is proved.

Next, suppose $R(\tau)$ is the covariance function of some q.m. continuous and wide-sense stationary process. That is, $R(t - s)$ is a continuous nonnegative definite function on $(-\infty, \infty) \times (-\infty, \infty)$. Now, define a sequence of continuous covariance functions $R_n(\tau)$ by

$$R_n(\tau) = \begin{cases} \left(1 - \dfrac{|\tau|}{2n}\right) R(\tau) & |\tau| \leq 2n \\ 0 & |\tau| > 2n \end{cases} \quad (5.25)$$

The fact that $R_n(\tau)$ is a nonnegative definite function follows from the fact that $\max(0, 1 - |\tau|/2n)$ is a nonnegative definite function. Now, clearly $R_n \in L_1 \cap C_0$. Therefore, there corresponds a sequence of spectral-density functions $\{S_n\}$ defined by

$$\begin{aligned} S_n(\nu) &= \int_{-\infty}^{\infty} R_n(\tau) e^{-i2\pi\nu\tau} \, d\tau \\ &= \int_{-2n}^{2n} \left(1 - \frac{|\tau|}{2n}\right) R(\tau) e^{-i2\pi\nu\tau} \, d\tau \end{aligned} \quad (5.26)$$

Now, let f and $\hat{f}$ be functions in $\mathcal{S}$ which are Fourier transform pairs,

$$\hat{f}(x) = \int_{-\infty}^{\infty} e^{-i2\pi xy} f(y) \, dy$$
$$f(y) = \int_{-\infty}^{\infty} e^{i2\pi xy} \hat{f}(x) \, dx$$

Define a linear functional ρ on $\mathcal{S}$ by

$$\rho(f) = \int_{-\infty}^{\infty} R(\tau) \hat{f}(\tau) \, d\tau \quad (5.27)$$

Using the dominated-convergence theorem, we have

$$\begin{aligned} |\rho(f)| &= \lim_{n \to \infty} \left| \int_{-\infty}^{\infty} R_n(\tau) \hat{f}(\tau) \, d\tau \right| \\ &= \lim_{n \to \infty} \left| \int_{-\infty}^{\infty} S_n(\nu) f(\nu) \, d\nu \right| \\ &\leq \sup_{\nu} |f(\nu)| \lim_{n \to \infty} \int_{-\infty}^{\infty} S_n(\nu) \, d\nu \\ &= \sup_{\nu} |f(\nu)| R(0) \end{aligned}$$

This means that ρ is a linear functional defined on a set $\mathcal{S}$ dense in C_0 and bounded with respect to the supremum norm. Thus, ρ can be extended to a bounded-linear functional on C_0, and the Riesz representation theorem [see, e.g., Rudin, 1966, p. 131] states that there exists a bounded σ additive set function F defined on the Borel sets such that

$$\rho(f) = \int_{-\infty}^{\infty} f(\nu) F(d\nu) \tag{5.28}$$

Because R is nonnegative definite, F is nonnegative and is thus a finite Borel measure. Comparing (5.27) and (5.28), we conclude that

$$\int_{-\infty}^{\infty} R(\tau) \hat{f}(\tau) \, d\tau = \int_{-\infty}^{\infty} f(\nu) F(d\nu) \qquad f \in \mathcal{S} \tag{5.29}$$

Now, for a fixed t define

$$f_n(\nu) = \exp\left[-\frac{1}{2}\frac{(2\pi\nu)^2}{n}\right] e^{i2\pi\nu t} \tag{5.30a}$$

then

$$\hat{f}_n(\tau) = \sqrt{\frac{n}{2\pi}} e^{-\frac{1}{2}n(t-\tau)^2} \tag{5.30b}$$

It is easy to verify that $f_n, \hat{f}_n$ are in $\mathcal{S}$. If R is continuous, then

$$\int_{-\infty}^{\infty} R(\tau) \hat{f}_n(\tau) \, d\tau \xrightarrow[n\to\infty]{} R(t)$$

and

$$\int_{-\infty}^{\infty} f_n(\nu) F(d\nu) \xrightarrow[n\to\infty]{} \int_{-\infty}^{\infty} e^{i2\pi\nu t} F(d\nu)$$

which proves the second half of Proposition 5.1. ∎

Proposition 5.1 gives an expression of a covariance function in terms of the corresponding spectral measure (alternatively, the spectral-distribution function), but gives no inversion formula expressing the spectral measure F in terms of the covariance function R. Actually, the proof already contains the heart of both the direct and the inversion formulas in (5.29). We need only to choose an approximating sequence of f's or $\hat{f}$'s. For example, suppose $b > a$ are two continuity points of F, that is, a and b are such that

$$F([a,b]) = F((a,b)) = F([a,b)) = F((a,b])$$

Denote the common value of these four quantities by F_{ab}, then F_{ab} can be found from (5.29) by approximating the indicator function of the interval a to b by a sequence in $\mathcal{S}$. Specifically, let

$$\begin{aligned} f_n(\nu) &= \sqrt{\frac{n}{2\pi}} \int_a^b e^{-\frac{1}{2}n(\nu-\nu')^2} \, d\nu' \\ \hat{f}_n(\tau) &= \exp\left[-\frac{1}{2}\frac{(2\pi\tau)^2}{n}\right] \frac{e^{-i2\pi\tau b} - e^{-i2\pi\tau a}}{-2\pi i\tau} \end{aligned} \tag{5.31}$$

6. SPECTRAL REPRESENTATION

Then, from (5.29) we have

$$F_{ab} = \lim_{n \to \infty} \int_{-\infty}^{\infty} f_n(\nu) F(d\nu)$$
$$= \lim_{n \to \infty} \int_{-\infty}^{\infty} \hat{f}_n(\tau) R(\tau) \, d\tau$$

Thus, we have in final form the inversion formula

$$F_{ab} = \lim_{n \to \infty} \int_{-\infty}^{\infty} R(\tau) \frac{e^{-i2\pi\tau b} - e^{-i2\pi\tau a}}{-2\pi i \tau} \exp\left[-\frac{1}{2} \frac{(2\pi\tau)^2}{n}\right] d\tau \qquad (5.32)$$

Equation (5.32) holds only at continuity points of F. However, because $R(0) = F((-\infty, \infty))$ is finite, the number of discontinuities of F which have jumps greater than $1/n$ is, at most, $nR(0)$. Hence, the discontinuity points (mass points) of F are, at most, countable. Thus, the continuity points of F are always dense in $(-\infty, \infty)$. It follows that F is completely determined by (5.32).

We can now repeat the arguments relating to linear time-invariant filtering operations on a wide-sense stationary process. Instead of (5.18), we have for the more general case

$$EY_t \bar{Y}_s = \int_{-\infty}^{\infty} e^{i2\pi\nu(t-s)} |\hat{h}(\nu)|^2 F(d\nu)$$

where Y_t is the output of a linear time-invariant filter and F is the spectral measure of the input. We postpone a detailed discusson of this point until later. First, we need to develop a spectral representation, not just for the covariance function as is given by Proposition 5.1, but for the process itself.

6. SPECTRAL REPRESENTATION

Let $\{X_t, -\infty < t < \infty\}$ be a q.m. continuous and wide-sense stationary process. If X_t had a Fourier integral, then the inversion formula would immediately give a representation of X_t as a linear combination of sinusoids. However, X_t does not have a Fourier integral. That is, $\int_{-\infty}^{\infty} e^{-i2\pi\nu t} X_t \, dt$ fails to exist as a q.m. integral or any other kind of integral. To obtain a spectral representation, we introduce a kind of integrated Fourier integral $\{\hat{X}_\lambda, -\infty < \lambda < \infty\}$ as follows:

$$\lim_{a \to -\infty} \text{in q.m. } \hat{X}_a = 0 \qquad (6.1a)$$

If a and b are continuity points of the spectral measure F, we set

$$\hat{X}_b - \hat{X}_a = \int_{-\infty}^{\infty} \left(\int_a^b e^{-i2\pi\nu t} \, d\nu\right) X_t \, dt \qquad (6.1b)$$

At discontinuity points of F, $\hat{X}_\lambda$ is defined so that $\{\hat{X}_\lambda, -\infty < \lambda < \infty\}$ is q.m. continuous from the left (6.1c)

Proposition 6.1. Defined as above, the process $\{\hat{X}_\lambda, -\infty < \lambda < \infty\}$ satisfies

$$E(\hat{X}_b - \hat{X}_a)\overline{(\hat{X}_d - \hat{X}_c)} = F([a,b) \cap [c,d)) \tag{6.2}$$

so that $\{\hat{X}_\lambda, -\infty < \lambda < \infty\}$ is a **process with orthogonal increments**.

Remark: Denoting $d\hat{X}_\lambda = \hat{X}_{\lambda+d\lambda} - \hat{X}_\lambda$ and $F(d\lambda) = F([\lambda, \lambda + d\lambda))$, we can write (6.2) in a differential form as

$$E\,d\hat{X}_\lambda\,\overline{d\hat{X}_\mu} = \delta_{\lambda\mu} F(d\lambda) \qquad \delta_{\lambda\mu} = \begin{cases} 1, \lambda = \mu \\ 0, \lambda \neq \mu \end{cases} \tag{6.3}$$

The process $\{\hat{X}_\lambda, -\infty < \lambda < \infty\}$ will be called the **spectral process** of $\{X_t, -\infty < t < \infty\}$. We shall show that

$$X_t = \int_{-\infty}^{\infty} e^{i2\pi\lambda t}\,d\hat{X}_\lambda$$

which is the spectral-representation formula that we are after. However, first we have to define integrals such as $\int_{-\infty}^{\infty} f(\lambda)\,d\hat{X}_\lambda$.

Definition. Let $\{\hat{X}_\lambda, -\infty < \lambda < \infty\}$ be a process with orthogonal increments which is q.m. continuous from the left. Let

$E\,d\hat{X}_\lambda\,\overline{d\hat{X}_\mu} = \delta_{\lambda\mu} F(d\lambda)$.

(a) If $f = I_{ab}$, the indicator function of $[a,b)$, we set

$$\int_{-\infty}^{\infty} f(\lambda)\,d\hat{X}_\lambda = \hat{X}_b - \hat{X}_a \tag{6.4}$$

(b) If $f = \sum_{\nu=1}^{n} a_\nu f_\nu$, we require

$$\int_{-\infty}^{\infty} f(\lambda)\,d\hat{X}_\lambda = \sum_{\nu=1}^{n} a_\nu \int_{-\infty}^{\infty} f_\nu(\lambda)\,d\hat{X}_\lambda \tag{6.5}$$

(c) If $\int_{-\infty}^{\infty} |f_n(\lambda) - f(\lambda)|^2 F(d\lambda) \xrightarrow[n \to \infty]{} 0$, we require

$$\int_{-\infty}^{\infty} f(\lambda)\,d\hat{X}_\lambda = \lim_{n \to \infty} \text{in q.m.} \int_{-\infty}^{\infty} f_n(\lambda)\,d\hat{X}_\lambda \tag{6.6}$$

It is clear that $\int_{-\infty}^{\infty} f(\lambda)\,d\hat{X}_\lambda$ is well defined for every f for which there exists a sequence of step functions $\{f_n\}$ (linear combinations of indicator

6. SPECTRAL REPRESENTATION

functions) such that

$$\int_{-\infty}^{\infty} |f_n(\lambda) - f(\lambda)|^2 F(d\lambda) \xrightarrow[n \to \infty]{} 0 \tag{6.7}$$

Although we won't prove it, the class of all such f is precisely $L_2(F)$, that is, the class of functions satisfying $\int_{-\infty}^{\infty} |f(\lambda)|^2 F(d\lambda) < \infty$. For a continuous function f in $L_2(F)$, a suitable approximating sequence of step functions can be constructed by sampling f at continuity points of F. For an arbitrary f in $L_2(F)$, the construction of an approximating sequence is somewhat more complicated [Doob, 1953, pp. 439–441].

Proposition 6.2. Let $\{\hat{X}_\lambda, -\infty < \lambda < \infty\}$ be a process with orthogonal increments assumed to be q.m. continuous from the left. Let F be a finite Borel measure so that $E\hat{X}_\lambda \overline{\hat{X}}_\mu = \delta_{\lambda\mu} F(d\lambda)$. Then,

$$E\left[\int_{-\infty}^{\infty} f(\lambda) \, d\hat{X}_\lambda\right] \overline{\left[\int_{-\infty}^{\infty} g(\lambda) \, d\hat{X}_\lambda\right]}$$
$$= \int_{-\infty}^{\infty} f(\lambda) \, \bar{g}(\lambda) F(d\lambda) \qquad f, g \in L_2(F) \tag{6.8}$$

Proof: Equation (6.8) is certainly true if f and g are step functions. Let $\{f_n\}$ and $\{g_n\}$ be approximating sequences of step functions for f and g, respectively, and write $A(f) = \int_{-\infty}^{\infty} f(\lambda) \, d\hat{X}_\lambda$. Then,

$$EA(f)\overline{A(g)} = E[A(f - f_n) + A(f_n)]\overline{[A(g - g_n) + A(g_n)]}$$
$$= \int_{-\infty}^{\infty} f_n(\lambda)\overline{g_n(\lambda)} F(d\lambda) + EA(f - f_n)\overline{A(g - g_n)}$$
$$+ EA(f_n)\overline{A(g - g_n)} + EA(f - f_n)\overline{A(g_n)}$$

Therefore,

$$\left| EA(f)\overline{A(g)} - \int_{-\infty}^{\infty} f_n(\lambda)\overline{g_n}(\lambda) F(d\lambda) \right|$$
$$\leq \sqrt{E|A(f - f_n)|^2} \sqrt{E|A(g - g_n)|^2} + \sqrt{E|A(f_n)|^2} \sqrt{E|A(g - g_n)|^2}$$
$$+ \sqrt{E|A(f - f_n)|^2} \sqrt{E|A(g_n)|^2}$$

and (6.8) follows by letting $n \to \infty$. ∎

Proposition 6.3. Let $\{X_t, -\infty < t < \infty\}$ be a q.m. continuous and wide-sense stationary process, and let $\{\hat{X}_\lambda, -\infty < \lambda < \infty\}$ be its spectral process as defined by (6.1). Then

$$\int_{-\infty}^{\infty} f(\lambda) \, d\hat{X}_\lambda = \int_{-\infty}^{\infty} \hat{f}(t) X_t \, dt \qquad f \in \mathcal{S} \tag{6.9}$$

where $\hat{f}$ denotes the Fourier transform of f and $\mathcal{S}$, as usual, denotes the space of infinitely differentiable functions of rapid descent.

Proof: The definition of $\hat{X}_\lambda$ as given by (6.1) yields

$$\int_{-\infty}^{\infty} I_{ab}(\lambda)\, d\hat{X}_\lambda = \int_{-\infty}^{\infty} \hat{I}_{ab}(t) X_t\, dt$$

where I_{ab} denotes the indicator function of $[a,b)$. Let $\{f_n\}$ be a sequence of step functions obtained by sampling f at continuity points of the spectral measure F. As the sampling points become dense in $(-\infty, \infty)$, f_n converges to f uniformly and in L_1 metric. Therefore, $\{\hat{f}_n\}$ converges to $\hat{f}$ uniformly. Now, for each n,

$$\int_{-\infty}^{\infty} f_n(\lambda)\, d\hat{X}_\lambda = \int_{-\infty}^{\infty} \hat{f}_n(t) X_t\, dt$$

Therefore, if $f \in \mathcal{S}$,

$$E\left| \int_{-\infty}^{\infty} f(\lambda)\, d\hat{X}_\lambda - \int_{-\infty}^{\infty} \hat{f}(t) X_t\, dt \right|^2 \leq 2E\left| \int_{-\infty}^{\infty} [f(\lambda) - f_n(\lambda)]\, d\hat{X}_\lambda \right|^2$$
$$+ 2E\left| \int_{-\infty}^{\infty} [\hat{f}(t) - \hat{f}_n(t)] X_t\, dt \right|^2 = 2\int_{-\infty}^{\infty} |f(\lambda) - f_n(\lambda)|^2 F(d\lambda)$$
$$+ 2\iint_{-\infty}^{\infty} R(t-s)[\hat{f}(t) - \hat{f}_n(t)]\overline{[\hat{f}(s) - \hat{f}_n(s)]}\, dt\, ds$$

The first of these integrals goes to zero as $n \to \infty$ by virtue of the construction of f_n. The second integral also goes to zero, because

$$\int_{T}^{T} [\hat{f}(t) - \hat{f}_n(t)]e^{i2\pi\nu t}\, dt \xrightarrow[T\to\infty]{} f(\nu) - f_n(\nu)$$

and the convergence is bounded. Hence, from (5.21), we have

$$\lim_{T\to\infty} \iint_{-T}^{T} R(t-s)[\hat{f}(t) - \hat{f}_n(t)]\overline{[\hat{f}(s) - \hat{f}_n(s)]}\, dt\, ds$$
$$= \int_{-\infty}^{\infty} |f(\nu) - f_n(\nu)|^2 F(d\nu) \xrightarrow[n\to\infty]{} 0 \quad \blacksquare$$

Corollary. Under the hypotheses of Proposition 6.3,

$$\int_{-\infty}^{\infty} f(\lambda)\bar{g}(\lambda)\, F(d\lambda) = \iint_{-\infty}^{\infty} \hat{f}(t)\bar{\hat{g}}(s) R(t-s)\, dt\, ds \qquad f, g \in \mathcal{S} \quad (6.10)$$

Remark: Note the similarity between (6.10) and (5.29). Indeed, (5.29) can be obtained from (6.10) by using

$$g_n(\lambda) = \exp\left(-\frac{1}{2}\frac{\lambda^2}{n}\right)$$

$$\hat{g}_n(s) = \sqrt{\frac{n}{2\pi}} \exp\left(-\frac{n}{2} s^2\right)$$

and letting $n \to \infty$.

6. SPECTRAL REPRESENTATION

Proposition 6.4 (Spectral Representation). A q.m. continuous process $\{X_t, -\infty < t < \infty\}$ is wide-sense stationary if and only if there exists a process with orthogonal increments $\{\hat{X}_\lambda, -\infty < \lambda < \infty\}$ such that $E|\hat{X}_\lambda|^2$ is bounded and

$$X_t = \int_{-\infty}^{\infty} e^{i2\pi\lambda t} \, d\hat{X}_\lambda \tag{6.11}$$

Proof: If $\{\hat{X}_\lambda, -\infty < \lambda < \infty\}$ is as described, we can always assume that it is continuous from the left. Then,

$$F([a,b)) = E|\hat{X}_b - \hat{X}_a|^2$$

defines a finite Borel measure, and $E \, d\hat{X}_\lambda \, \overline{d\hat{X}_\mu} = \delta_{\lambda\mu} F(d\lambda)$. Since

$$E \left(\int_{-\infty}^{\infty} e^{i2\pi\lambda t} \, dX_\lambda \right) \overline{\left(\int_{-\infty}^{\infty} e^{i2\pi\nu s} \, d\hat{X}_\nu \right)} = \int_{-\infty}^{\infty} e^{i2\pi\lambda(t-s)} F(d\lambda)$$

the process defined by $\int_{-\infty}^{\infty} e^{i2\pi\lambda t} \, d\hat{X}_\lambda$ must be q.m. continuous and wide-sense stationary by virtue of Bochner's theorem (Proposition 5.1).

Conversely, let $\{X_t, -\infty < t < \infty\}$ be q.m. continuous and wide-sense stationary and define $\hat{X}_\lambda$ by (6.1). Then, (6.11) follows from (6.9) by using the familiar approximations

$$f_n(\lambda) = e^{i2\pi\lambda t_0} \exp\left(\frac{1}{2} \frac{\lambda^2}{n}\right)$$

$$\hat{f}_n(t) = \sqrt{\frac{n}{2\pi}} \exp\left[-\frac{1}{2} n(t-t_0)^2\right]$$

and letting $n \to \infty$. ∎

Equation (6.11) finally provides us with a representation of wide-sense stationary processes as a linear combination of sinusoids. The spectral process $\{\hat{X}_\lambda, -\infty < \lambda < \infty\}$ also provides a complete characterization of the Hilbert space $\mathcal{H}_X$ generated by the process $\{X_t, -\infty < t < \infty\}$, and a complete characterization of every process that can be said to be derived by a time-invariant linear operation on $\{X_t, -\infty < t < \infty\}$.

Proposition 6.5. Let $\{X_t, -\infty < t < \infty\}$ be a q.m. continuous and wide-sense stationary process with spectral process $\{\hat{X}_\lambda, -\infty < \lambda < \infty\}$ and spectral measure F. Let $\mathcal{H}_X$ be the Hilbert space generated by $\{X_t, -\infty < t < \infty\}$. Then, a random variable Y belongs to $\mathcal{H}_X$ if and only if there exists $\eta \in L_2(F)$ such that

$$Y = \int_{-\infty}^{\infty} \eta(\lambda) \, d\hat{X}_\lambda \tag{6.12}$$

Proof: The proof that every Y of the form of (6.12) is in $\mathcal{H}_X$ is elementary. We shall only consider the other half. First, suppose Y is of the form $\sum_{\nu=1}^{n} a_\nu X_{t_\nu}$; then from (6.11), we have

$$Y = \int_{-\infty}^{\infty} \left(\sum_{\nu=1}^{n} a_\nu e^{i2\pi\lambda t_\nu} \right) d\hat{X}_\lambda$$

hence (6.12) holds. In general, $Y \in \mathcal{H}_X$ is the q.m. limit of a sequence $\{Y_n\}$, each of which is a finite sum. Hence

$$Y = \lim_{n \to \infty} \text{in q.m.} \quad Y_n = \lim_{n \to \infty} \text{in q.m.} \quad \int_{-\infty}^{\infty} \eta_n(\lambda) \, d\hat{X}_\lambda$$

Since $\{Y_n\}$ is q.m. convergent, it is also mutually q.m. convergent, and from (6.8) we have

$$E|Y_m - Y_n|^2 = \int_{-\infty}^{\infty} |\eta_m(\lambda) - \eta_n(\lambda)|^2 F(d\lambda) \xrightarrow[m,n \to \infty]{} 0$$

This means that $\{\eta_n\}$ is a Cauchy sequence in $L_2(F)$, and the completeness of $L_2(F)$ implies the existence of $\eta \in L_2(F)$ such that

$$\int_{-\infty}^{\infty} |\eta_n(\lambda) - \eta(\lambda)|^2 F(d\lambda) \xrightarrow[n \to \infty]{}$$

But this is equivalent to saying that $\int_{-\infty}^{\infty} \eta_n(\lambda) \, d\hat{X}_\lambda$ converges in q.m. to $\int_{-\infty}^{\infty} \eta(\lambda) \, d\hat{X}_\lambda$. The proof is complete. ∎

Proposition 6.5 provides an explicit representation for elements of $\mathcal{H}_X$ in terms of functions in $L_2(F)$. Indeed, it provides a one-to-one mapping of $L_2(F)$ onto $\mathcal{H}_X$ which preserves the inner products, that is,

$$EY_1 \bar{Y}_2 = \int_{-\infty}^{\infty} \eta_1(\lambda) \bar{\eta}_2(\lambda) F(d\lambda) \tag{6.13}$$

That is, (6.12) is an isometry between $\mathcal{H}_X$ and $L_2(F)$. Equation (6.12) is extremely useful in representing an arbitrary linear operation on a wide-sense stationary process.

Proposition 6.6. Let $\{X_t, -\infty < t < \infty\}$ be as in Proposition 6.5. Let $\{U_t, -\infty < t < \infty\}$ be the corresponding translation group of unitary operators as defined by (5.3). Suppose $\{Y_t, -\infty < t < \infty\}$ is a process such that for each t, $Y_t \in \mathcal{H}_X$. Then, the following conditions are equivalent:

(a) There exists $\eta \in L_2(F)$ such that

$$Y_t = \int_{-\infty}^{\infty} e^{i2\pi\lambda t} \eta(\lambda) \, d\hat{X}_\lambda \qquad t \in (-\infty, \infty) \tag{6.14}$$

6. SPECTRAL REPRESENTATION

(b) For each $t \in (-\infty, \infty)$,

$$Y_t = U_t Y_0 \tag{6.15}$$

(c) For arbitrary t and s in $(-\infty, \infty)$,

$$EY_t X_s = EY_0 X_{s-t} \tag{6.16}$$

Proof: It is nearly obvious that (a) implies (c). By direct computation we have

$$EY_t \bar{X}_s = \int_{-\infty}^{\infty} e^{i2\pi\lambda(t-s)} \eta(\lambda) F(d\lambda) = EY_0 \bar{X}_{s-t}$$

To prove that (a) implies (b), we make use of the fact that

$$U_t \, d\hat{X}_\lambda = e^{i2\pi\lambda t} \, d\hat{X}_\lambda \tag{6.17}$$

More precisely, this means that

$$E|U_t(\hat{X}_{\lambda+\epsilon} - \hat{X}_\lambda) - e^{i2\pi\lambda t}(\hat{X}_{\lambda+\epsilon} - \hat{X}_\lambda)|^2 \xrightarrow[\epsilon \to 0]{} 0$$

Finally, if (6.14) holds, then we can write

$$U_t Y_0 = \int_{-\infty}^{\infty} \eta(\lambda) U_t \, d\hat{X}_\lambda = \int_{-\infty}^{\infty} e^{i2\pi\lambda t} \eta(\lambda) \, d\hat{X}_\lambda = Y_t$$

This step can be made precise by approximating the stochastic integral in (6.14) by sums.

Next, we prove that each of (b) and (c) implies (a). Suppose (b) is satisfied. Because of Proposition 6.4, we can write

$$Y_0 = \int_{-\infty}^{\infty} \eta(\lambda) \, d\hat{X}_\lambda$$

Thus, (6.14) follows upon using (6.17). To prove that (c) implies (a), we first note that from Proposition 6.4 for each t there exists $\eta(\cdot, t) \in L_2(F)$ such that

$$Y_t = \int_{-\infty}^{\infty} \eta(\lambda, t) \, d\hat{X}_\lambda$$

If (c) is satisfied, then from (6.16),

$$EY_t \bar{X}_s = \int_{-\infty}^{\infty} \eta(\lambda, t) e^{-i2\pi\lambda s} F(d\lambda)$$
$$= EY_0 \bar{X}_{s-t} = \int_{-\infty}^{\infty} \eta(\lambda, 0) e^{i2\pi\lambda(t-s)} F(d\lambda)$$

This means that for arbitrary t, s in $(-\infty, \infty)$, we have

$$\int_{-\infty}^{\infty} e^{-i2\pi\lambda s} [\eta(\lambda, t) - e^{i2\pi\lambda t} \eta(\lambda, 0)] F(d\lambda) = 0$$

If we denote $\tilde{Y}_t = \int_{-\infty}^{\infty} e^{i2\pi\lambda t}\eta(\lambda,0)\, d\hat{X}_\lambda$, then the above equation is equivalent to

$$E(Y_t - \tilde{Y}_t)\bar{X}_s = 0 \qquad t,s \in (-\infty,\infty)$$

Since $(Y_t - \tilde{Y}_t)$, being in $\mathcal{3C}_X$, is the q.m. limit of a sequence of finite sums of the form $\sum_\nu a_\nu X_{t_\nu}$, we have $E|Y_t - \tilde{Y}_t|^2 = 0$. The proof is now complete. ∎

Remark: The process $\{Y_t, -\infty < t < \infty\}$ is necessarily wide-sense stationary. More than that, $\{X_t, Y_t, -\infty < t < \infty\}$ may be said to be **jointly stationary** (wide-sense) in the sense that every linear combination $\alpha X_t + \beta Y_t$ defines a wide-sense stationary process. On the other hand, suppose that for every t in $(-\infty,\infty)$, $Z_t \in \mathcal{3C}_X$ and $\{Z_t, -\infty < t < \infty\}$ is wide-sense stationary. The process $\{Z_t, -\infty < t < \infty\}$ does not necessarily satisfy the conditions (6.14) to (6.16) and is not necessarily jointly stationary with $\{X_t, -\infty < t < \infty\}$ (see Exercise 12).

If $h(t), -\infty < t < \infty$, is a function such that its Fourier transform $\hat{h}$ is in $L_2(F)$, then

$$\begin{aligned} Y_t &= \int_{-\infty}^{\infty} h(t-\tau)X_\tau\, d\tau \\ &= \int_{-\infty}^{\infty} \hat{h}(\lambda)e^{i2\pi\lambda t}\, d\hat{X}_\lambda \end{aligned} \qquad (6.18)$$

Therefore, every process which can be expressed as a convolution of an impulse response and a wide-sense stationary process has a spectral representation of the form (6.14), and η is just the transfer function. However, not every process of the form (6.14) can be represented as a convolution. Hence, (6.14) is a more general representation. We shall call every process of the form (6.14) a process derivable by a linear and time-invariant filtering on the X process. Naturally, η will be referred to as the transfer function of the filter. We note that if $\{Y_t, -\infty < t < \infty\}$ is given by (6.14), then its spectral process $\{\tilde{Y}_\lambda, -\infty < \lambda < \infty\}$ is defined by

$$d\hat{Y}_\lambda = \eta(\lambda)\, d\hat{X}_\lambda \qquad (6.19)$$

The Y process is automatically wide-sense stationary and q.m. continuous with

$$EY_t\bar{Y}_s = \int_{-\infty}^{\infty} e^{i2\pi\lambda(t-s)}|\eta(\lambda)|^2 F(d\lambda) \qquad (6.20)$$

This means that

$$F_Y(d\lambda) = |\eta(\lambda)|^2 F(d\lambda) \qquad (6.21)$$

Equation (6.21) generalizes (5.19). Equation (6.19) should be considered the random-process counterpart of the familiar formula relating the Fourier transforms of the input and output of a linear time-invariant filter.

It may be useful to consider some examples. First, if $\{X_t, -\infty < t < \infty\}$ is a wide-sense stationary process, then its q.m. derivative as defined by (3.5) can be expressed as

$$\dot{X}_t = \int_{-\infty}^{\infty} (i2\pi\lambda)e^{i2\pi\lambda t} \, d\hat{X}_\lambda \tag{6.22}$$

whenever it exists. It is rather obvious from (6.22) that the q.m. derivative exists if and only if $\int_{-\infty}^{\infty} |\lambda|^2 F(d\lambda) < \infty$. As a second example, consider the process $\{Y_t, -\infty < t < \infty\}$ defined by

$$Y_t = \int_{-\infty}^{\infty} (-i \operatorname{sgn} \lambda) e^{i2\pi\lambda t} \, d\hat{X}_\lambda \tag{6.23}$$

where $\operatorname{sgn} \lambda = 1$ or -1 according as $\lambda \geq 0$ or $\lambda < 0$. Thus defined, Y_t is known as the **Hilbert transform** of X_t. The Y process has exactly the same covariance function as the X process. Also, since

$$X_t + iY_t = 2\int_0^{\infty} e^{i2\pi\lambda t} \, d\hat{X}_\lambda$$

we can say that $X_t + iY_t$ has no negative frequencies. Hilbert transforms will be made use of again in the next section.

7. LOWPASS AND BANDPASS PROCESSES

Definition. A wide-sense stationary and q.m. continuous process $\{X_t, -\infty < t < \infty\}$ is said to be **bandlimited** to frequency W if its spectral measure F satisfies

$$F((-\infty, -W]) = 0 = F([W, \infty)) \tag{7.1}$$

In other words, the X process has no average power for frequencies $\leq -W$ and $\geq W$. We note that the points $-W$ and W are required to be continuity points of F. A process bandlimited to frequency W can always be written as

$$X_t = \int_{-W}^{W} e^{i2\pi\nu t} \, d\hat{X}_\nu \tag{7.2}$$

where it makes no difference if the limits are replaced by $W \pm 0$ and $-W \pm 0$. Equation (7.2) results from the fact that

$$E\left|X_t - \int_{-W+0}^{W-0} e^{i2\pi\nu t} \, d\hat{X}_\nu\right|^2 = E\left|\int_{|\nu| \geq W} e^{i2\pi\nu t} \, d\hat{X}_\nu\right|^2$$
$$= F((-\infty, -W]) + F([W, \infty)) = 0$$

The concept of a bandlimited process is an important one in practice, because such a process is specified for all t by its sampled values spaced $1/2W$ apart in a formula which is known as the **sampling theorem**.

Proposition 7.1 (Sampling Theorem). Let $\{X_t, -\infty < t < \infty\}$ be a process bandlimited to frequency W. Then

$$X_t = \lim_{N \to \infty} \text{in q.m.} \sum_{k=-N}^{N} X_{\alpha + k/2W} \frac{\sin 2\pi W(t - \alpha - k/2W)}{2\pi W(t - \alpha - k/2W)} \quad (7.3)$$

for all $t \in (-\infty, \infty)$, where α is an arbitrary constant.

Remark: Proposition 7.1 implies that $\mathcal{H}_X$ has a countable basis, namely $\{X_{\alpha + k/2W}, k = 0, \pm 1, \pm 2, \ldots\}$.

Proof: We begin with the representation (7.2) and for emphasis rewrite it as

$$X_t = \int_{-W+0}^{W-0} e^{i2\pi \nu t} \, d\hat{X} \quad (7.4)$$

For a fixed t, the function $f(\nu) = e^{i2\pi \nu t}$ is square integrable over $(-W, W)$. Since $\{\varphi^k(\nu) = e^{i2\pi \nu k/2W}, k = 0, \pm 1, \pm 2, \ldots\}$ forms an orthonormal basis in $L_2(-W, W)$, we have

$$e^{i2\pi \nu t} = \lim_{N \to \infty} \sum_{k=-N}^{N} e^{i2\pi \nu k/2W} \frac{1}{2W} \int_{-W}^{W} e^{i2\pi \nu(t - k/2W)} \, d\nu$$

where l.i.m. denotes limit in L_2 mean. The Fourier series

$$\sum_{k=-N}^{N} e^{i2\pi \nu k/2W} \frac{1}{2W} \int_{-W}^{W} e^{i2\pi \nu(t - k/2W)} \, d\nu$$

$$= \sum_{k=-N}^{N} e^{i2\pi \nu k/2W} \frac{\sin 2\pi W(t - k/2W)}{2\pi W(t - k/2W)}$$

also converges pointwise to $e^{i2\pi \nu t}$ for $\nu \in (-W, W)$, but not at the endpoints $\nu = \pm W$. At the endpoints, it converges to $\frac{1}{2}(e^{i2\pi W t} + e^{-i2\pi W t}) = \cos 2\pi W t$. Therefore,

$$E \left| X_t - \sum_{k=-N}^{N} X_{k/2W} \frac{\sin 2\pi W(t - k/2W)}{2\pi W(t - k/2W)} \right|^2$$

$$= E \left| \int_{-W+0}^{W-0} \left[e^{i2\pi \nu t} - \sum_{k=-N}^{N} e^{i2\pi \nu k/2W} \frac{\sin 2\pi W(t - k/2W)}{2\pi W(t - k/2W)} \right] d\hat{X}_\lambda \right|^2$$

$$= \int_{-W+0}^{W-0} \left| e^{i2\pi \nu t} - \sum_{k=-N}^{N} e^{i2\pi \nu k/2W} \frac{\sin 2\pi W(t - k/2W)}{2\pi W(t - k/2W)} \right|^2 F(d\lambda) \xrightarrow[N \to \infty]{} 0$$

7. LOWPASS AND BANDPASS PROCESSES

which proves the theorem for $\alpha = 0$. For $\alpha \neq 0$, we note that $Y_t = X_{t+\alpha}$ is also bandlimited to W and (7.3) is merely the expansion for $Y_{t-\alpha}$. ∎

Remark: The proof fails if F merely satisfies $F((-\infty, -W)) = 0$ and $F((W, \infty)) = 0$ but has a finite mass at one or both of the points $\pm W$. Then, we would have to write

$$X_t = \int_{-W-0}^{W+0} e^{i2\pi \nu t} \, d\hat{X}_\nu$$

and would find

$$E \left| X_t - \sum_{k=-N}^{N} X_{k/2W} \frac{\sin 2\pi W(t - k/2W)}{2\pi W(t - k/2W)} \right|^2$$

$$\xrightarrow[N \to \infty]{} \sin^2 2\pi W t (F^+ + F^-) \quad (7.5)$$

where F^+ and F^- are, respectively, the mass of F at $+W$ and $-W$. As an example, suppose $X_t = \cos(2\pi W t + \varphi)$ where φ is a random variable uniformly distributed in $[0, 2\pi)$. Then, $\{X_t, -\infty < t < \infty\}$ is a wide-sense stationary and q.m. continuous process with

$$EX_t X_s = R(t - s) = \tfrac{1}{2} \cos 2\pi W(t - s)$$

Therefore, F is concentrated at $\pm W$, assigning a mass of $\tfrac{1}{4}$ to each point. The sampling theorem obviously fails, because in this case,

$$X_{\alpha + k/2W} = (-1)^k X_\alpha \qquad k = 0, \pm 1, \ldots$$

which means that for every t, the sum

$$\sum_k X_{\alpha + k/2W} \frac{\sin 2\pi W(t - \alpha - k/2W)}{2\pi W(t - \alpha - k/2W)}$$

is proportional to X_α. On the other hand $X_{\alpha + 1/4W}$ is orthogonal to X_α. So, $X_{\alpha + 1/4W}$ obviously cannot be represented by the sampling formula.

In communication problems, we frequently encounter the so-called bandpass processes. Here, we define a **bandpass process** with **center frequency** W_0 and **bandwidth** $2W$ ($W_0 > W$) as q.m. continuous and a wide-sense stationary process with

$$F((-\infty, -W_0 - W]) = F([-W_0 + W, W_0 - W])$$
$$= F([+W_0 + W, \infty)) = 0 \quad (7.6)$$

In other words, the average power is concentrated in the frequency ranges $(-W_0 - W, -W_0 + W)$ and $(W_0 - W, W_0 + W)$. Of course, a bandpass process is bandlimited to frequency $W_0 + W$. Therefore, the sampling theorem can be applied. However, a straightforward application of the sampling theorem requires a sampling rate of $2W_0 + 2W$, which does not reflect the fact that there is no average power in the frequency range $[-W_0 + W, W_0 - W]$. From the point of view of bandwidth, a bandpass has the same bandwidth as a process bandlimited to $2W$. Thus, we should expect a sampling rate of $4W$ rather than $2(W_0 + W)$. In practice W_0 is much greater than W, and there is a substantial difference between these two rates of sampling. It turns out that a bandpass process with bandwidth $2W$ can be represented in terms of the samples of itself and the Hilbert transform, each being sampled at a rate of $2W$ to give a combined rate of $4W$. Let $\{X_t, -\infty < t < \infty\}$ be a bandpass process with center frequency W_0 and bandwidth $2W$. Since $W_0 > W$, $\nu = 0$ is a continuity point of F. We define the Hilbert transform by

$$Y_t = \int_{-\infty}^{\infty} (-i \operatorname{sgn} \nu) e^{i2\pi\nu t} \, d\hat{X}_\nu \tag{7.7}$$

where $\operatorname{sgn} \nu = 1$ or -1 according as $\nu > 0$ or $\nu < 0$. The value of $\operatorname{sgn} 0$ is immaterial, since (7-7) can be rewritten as

$$Y_t = i \int_{-W_0-W}^{-W_0+W} e^{i2\pi\nu t} \, d\hat{X}_\nu - i \int_{+W_0-W}^{W_0+W} e^{i2\pi\nu t} \, d\hat{X}_\nu \tag{7.8}$$

Again, we note that the limits in the integrals in (7.8) can be replaced by their limits from the left or the right without any material effect. The process $\{X_t, -\infty < t < \infty\}$ can be written as

$$X_t = \int_{-W_0-W}^{-W_0+W} e^{i2\pi\nu t} \, d\hat{X}_\nu + \int_{W_0-W}^{W_0+W} e^{i2\pi\nu t} \, d\hat{X}_\nu \tag{7.9}$$

It follows that we can write

$$\begin{aligned} X_t + iY_t &= 2 \int_{W_0-W}^{W_0+W} e^{i2\pi\nu t} \, d\hat{X}_\nu \\ &= 2e^{i2\pi W_0 t} \int_{-W}^{W} e^{i2\pi\nu t} \, d\hat{X}_{\nu+W_0} \\ &= 2e^{i2\pi W_0 t} \xi_t \end{aligned} \tag{7.10}$$

$$\begin{aligned} X_t - iY_t &= 2e^{-i2\pi W_0 t} \int_{-W}^{W} e^{i2\pi\nu t} \, d\hat{X}_{\nu-W_0} \\ &= 2e^{-i2\pi W_0 t} \eta_t \end{aligned} \tag{7.11}$$

It is clear that both $\{\xi_t, -\infty < t < \infty\}$ and $\{\eta_t, -\infty < t < \infty\}$ are bandlimited to W. Hence, by applying (7.3) to ξ_t and η_t, we get

$$\begin{aligned} X_t &= e^{-i2\pi W_0 t} \eta_t + e^{i2\pi W_0 t} \xi_t \\ &= \sum_{n=-\infty}^{\infty} (\eta_{\alpha+k/2W} e^{-i2\pi W_0 t} + \xi_{\alpha+k/2W} e^{i2\pi W_0 t}) \frac{\sin 2\pi W(t - k/2W - \alpha)}{2\pi W(t - k/2W - \alpha)} \end{aligned}$$

Reexpressing $\eta_{\alpha+k/2W}$ and $\xi_{\alpha+k/2W}$ in terms of $X_{\alpha+k/2W}$ and $Y_{\alpha+k/2W}$, we get the following result.

8. WHITE NOISE AND WHITE-NOISE INTEGRALS

Proposition 7.2. Let $\{X_t, -\infty < t < \infty\}$ be a bandpass process with center frequency W_0 and bandwidth $2W$. Let Y_t be its Hilbert transform defined by (7.7). Then

$$X_t = \lim_{N\to\infty} \text{in q.m.} \sum_{k=-N}^{N} \frac{\sin 2\pi W(t - k/2W - \alpha)}{2\pi W(t - k/2W - \alpha)}$$
$$[X_{k/2W+\alpha} \cos 2\pi W_0(t - k/2W - \alpha)$$
$$- Y_{k/2W+\alpha} \sin 2\pi W_0(t - k/2W - \alpha)] \quad (7.12)$$

Equation (7.12) involves both X_t and Y_t being sampled at $2W$, giving a total sampling rate of $4W$.

It is interesting to note that

$$|X_t|^2 + |Y_t|^2 = |\xi_t|^2 + |\eta_t|^2 \quad (7.13)$$

Since ξ_t and η_t are both bandlimited to W, we can expect $|\xi_t|^2 + |\eta_t|^2$ to be relatively slowly varying when $W \ll W_0$. Therefore, $|X_t|^2 + |Y_t|^2$ is also slowly varying (relative to a sinusoid with frequency W_0). On the other hand, most of the average power of X_t is concentrated near $\pm W_0$, so X_t itself is rapidly varying. If X_t is real valued, then it can be written as

$$X_t = \sqrt{X_t^2 + Y_t^2} \cos[2\pi W_0 t + \theta(t)]$$

where both $\sqrt{X_t^2 + Y_t^2}$ and $\theta(t)$ are slowly varying. Thus, $\sqrt{X_t^2 + Y_t^2}$ has the interpretation of the envelope and $\theta(t)$ the phase of a sinusoid being slowly modulated in both its amplitude and phase. Thus, we see that the Hilbert transform plays a rather important role in bandpass processes. It permits the envelope of such a process to be expressed simply.

8. WHITE NOISE AND WHITE-NOISE INTEGRALS

A **white noise** is usually described as a wide-sense stationary process with a spectral-density function given by

$$S(\nu) = S_0 \quad \text{for all } \nu \quad (8.1)$$

Since for this case $\int_{-\infty}^{\infty} S(\nu) \, d\nu = R(0) = \infty$, a white noise is not really a second-order process at all, and the spectral-density function is not well defined. However, if we proceed heuristically, (8.1) suggests that $R(\tau)$ must be given by

$$R(\tau) = EX_{t+\tau}\bar{X}_t = \delta(\tau)S_0$$

where $\delta(\tau)$ is the Dirac δ function. The reason for this is that

$$S(\nu) = \int_{-\infty}^{\infty} e^{i2\pi\nu\tau} R(\tau) \, d\tau$$

would then give (8.1). These considerations are purely formal and require elaboration and substantiation. At the outset, we should distinguish between the problem of handling white noise in a mathematically consistent way and the problem of interpreting white noise as an abstraction of physical phenomena. As far as the calculus of white noise is concerned, the problem is not difficult, at least for linear problems. Nonlinear problems involving Gaussian white noise are substantially more complex and will be dealt with in a later chapter. The principal tool that we shall use in establishing a self-consistent calculus for white noise is the second-order stochastic integral that we introduced in Sec. 6 in connection with spectral representation. There remains the problem of interpretation.

Since $R(0) = \infty$ implies an infinite average power, a white noise cannot be a physical process. If a white noise is not a physical process, and if it leads to mathematical complications, then why is it used at all? First, even though the calculus of white noise requires justification, once justified it leads to a tremendous analytical simplification in many problems. Secondly, many processes that one encounters in practice are well approximated by white noise, but this statement requires amplification. Because a white noise is not a second-order process (indeed, it is not a stochastic process at all!), no sequence of processes $\{X_n(t), t \in (-\infty, \infty)\}$ which is q.m. convergent for each t can converge to a white noise. The way out of this difficulty is to recall that just as a δ function is never used outside of an integral, the same is true with white noise.

Definition. A sequence of q.m. continuous processes $\{X_t^{(n)}, t \in (-\infty, \infty)\}$ is said to "converge to a white noise" if
  (a) For each $f \in L_2$, $\{X_n(f)\}$ is a q.m. convergent sequence, where $X_n(f)$ is defined by

$$X_n(f) = \int_{-\infty}^{\infty} f(t) X_t^{(n)} dt \qquad (8.2)$$

  (b) There exists a positive constant S_0 such that

$$\lim_{n \to \infty} E X_n(f) \overline{X_n(g)} = S_0 \int_{-\infty}^{\infty} f(t) \bar{g}(t) dt \qquad (8.3)$$

This definition helps to make clear the idea that a process $\{X_t, -\infty < t < \infty\}$ is approximately a white noise. What we really mean is that for all function f that we are concerned with, the quantity $EX(f)\overline{X(f)}$ is very nearly equal to $S_0 \int_{-\infty}^{\infty} |f(t)|^2 dt$.

Suppose $\{X_t^{(n)}, t \in (-\infty, \infty)\}$ is a sequence of processes converging to a white noise. By definition, for every $f \in L_2$, there exists a second-

8. WHITE NOISE AND WHITE-NOISE INTEGRALS

order random variable $X(f)$ such that

$$EX(f)\overline{X(f)} = S_0 \int_{-\infty}^{\infty} |f(t)|^2 \, dt \tag{8.4}$$

It is common practice to write $X(f)$ as

$$X(f) = \int_{-\infty}^{\infty} f(t) X_t \, dt \tag{8.5}$$

where X_t is a white noise. We shall do so on many occasions. It should not be forgotten, however, that the right-hand side of (8.5) is nothing more than a symbolic way of writing $X(f)$, and there exists no stochastic process X_t for which the right-hand side of (8.5) is an integral. Although (8.5) is merely formal, $X(f)$ does admit a representation as a second-order stochastic integral as is indicated by the following proposition.

Proposition 8.1. Let $\{X_t^{(n)}, -\infty < t < \infty\}$ be a sequence of q.m. continuous converging to a white noise. Let

$$X(f) = \lim_{n\to\infty} \text{in q.m.} \int_{-\infty}^{\infty} f(t) X_t^{(n)} \, dt \qquad f \in L_2 \tag{8.6}$$

Then

$$X(f) = \int_{-\infty}^{\infty} f(t) \, dZ_t \tag{8.7}$$

where $\{Z_t, -\infty < t < \infty\}$ is a process with orthogonal increments defined by

$$Z_t = \lim_{n\to\infty} \text{in q.m.} \int_0^t X_s^{(n)} \, ds \tag{8.8}$$

Proof: Define $\{Z_t, -\infty < t < \infty\}$ by (8.8). If we denote the indicator function of $[a,b]$ by I_{ab}, then we can write

$$Z_b - Z_a = \lim_{n\to\infty} \text{in q.m.} \, X_n(I_{ab}) \tag{8.9}$$

From (8.3) we have

$$E(Z_b - Z_a)(Z_d - Z_c) = S_0 \int_{-\infty}^{\infty} I_{ab}(t) I_{cd}(t) \, dt \tag{8.10}$$

Hence, $\{Z_t, -\infty < t < \infty\}$ has orthogonal increments and

$$E \, dZ_t \, \overline{dZ_s} = S_0 \, \delta_{ts} \, dt \tag{8.11}$$

Define the second-order stochastic integral $\int_{-\infty}^{\infty} f(t) \, dZ_t$ as in Sec. 6 for $f \in L_2$. If f is a step function, it is obvious from (8.9) that

$$X(f) = \int_{-\infty}^{\infty} f(t) \, dZ_t$$

If $f \in L_2$ is not a step, there exists a sequence of step functions converging to f in L_2 distance. Therefore

$$X(f) = \int_{-\infty}^{\infty} f_n(t)\, dZ_t + X(f - f_n) \xrightarrow[n \to \infty]{} \int_{-\infty}^{\infty} f(t)\, dZ_t + X(0)$$

$$= \int_{-\infty}^{\infty} f(t)\, dZ_t$$

which proves the theorem. ∎

Manipulations of **white-noise integrals** $\int_{-\infty}^{\infty} X_t f(t)\, dt$ are justified by replacing it by the stochastic integral $\int_{-\infty}^{\infty} f(t)\, dZ_t$. In the future, when we speak of being given a white noise $\{X_t, -\infty < t < \infty\}$, we shall take it to mean that we are given a process with orthogonal increments $\{Z_t, -\infty < t < \infty\}$ which satisfies $E\, dZ_t\, \overline{dZ_s} = S_0\, \delta_{ts}\, dt$. From now on we shall also omit the constant S_0 which is irrelevant to most considerations. A most interesting property of a white noise is that its Fourier transform is again a white noise. Of course, Fourier transform has to be defined properly here.

Proposition 8.2. Let $\{X_t, -\infty < t < \infty\}$ be a white noise. Then there exists a second white noise $\{\hat{X}_\nu, -\infty < \nu < \infty\}$ such that

$$\int_{-\infty}^{\infty} \hat{h}(t) X_t\, dt = \int_{-\infty}^{\infty} h(\nu) \hat{X}_\nu\, d\nu \tag{8.12}$$

for all $h, \hat{h} \in L_2$.

Remark: We repeat once again that the integrals in (8.12) are merely symbolic representations of $\int_{-\infty}^{\infty} \hat{h}(t)\, dZ_t$ and $\int_{-\infty}^{\infty} h(\nu)\, d\hat{Z}_\nu$.

Proof: Define $\{\hat{Z}_\nu, -\infty < \nu < \infty\}$ as follows:
 (a) $\hat{Z}_0 = 0$
 (b) $\hat{Z}_b - \hat{Z}_a = \int_{-\infty}^{\infty} \hat{I}_{ab}(t)\, dZ_t$
Then,

$$E(\hat{Z}_b - \hat{Z}_a)\overline{(\hat{Z}_d - \hat{Z}_c)} = \int_{-\infty}^{\infty} \hat{I}_{ab}(t) \overline{\hat{I}_{cd}}(t)\, dt$$

$$= \int_{-\infty}^{\infty} I_{ab}(\nu) \bar{I}_{cd}(\nu)\, d\nu$$

Hence, $E\, d\hat{Z}_\nu\, \overline{d\hat{Z}_\mu} = \delta_{\nu\mu}\, d\nu$, and $\{\hat{Z}_\nu, -\infty < \nu < \infty\}$ has the same second-order properties as $\{Z_t, -\infty < t < \infty\}$. For an arbitrary $h \in L_2$, by approximating h by step functions in the familiar way, we find

$$\int_{-\infty}^{\infty} \hat{h}(t)\, dZ_t = \int_{-\infty}^{\infty} h(\nu)\, d\hat{Z}_\nu \tag{8.13}$$

proving the theorem. ∎

8. WHITE NOISE AND WHITE-NOISE INTEGRALS

The spectral process of a white noise can now be used to define linear time-invariant filtering operations on white noise. Let $\eta \in L_2$, and let $\{X_t, -\infty < t < \infty\}$ be a white noise. Define

$$Y_t = \int_{-\infty}^{\infty} \eta(\nu) e^{i2\pi\nu t} \hat{X}_\nu \, d\nu \tag{8.14}$$

that is,

$$Y_t = \int_{-\infty}^{\infty} \eta(\nu) e^{i2\pi\nu t} \, d\hat{Z}_\nu$$

Then, Y_t is a q.m. continuous and wide-sense stationary process with a spectral-density function given by

$$S_Y(\nu) = |\eta(\nu)|^2 \tag{8.15}$$

Equation (8.15) justifies the interpretation that a white noise has a spectral-density function which is equal to a constant for all frequencies. This is because if we filter a white noise $\{X_t, -\infty < t < \infty\}$ with an ideal filter having a transfer function

$$\eta(\nu) = \begin{cases} 1 & a \leq \nu \leq b \\ 0 & \text{otherwise} \end{cases}$$

then the total average output power is given by

$$E|Y_t|^2 = E \left| \int_a^b e^{i2\pi\nu t} \hat{X}_\nu \, d\nu \right|^2$$
$$= \int_a^b d\nu = b - a$$

Equation (8.15) also has another interpretation. It suggests that every process with a spectral density $S(\nu)$ can be represented as white noise filtered by a time-invariant linear filter whose transfer function $\hat{h}$ satisfies $|\hat{h}(\nu)|^2 = S(\nu)$. This interpretation plays an important role in Wiener filtering problems. Closely related to linear time-invariant filtering is the idea of a differential equation driven by a white noise. Consider the following example. Suppose X_t is a white noise and Y_t satisfies

$$\dot{Y}_t = -Y_t + X_t \tag{8.16}$$

A good guess is that a solution of this equation is given by

$$Y_t = \int_{-\infty}^{\infty} \frac{1}{1 + i2\pi\nu} e^{i2\pi\nu t} \hat{X}_\nu \, d\nu \tag{8.17}$$

Before we can ascertain this, we need to give a precise interpretation to an equation such as (8.16).

Even if X_t is not a white noise, but a q.m. continuous process, an equation like (8.16) still cannot be interpreted as an ordinary differential equation involving sample functions of the two processes without assumptions such as separability, sample differentiability, etc. However, these assumptions would not be necessary if we interpret (8.16) as an integral equation. More generally, let $\{X_t, -\infty < t < \infty\}$ be a white noise with $\{Z_t, -\infty < t < \infty\}$ as the corresponding process with orthogonal increments. A second-order process $\{Y_t, t \in [a,b]\}$ is said to be the solution of the differential equation

$$\dot{Y}_t = \alpha(t) Y_t + \beta(t) X_t \tag{8.18}$$

if it satisfies

$$Y_t = Y_a + \int_a^t \alpha(s) Y_s \, ds + \int_a^t \beta(s) \, dZ_s \qquad a \leq t \leq b \tag{8.19}$$

Proposition 8.3. Let α and $\beta \in L_2(a,b)$. Then, (8.19) has one and only one solution with the same initial condition Y_a, provided that $E|Y_a|^2 < \infty$.

Remark: The proof will be omitted, because it is identical to the corresponding proof for stochastic differential equations of the Ito type. At this point we merely point out the following facts concerning the solution of (8.19).

(a) If we set $Y_t^{(0)} = Y_a$, $a \leq t \leq b$, and

$$Y_t^{(n+1)} = \int_a^t \alpha(s) Y_s^{(n)} \, ds + \int_a^t \beta(s) \, dZ_s$$

then $\{Y_t^{(n)}, a \leq t \leq b, n = 0, 1, \ldots\}$ converges in q.m. to the unique solution of (8.19).

(b) The solution of (8.19) is q.m. continuous on $[a,b]$.

If $\{X_t^{(n)}, -\infty < t < \infty, n = 1, 2, \ldots\}$ is a sequence of q.m. continuous processes, then for each n, we can consider the differential equation

$$\dot{Y}_t^{(n)} = \alpha(t) Y_t^{(n)} + \beta(t) X_t^{(n)} \tag{8.20}$$

where $\dot{Y}_t^{(n)}$ is the q.m. derivative of $Y_t^{(n)}$. Equation (8.20) is equivalent to the integral equation

$$Y_t^{(n)} = Y_a^{(n)} + \int_a^t \alpha(s) Y_s^{(n)} \, ds + \int_a^t \beta(s) X_s^{(n)} \, ds$$

where the integrals are q.m. integrals. Suppose $Y_a^{(n)} \xrightarrow[n \to \infty]{q.m.} Y_a$, and $\{X_t^{(n)}\}$ converges to a white noise X_t. Then we can show that $\{Y_t^{(n)}, a \leq t \leq b\}$ converges to a q.m. continuous process $\{Y_t, a \leq t \leq b\}$ for any b such that $\alpha, \beta \in L_2(a,b)$. Further, $\{Y_t, a \leq t \leq b\}$ satisfies the equation

$$Y_t = Y_a + \int_a^t \alpha(s) Y_s \, ds + \int_a^t \beta(s) X_s \, ds \qquad a \leq t \leq b$$

8. WHITE NOISE AND WHITE-NOISE INTEGRALS

These considerations justify the use of (8.19), even when the driving force is "not quite white." Finally, we note that for our earlier example, we can easily show that

$$Y_t = \int_{-\infty}^{\infty} \frac{1}{1 + i2\pi\nu} e^{i2\pi\nu t} \, d\hat{Z}_\nu \tag{8.21}$$

is the unique solution to the equation

$$Y_t = Y_a - \int_a^t Y_s \, ds + (Z_t - Z_a) \qquad t \geq a \tag{8.22}$$

corresponding to the initial condition

$$Y_a = \int_{-\infty}^{\infty} \frac{1}{1 + i2\pi\nu} e^{i2\pi\nu a} \, d\hat{Z}_\nu \tag{8.23}$$

For an arbitrary initial condition Y_a, the general solution of (8.22) is of the form

$$Y_t = Y_a e^{-(t-a)} + \int_a^t e^{-(t-s)} \, dZ_s \tag{8.24}$$

Hence, if $E|Y_a|^2$ stays finite as $a \to -\infty$, then

$$Y_t \xrightarrow[a \to -\infty]{\text{q.m.}} \int_{-\infty}^t e^{-(t-s)} \, dZ_s$$

By Proposition 8.2,

$$\int_{-\infty}^t e^{-(t-s)} \, dZ_s = \int_{-\infty}^{\infty} \frac{1}{2\pi i\nu + 1} e^{i2\pi\nu t} \, d\hat{Z}_\nu$$

Therefore, under rather general conditions, every solution of (8.22) approaches (8.21) as the initial time $a \to -\infty$.

It is easy to generalize the preceding considerations to vector differential equations

$$\dot{\mathbf{Y}}_t = \mathbf{A}(t)\mathbf{Y}_t + \mathbf{B}(t)\mathbf{X}_t$$

where $\mathbf{A}$ and $\mathbf{B}$ are now matrices and $\mathbf{Y}_t$ and $\mathbf{X}_t$ are column vectors of possibly different dimensions whose components are, respectively, second-order processes and orthogonal white-noise processes. Proposition 8.3 is easily generalized to this case. The required conditions on A and B are now

$$\sum_{i,j} \int_a^b |A_{ij}(t)|^2 \, dt < \infty$$

$$\sum_{i,j} \int_a^b |B_{ij}(t)|^2 \, dt < \infty$$

9. LINEAR PREDICTION AND FILTERING

Linear prediction and filtering are special cases of a general estimation problem that can be formulated as follows: Let $\{X_t, t \in T\}$ be a second-order process and let Y be a second-order random variable. Let $\mathcal{H}_X$ denote Hilbert space generated by $\{X_t, t \in T\}$. A random variable $\tilde{Y}$ is said to be a **linear least-squares estimator** of Y given $\{X_t, t \in T\}$ if (1) $\tilde{Y} \in \mathcal{H}_X$, and (2)

$$E|\tilde{Y} - Y|^2 = \min_{Z \in \mathcal{H}_X} E|Z - Y|^2 \tag{9.1}$$

Filtering and prediction problems are estimation problems involving two second-order processes $\{X_t\}$ and $\{Y_t\}$, where for each t we want to find the linear least-squares estimator of Y_t given $\{X_s, s \leq t\}$. A pure prediction problem is one where $Y_t = X_{t+\alpha}$ for some positive α. In practice, the parameter space for the two processes is usually the same, the most common cases being $[0, \infty)$ or $(-\infty, \infty)$. The fact that an estimator is required for each t makes it especially important that the implementation of the estimator be simple. Both the Wiener filter and Kalman-Bucy filter, in different circumstances, achieve this goal extremely well. This fact is responsible for the importance and the widespread use of these filters. Before discussing these problems in detail, we first prove the following important characterization of linear least-squares estimators.

Proposition 9.1. Let $\tilde{Y} \in \mathcal{H}_X$. Then, $\tilde{Y}$ is a linear least-squares estimator of Y given $\{X_t, t \in T\}$ if and only if

$$E(\tilde{Y} - Y)\bar{X}_t = 0 \quad \text{for every } t \in T \tag{9.2}$$

and equivalently,

$$E(\tilde{Y} - Y)\bar{Z} = 0 \quad \text{for every } Z \in \mathcal{H}_X$$

Proof: For any $Z \in \mathcal{H}_X$, $\tilde{Y} - Z$ is again in $\mathcal{H}_X$. Hence, if (9.2) holds, then

$$\begin{aligned} E|Z - Y|^2 &= E|Z - \tilde{Y} + \tilde{Y} - Y|^2 \\ &= E|Z - \tilde{Y}|^2 + E|\tilde{Y} - Y|^2 \end{aligned} \tag{9.3}$$

Therefore, (9.1) is satisfied.

Conversely, suppose $\tilde{Y}$ satisfies (9.1). Set

$$Z = Y - \frac{E(\tilde{Y} - Y)\bar{X}_t}{E|X_t|^2} X_t$$

Then, $Z \in \mathcal{H}_X$, and from (9.1), we have

$$E|Z - Y|^2 - E|\tilde{Y} - Y|^2 \geq 0$$

9. LINEAR PREDICTION AND FILTERING

However, by direct computation, we find

$$E|Z - Y|^2 = E|Y - \tilde{Y}|^2 - \frac{|E(\tilde{Y} - Y)X_t|^2}{E|X_t|^2} \geq E|Y - \tilde{Y}|^2$$

Therefore, (9.2) follows.

Remark: Equation (9.3) proves the uniqueness of linear least-squares estimators. If Z and $\tilde{Y}$ are both linear least-squares estimators of Y given $\{X_t, t \in T\}$, then $E|Z - Y|^2 = E|\tilde{Y} - Y|^2$, so (9.3) implies $E|Z - Y|^2 = 0$. Hence, we shall refer to *the* linear least-squares estimator of Y given $\{X_t, t \in T\}$. Furthermore, from Proposition 9.1 we see that the linear least-squares estimator of Y given $\{X_t, t \in T\}$ is the projection of Y into the smallest Hilbert space $\mathcal{H}_X$ spanned by $\{X_t, t \in T\}$. We shall denote this projection by a more explicit notation

$$\hat{E}(Y|\mathcal{H}_X)$$

The Wiener theory of filtering and prediction is distinguished by two facts. (1) It deals with wide-sense stationary processes with well-defined spectral-density functions, and (2) the given information on which the estimator is to be constructed is the infinite past. These conditions will be made more precise later. Under these conditions, it is often possible to express the estimator as the output of a linear time-invariant filtering operation on the observed process. This is an extremely convenient form for the solution from the point of view of implementation.

Let $\{X_t, -\infty < t < \infty\}$ and $\{Y_t, -\infty < t < \infty\}$ be wide-sense stationary processes. We assume that there exist bounded functions S_x, S_y, and S_{xy} such that

$$EX_{t+\tau}\bar{X}_t = \int_{-\infty}^{\infty} S_x(\nu) e^{i2\pi\nu\tau} d\nu \tag{9.4a}$$

$$EX_{t+\tau}\bar{Y}_t = \int_{-\infty}^{\infty} S_{xy}(\nu) e^{i2\pi\nu\tau} d\nu \tag{9.4b}$$

$$EY_t\bar{Y}_{t+\tau} = \int_{-\infty}^{\infty} S_y(\nu) e^{i2\pi\nu\tau} d\nu \tag{9.4c}$$

The spectral-density functions S_x and S_y are necessarily real and positive. The cross-spectral density S_{xy} need not be real. Under these assumptions X_t, Y_t are not only individually wide-sense stationary processes, but are in fact jointly wide-sense stationary in the sense that for arbitrary complex constants a and b, $Z_t = aX_t + bY_t$ is again a wide-sense stationary process. For such a process Z_t, we have

$$EZ_{t+\tau}\bar{Z}_t = \int_{-\infty}^{\infty} [|a|^2 S_x(\nu) + |b|^2 S_y(\nu) + a\bar{b} S_{xy}(\nu) \\ + \bar{a}b \bar{S}_{xy}(\nu)] e^{i2\pi\nu\tau} d\nu \tag{9.5}$$

In other words, the spectral-density function of $\{Z_t, -\infty < t < \infty\}$ is given by

$$S_z(\nu) = |a|^2 S_x(\nu) + |b|^2 S_y(\nu) + a\bar{b}S_{xy}(\nu) + \bar{a}b\bar{S}_{xy}(\nu) \tag{9.6}$$

It is obvious that S_z must be real and nonnegative for arbitrary a and b. A necessary and sufficient condition for this is that for every ν,

$$S_x(\nu)S_y(\nu) \geq |S_{xy}(\nu)|^2 \tag{9.7}$$

A proof of this fact is easily constructed by noting the fact that (9.6) is a quadratic form involving the Hermitian matrix

$$\mathbf{S}(\nu) = \begin{bmatrix} S_x(\nu) & S_{xy}(\nu) \\ \bar{S}_{xy}(\nu) & S_y(\nu) \end{bmatrix} \tag{9.8}$$

Hence, for $S_z(\nu)$ to be nonnegative for every ν and a and b, it is necessary and sufficient that $\mathbf{S}(\nu)$ be nonnegative definite for every ν. This condition is equivalent to (9.7).

For such a pair of processes $\{X_t, Y_t, -\infty < t < \infty\}$, we shall consider the problem of finding the linear least-squares estimator of Y_t given $\{X_s, -\infty < s < t\}$. If we denote by $\mathcal{H}_X{}^t$ the Hilbert space spanned by $\{X_s, -\infty \leq s < t\}$, the problem is to find the projection $\hat{E}(Y_t|\mathcal{H}_X{}^t)$. We recall that $\hat{E}(Y_t|\mathcal{H}_X{}^t)$ is characterized by two properties: (1) $\hat{E}(Y_t|\mathcal{H}_X{}^t)$ belongs to $\mathcal{H}_X{}^t$ and (2) $Y_t - \hat{E}(Y_t|\mathcal{H}_X{}^t)$ is orthogonal to $\mathcal{H}_X{}^t$. A solution to this problem in the form of a filtering operation will be constructed using the ideas of white noise and white-noise integrals that we introduced in the last section. First, we state the following fundamental result of Paley and Wiener [1934, chap. 1].

Proposition 9.2. Let $S \in L_1$ be a real nonnegative function such that

$$\int_{-\infty}^{\infty} \frac{|\ln S(\nu)|}{1+\nu^2} d\nu < \infty \tag{9.9}$$

Then there exists a function $\hat{h} \subset L_2$ such that

$$|\hat{h}(\nu)|^2 = S(\nu) \tag{9.10a}$$
$$h(t) = \int_{-\infty}^{\infty} e^{i2\pi\nu t}\hat{h}(\nu)\,d\nu = 0 \quad t < 0 \tag{9.10b}$$
$$\hat{h}(u+iv) = \int_{-\infty}^{\infty} e^{i2\pi(u+iv)t}h(t)\,dt \neq 0 \quad v > 0 \tag{9.10c}$$

Remark: Condition (9.10b) shows that $\hat{h}$ is the transfer function of a **nonanticipative filter** (physically realizable filter). Condition (9.10c) is known as the **minimum-phase condition** in circuit theory. Roughly speaking, it means that $1/\hat{h}$ is again nonanticipative. However, $1/\hat{h}$

9. LINEAR PREDICTION AND FILTERING

has no inverse Fourier transform, so the nonanticipative property is a little more difficult to define.

A complete proof is somewhat complicated and will be omitted. Instead, we shall outline a procedure by which the desired $\hat{h}$ can be found. First, consider the transformation

$$\nu = -\tan\frac{\theta}{2} \tag{9.11}$$

Then, (9.9) becomes

$$\frac{1}{2}\int_{-\pi}^{\pi}\left|\ln S\left(-\tan\frac{\theta}{2}\right)\right|d\theta < \infty$$

so that $\ln S(-\tan\theta/2)$ has a Fourier series

$$\ln S\left(-\tan\frac{\theta}{2}\right) = \sum_{n=-\infty}^{\infty} a_n e^{in\theta} \tag{9.12}$$

where a_n are given by

$$a_n = \frac{1}{2\pi}\int_{-\pi}^{\pi} e^{-in\theta}\ln S\left(-\tan\frac{\theta}{2}\right)d\theta \tag{9.13}$$

With the correspondence (9.11), we have

$$e^{i\theta} = \frac{1-i\nu}{1+i\nu} \tag{9.14}$$

so that (9.12) and (9.13) become

$$\ln S(\nu) = \sum_{n=-\infty}^{\infty} a_n\left(\frac{1-i\nu}{1+i\nu}\right)^n \tag{9.15}$$

and

$$a_n = \frac{1}{\pi}\int_{-\infty}^{\infty}\frac{1}{1+\nu^2}\left(\frac{1+i\nu}{1-i\nu}\right)^n \ln S(\nu)\,d\nu \tag{9.16}$$

If $|\hat{h}|^2 = S$, then

$$\ln S(\nu) = \ln \hat{h} + \ln \bar{\hat{h}} \tag{9.17}$$

We now identify

$$\hat{h}(\nu) = \exp\left[\frac{a_0}{2} + \sum_{n=1}^{\infty} a_n\left(\frac{1-i\nu}{1+i\nu}\right)^n\right] \tag{9.18}$$

Because $a_{-n} = \bar{a}_n$, (9.17) is satisfied, hence also (9.10a). Conditions (9.10b) and (9.10c) follow from the fact that

$$f(u+iv) = \frac{a_0}{2} + \sum_{n=1}^{\infty} a_n \left[\frac{1-i(u+iv)}{1+i(u+iv)}\right]^n \tag{9.19}$$

is analytic for $v < 0$. Hence, $\hat{h}(u+iv)$ is analytic for $v < 0$, and (9.10b) follows (more or less) from contour integration closing the contour in the lower half plane. Condition (9.10c) follows from the fact that if $f(z)$ is analytic, then $e^{f(z)}$ has no zero. We call the above procedure the **spectral-factorization** procedure, since S is usually identified with a spectral-density function.

Let $\{X_t, -\infty < t < \infty\}$ have a spectral-density function S_x, and let $\hat{h}$ be obtained by factoring S_x so that (9.10) is satisfied with S_x replacing S in (9.10a). Then, in view of our discussion on (8.15), $\{X_t, -\infty < t < \infty\}$ can be regarded as the output of filtering a white noise $\{\zeta_t, -\infty < t < \infty\}$ with a nonanticipative filter having transfer function $\hat{h}$. Condition (9.10c) means that the white noise $\{\zeta_t, -\infty < t < \infty\}$ can in turn be obtained by filtering $\{X_t, -\infty < t < \infty\}$ by $1/\hat{h}$, which is also nonanticipative. A more precise statement of these results can be made in terms of the process with orthogonal increments $\{Z_t, -\infty < t < \infty\}$ corresponding to $\{\zeta_t, -\infty < t < \infty\}$. Condition (9.10b) then implies that there exists a process $\{Z_t, -\infty < t < \infty\}$ with $Z_0 = 0$ and $E\,dZ_t\,\overline{dZ_s} = \delta_{ts}\,dt$ such that

$$\begin{aligned}X_t &= \int_{-\infty}^{\infty} h(t-s)\,dZ_s \\ &= \int_{-\infty}^{\infty} e^{i2\pi\nu t}\hat{h}(\nu)\,d\hat{Z}_\nu\end{aligned} \tag{9.20}$$

Condition (9.10c) implies that for each t

$$Z_t \in \mathcal{H}_X{}^t \tag{9.21}$$

We can express Z_t more explicitly in terms of the spectral process $\{\hat{X}_\nu, -\infty < \nu < \infty\}$ as

$$Z_t = \int_{-\infty}^{\infty} \frac{1}{\hat{h}(\nu)} \left(\int_0^t e^{i2\pi\nu s}\,ds\right) d\hat{X}_\nu \tag{9.22}$$

which points out even more clearly that $\zeta_t\,(=\dot{Z}_t)$ is obtained by filtering X_t by $1/\hat{h}$.

We shall now give the main result of the Wiener theory of filtering as follows [Wiener, 1949].

Proposition 9.3. Let $\{X_t, Y_t, -\infty < t < \infty\}$ be a pair of wide-sense stationary processes satisfying (9.4). Let $\hat{h}$ be obtained by factoring S_x so that (9.10a) to (9.10c) are satisfied. Let $\{Z_t, -\infty < t < \infty\}$ be

9. LINEAR PREDICTION AND FILTERING

as in (9.20) to (9.22). Then,

$$\hat{E}(Y_t|\mathcal{3C}_X{}^t) = \int_{-\infty}^{t} g(t-s)\, dZ_s \tag{9.23}$$

where g is given by

$$g(t) = \int_{-\infty}^{\infty} e^{i2\pi \nu t}\, \overline{\frac{S_{xy}(\nu)}{\hat{h}(\nu)}}\, d\nu \qquad -\infty < t < \infty \tag{9.24}$$

Remark: Because $|S_{xy}/\hat{h}|^2 = |S_{xy}|^2/S_x$ is bounded by S_y (from (9.7)), the integral in (9.24) is well defined as a limit in L_2 mean.

Proof: To prove (9.23), we need to verify the two conditions:

$$\int_{-\infty}^{t} g(t-s)\, dZ_s \in \mathcal{3C}_X{}^t \tag{9.25a}$$

$$EY_t\bar{X}_\tau = E\left[\int_{-\infty}^{t} g(t-s)\, dZ_s\right]\bar{X}_\tau \qquad \tau \le t \tag{9.25b}$$

To verify (9.25a), we note that $\int_{-\infty}^{t} g(t-s)\, dZ_s$ is in $\mathcal{3C}_Z{}^t$, where $\mathcal{3C}_Z{}^t$ denotes the Hilbert space spanned by $\{Z_s, -\infty < s \le t\}$. Since for each t, $Z_t \in \mathcal{3C}_X{}^t$, so $\mathcal{3C}_Z{}^t \subset \mathcal{3C}_X{}^t$, verifying (9.25a). To verify (9.25b), we note that

$$\int_{-\infty}^{\infty} g(t-s)\, dZ_s = \int_{-\infty}^{t} g(t-s)\, dZ_s + \int_{t}^{\infty} g(t-s)\, dZ_s$$

$$= \int_{-\infty}^{\infty} \overline{[S_{xy}(\nu)/\hat{h}(\nu)]} e^{i2\pi \nu t}\, d\hat{Z}_\nu \tag{9.26}$$

Hence, from (9.20) we have

$$E\left[\int_{-\infty}^{\infty} g(t-s)\, dZ_s\right]\bar{X}_\tau = \int_{-\infty}^{\infty} \tilde{S}_{xy}(\nu) e^{i2\pi \nu(t-\tau)}\, d\nu$$

$$= EY_t\bar{X}_\tau \tag{9.27}$$

On the other hand, since

$$X_\tau = \int_{-\infty}^{\tau} h(\tau - s)\, dZ_s$$

and $E\, dZ_t\, \overline{dZ_s} = \delta_{ts}\, dt$, we have

$$E\left[\int_{t}^{\infty} g(t-s)\, dZ_s\right]\bar{X}_\tau = 0 \qquad \text{whenever } \tau \le t$$

Therefore,

$$E\left[\int_{-\infty}^{t} g(t-s)\, dZ_s\right]\bar{X}_\tau = E\left[\int_{-\infty}^{\infty} g(t-s)\, dZ_s\right]\bar{X}_\tau$$

$$= EY_t\bar{X}_\tau \qquad \tau \le t$$

which verifies (9.25b) and completes the proof. ∎

The solution (9.23) can be put in a more useful form by using (9.22). If we define

$$\hat{\gamma}(\nu) = \int_0^\infty g(t)e^{-i2\pi\nu t}\,dt \tag{9.28}$$

then we can write

$$E(Y_t|\mathcal{K}_X{}^t) = \int_{-\infty}^\infty \frac{\hat{\gamma}(\nu)}{\hat{h}(\nu)} e^{i2\pi\nu t}\,d\hat{X}_\nu \tag{9.29}$$

which is the output to a time-invariant linear filter with $\{X_t, -\infty < t < \infty\}$ as input and $\hat{\gamma}/\hat{h}$ as its transfer function. The filter is called the **Wiener filter.** We should note that $\hat{\gamma}$ is *not* the Fourier integral of g, since the lower limit in the integral in (9.28) is 0 and not $-\infty$, and $g(t)$ in general is not zero for $t < 0$.

For a general S_x, the spectral-factorization procedure that we outlined earlier (9.16 and 9.18) may not lead to a closed-form solution for $\hat{h}$. It is also somewhat complicated. The factorization problem becomes trivial if S_x is a rational function. If S_x is rational, then we can always write

$$S_x(\nu) = K^2 \frac{\prod_{k=1}^m (\nu - z_k)(\nu - \bar{z}_k)}{\prod_{k=1}^n (\nu - p_k)(\nu - \bar{p}_k)} \tag{9.30}$$

where every z_k and p_k have positive imaginary parts. Since $|\hat{h}|^2 = S_x$ and $\hat{h}(u + iv)$ has neither poles nor zeros for $v < 0$, $\hat{h}$ must be of the form

$$\hat{h}(\nu) = A \frac{\prod_{k=1}^m (\nu - z_k)}{\prod_{k=1}^n (\nu - p_k)} \tag{9.31}$$

where $|A|^2 = K^2$. The constant A can be determined from (9.18),

$$\hat{h}(-i) = e^{a_0/2} = A \frac{\prod_{k=1}^m (-i - z_k)}{\prod_{k=1}^n (-i - p_k)}$$

9. LINEAR PREDICTION AND FILTERING

and a_0 is given by

$$a_0 = \frac{1}{\pi} \int_{-\infty}^{\infty} \frac{1}{1+\nu^2} \ln S(\nu)\, d\nu$$

However, we note that any choice of A would yield an $\hat{h}$ satisfying (9.10). Equation (9.18) merely gives a specific one. When $\{X_t, -\infty < t < \infty\}$ is real, then $S_x(\nu)$ is an even function of ν, and the constant A in (9.31) can always be chosen so that the impulse response

$$h(t) = \int_{-\infty}^{\infty} e^{i2\pi \nu t} \hat{h}(\nu)\, d\nu$$

is real valued. This is a convenient choice. However, as far as the Wiener filtering problem is concerned, the choice of A is unimportant. It is fairly obvious that the solution (9.23) or (9.29) is independent of this choice.

For our first example, consider the following pure prediction problem: $S_x(\nu) = [1 + (2\pi\nu)^2]^{-1}$ and $Y_t = X_{t+\alpha}$, $\alpha > 0$. The factorization is trivial and we can take

$$\hat{h}(\nu) = \frac{1}{1 + i2\pi\nu}$$

From (9.24) $g(t)$ is found to be

$$g(t) = \int_{-\infty}^{\infty} e^{i2\pi\nu t} e^{i2\pi\nu\alpha} \frac{1}{1 + i2\pi\nu}\, d\nu$$
$$= \begin{cases} e^{-(t+\alpha)} & t > -\alpha \\ 0 & t < -\alpha \end{cases}$$

From (9.28) we find

$$\hat{\gamma}(\nu) = e^{-\alpha} \frac{1}{1 + i2\pi\nu}$$

so that

$$\hat{E}(Y_t | \mathcal{K}_x{}^t) = e^{-\alpha} \int_{-\infty}^{\infty} e^{i2\pi\nu t}\, d\hat{X}_\nu = e^{-\alpha} X_t$$

For this simple example, the predictor is nothing more than an attenuator.

As a second example, suppose

$$\mathbf{S}(\nu) = \begin{bmatrix} \frac{1}{1+(2\pi\nu)^2} + \frac{1}{4+(2\pi\nu)^2} & \frac{1}{1+(2\pi\nu)^2} \\ \frac{1}{1+(2\pi\nu)^2} & \frac{1}{1+(2\pi\nu)^2} \end{bmatrix}$$

Factorization of $S_x(\nu) = [1/1 + (2\pi\nu)^2] + [1/4 + (2\pi\nu)^2]$ yields

$$\hat{h}(\nu) = \frac{\sqrt{5} + \sqrt{2}\, i2\pi\nu}{(1 + i2\pi\nu)(2 + i2\pi\nu)}$$

This gives

$$\frac{S_{xy}(\nu)}{\hat{h}(\nu)} = \frac{2 + i2\pi\nu}{[\sqrt{5} + \sqrt{2}\,(i2\pi\nu)](1 - i2\pi\nu)}$$

Therefore, (9.24) can be computed to yield

$$g(t) = \int_{-\infty}^{\infty} e^{i2\pi\nu t} \frac{2 - i2\pi\nu}{(1 + i2\pi\nu)[\sqrt{5} - \sqrt{2}\,(i2\pi\nu)]}\, d\nu$$

$$= \int_{-\infty}^{\infty} \left(\frac{3}{\sqrt{5} + \sqrt{2}} \frac{1}{1 + i2\pi\nu} \right.$$

$$\left. + \frac{2 + \sqrt{5/2}}{1 + \sqrt{5/2}} \frac{1}{\sqrt{5} - \sqrt{2}\, i2\pi\nu} \right) e^{i2\pi\nu t}\, d\nu$$

$$= \begin{cases} \dfrac{3}{\sqrt{5} + \sqrt{2}} e^{-t} & t > 0 \\ \dfrac{2 - \sqrt{5/2}}{\sqrt{5} + \sqrt{2}} \exp\left(\sqrt{\dfrac{5}{2}}\, t\right) & t < 0 \end{cases}$$

and (9.28) gives

$$\hat{\gamma}(\nu) = \frac{3}{\sqrt{5} + \sqrt{2}} \frac{1}{1 + i2\pi\nu}$$

Finally, from (9.28) we get

$$\hat{E}(Y_t|\mathcal{H}_{X^t}) = \int_{-\infty}^{\infty} \frac{3}{\sqrt{5} + \sqrt{2}} \frac{2 + i2\pi\nu}{\sqrt{5} + \sqrt{2}\, i2\pi\nu} e^{i2\pi\nu t}\, d\hat{X}_\nu$$

$$= \frac{3}{2 + \sqrt{10}} \left[X_t + (2 - \sqrt{5/2}) \int_{-\infty}^{t} \exp \right.$$

$$\left. [-\sqrt{5/2}\,(t - s)] X_s\, ds \right] \quad (9.31a)$$

It is quite obvious that our formulation of the filtering problem can be extended to cover cases where both the process to be estimated and the observation process are vector valued. The general case involves a pair of vector-valued processes $\{\mathbf{X}_t, \mathbf{Y}_t, t \in T\}$, not necessarily of the same dimension, whose second-order properties are completely known. The problem is then to estimate $\mathbf{Y}_t$ given $\{\mathbf{X}_s, s \leq t\}$. In wide-sense stationary cases where spectral-density functions exist, one would expect that the Wiener theory can again be developed. If we denote by $\mathbf{A}^+$ the Hermitian adjoint of a matrix $\mathbf{A}$, we can define the matrix spectral-density function $\mathbf{S}_x$ by

$$E\mathbf{X}_{t+\tau}\mathbf{X}_\tau^+ = \int_{-\infty}^{\infty} \mathbf{S}_x(\nu) e^{i2\pi\nu t}\, d\nu$$

The key step in obtaining a solution to the Wiener filtering problem is again a spectral-factorization problem. But now, we have to obtain a factorization of the matrix-valued function $\mathbf{S}_x(\nu)$ into the form

$$\mathbf{S}_x(\nu) = \hat{\mathbf{h}}(\nu)\hat{\mathbf{h}}^+(\nu)$$

such that the matrix $\hat{\mathbf{h}}(\nu)$ satisfies conditions similar to (9.10b) and (9.10c). Except that instead of (9.10c), the determinant of $\hat{\mathbf{h}}(u+iv)$ is to have no zero for $v > 0$. The final solution can be expressed in a form which generalizes (9.29) as

$$\hat{E}(\mathbf{Y}_t | \mathcal{3C}_X{}^t) = \int_{-\infty}^{\infty} e^{i2\pi\nu t}[\hat{\mathbf{h}}^{-1}(\nu)\boldsymbol{\gamma}(\nu)]\, d\hat{\mathbf{X}}_\nu$$

where the matrix $\boldsymbol{\gamma}(\nu)$ is similarly defined, as in the scalar case. The matrix spectral-factorization problem is considerably more difficult than the scalar problem. For the rational case, a number of finite algorithms to achieve the factorization have been derived [Wiener and Masani, 1958; Wong and Thomas, 1961; Youla, 1961].

In some areas of application, the Wiener formulation of the filtering problem is not appropriate because of some of its inherent assumptions. Among these are the following: (1) wide-sense stationarity and existence of spectral densities; (2) the second-order properties of the processes $\{X_t, Y_t, -\infty < t < \infty\}$ are known, and no other information is known; (3) the estimator is to be based on the infinite past of the observation process. These limitations are removed in the formulation of the filtering problem due to Kalman and Bucy. Instead, they made other assumptions which are more natural in a great variety of applications. The form of the solution is also different. While in the Wiener theory, the final solution is in the form of a time-invariant linear and nonanticipative filter, the Kalman-Bucy theory yields a differential equation which is satisfied by the estimator. Implementation of the "filter" in feedback form is thus immediate [Kalman and Bucy, 1961].

The Kalman-Bucy filter problem is usually stated in vector form as follows: Let $\{\mathbf{X}_t, \mathbf{Y}_t, t \geq t_0\}$ be a pair of vector-valued second-order processes. The X process will be the observation process, and the Y process is to be estimated. While this notational convention is consistent with our earlier discussion, it is not universal. Often in the literature, the two letters X and Y are used in just the opposite way. The basic assumptions are the following: Throughout, **boldface** will be used to denote vectors and matrices, prime denotes transpose, + denotes Hermitian adjoint, and **I** denotes identity matrix.

1. The process to be estimated satisfies

$$\dot{\mathbf{Y}}_t = \mathbf{F}(t)\mathbf{Y}_t + \mathbf{A}(t)\boldsymbol{\zeta}_t \qquad t > t_0 \tag{9.32}$$

where $\boldsymbol{\zeta}_t$ has components that are orthogonal white-noise processes with

$$E\boldsymbol{\zeta}_t\boldsymbol{\zeta}_s^+ = \delta(t-s)\,\mathbf{I} \tag{9.33}$$

2. The observation process satisfies

$$\dot{\mathbf{X}}_t = \mathbf{H}(t)\mathbf{Y}_t + \mathbf{B}(t)\boldsymbol{\zeta}_t \qquad t > t_0 \tag{9.34}$$

Remarks:

(a) Both (9.32) and (9.34) are to be interpreted along the lines discussed in Sec. 8. We shall denote the process with orthogonal components which correspond to $\boldsymbol{\zeta}_t$ by $\mathbf{Z}_t$, so that formally $\dot{\mathbf{Z}}_t = \boldsymbol{\zeta}_t$. If $t_0 > -\infty$, it is convenient to set $\mathbf{Z}_{t_0} = 0$.

(b) We note that (9.34) is not really a differential equation, since $\mathbf{X}_t$ can be immediately expressed explicitly in terms of the $\mathbf{Y}$ and $\boldsymbol{\zeta}$ process by integrating (9.34). However, the problem would be no more general if we replace (9.34) by a linear differential equation in $\mathbf{X}_t$. Such an equation can always be changed into (9.34) by redefining the observation process.

(c) It is necessary to assume that the initial values $\mathbf{X}_{t_0}$ and $\mathbf{Y}_{t_0}$ are random variables orthogonal to $\{\mathbf{Z}_t,\, t \geq t_0\}$, in particular they can simply be constants.

As usual, let $\mathcal{3C}_X{}^t$ denote the smallest Hilbert space generated by $\{\mathbf{X}_\tau, t_0 \leq \tau \leq t\}$, and let $\hat{E}$ denote projection. The Kalman-Bucy filtering problem is to find $\hat{E}(\mathbf{Y}_s|\mathcal{3C}_X{}^t)$, and the main results can be summarized as follows.

Proposition 9.4. Let $\{\mathbf{X}_t, \mathbf{Y}_t, t \geq t_0\}$ satisfy (9.32) and (9.34). Let $\boldsymbol{\Phi}(s|t)$ be the unique solution of

$$\frac{d}{ds}\boldsymbol{\Phi}(s|t) = \mathbf{F}(s)\boldsymbol{\Phi}(s|t) \qquad s > t \tag{9.35}$$

with initial condition $\boldsymbol{\Phi}(t|t) = \mathbf{I}$. Let $\mathbf{A}(t)$ and $\mathbf{B}(t)$ in (9.32) and (9.33) be continuous functions on $[t_0, \infty)$.

9. LINEAR PREDICTION AND FILTERING

(a) For $s \geq t$,

$$\hat{E}(\mathbf{Y}_s|\mathcal{K}_X{}^t) = \boldsymbol{\Phi}(s|t)\hat{E}(\mathbf{Y}^t|\mathcal{K}_X{}^t) \tag{9.36}$$

(b) Let $\tilde{\mathbf{Y}}_t$ denote $\hat{E}(\mathbf{Y}_t|\mathcal{K}_X{}^t)$, then

$$\dot{\tilde{\mathbf{Y}}}_t = \mathbf{F}(t)\tilde{\mathbf{Y}}_t + \mathbf{K}(t)[\dot{\mathbf{X}}_t - \mathbf{H}(t)\tilde{\mathbf{Y}}_t] \tag{9.37}$$

(c) Let $\boldsymbol{\varepsilon}_t = \mathbf{Y}_t - \tilde{\mathbf{Y}}_t$ be the error vector, and let $\boldsymbol{\Sigma}(t)$ denote the covariance matrix

$$\boldsymbol{\Sigma}(t) = E\boldsymbol{\varepsilon}_t\boldsymbol{\varepsilon}_t{}^+ \tag{9.38}$$

Then,

$$\dot{\boldsymbol{\Sigma}}(t) = [\mathbf{A}(t) - \mathbf{K}(t)\mathbf{B}(t)][\mathbf{A}(t) - \mathbf{K}(t)\mathbf{B}(t)]^+ \\ + [\mathbf{F}(t) - \mathbf{K}(t)\mathbf{H}(t)]\boldsymbol{\Sigma}(t) + \boldsymbol{\Sigma}(t)[\mathbf{F}(t) - \mathbf{K}(t)\mathbf{H}(t)]^+ \tag{9.39}$$

and

$$\mathbf{K}(t)\mathbf{B}(t)\mathbf{B}^+(t) = \mathbf{A}(t)\mathbf{B}^+(t) + \boldsymbol{\Sigma}(t)\mathbf{H}^+(t) \tag{9.40}$$

Remarks:

(a) A complete proof is rather complicated, and will not be presented. Instead, we shall give a heuristic derivation.

(b) The continuity conditions on $\mathbf{A}(t)$ and $\mathbf{B}(t)$ are sufficient, but not necessary. However, some smoothness condition is needed. Unfortunately, this point is largely lost in our formal derivation.

(c) Equation (9.39) can be simplified somewhat by using (9.40). If $\mathbf{B}(t)\mathbf{B}^+(t)$ is invertible, these two equations can be combined to give a single equation in $\boldsymbol{\Sigma}(t)$, which is a nonlinear differential equation of the Riccati type.

(d) Once $\mathbf{K}$ is determined from (9.39) and (9.40), implementation of (9.37) in feedback form is immediate and yields a continuous estimate. Feedback implementation of (9.37) is often referred to as the **Kalman-Bucy filter**.

First, we derive (9.36) as follows: From (9.32) we can write for $s \geq t$,

$$\mathbf{Y}_s = \boldsymbol{\Phi}(s|t)\mathbf{Y}_t + \int_t^s \boldsymbol{\Phi}(s|t)\mathbf{A}(\tau)\,d\mathbf{Z}_\tau$$

Therefore,

$$\hat{E}(\mathbf{Y}_s|\mathcal{K}_X{}^t) = \boldsymbol{\Phi}(s|t)\tilde{\mathbf{Y}}_t + \int_t^s \boldsymbol{\Phi}(s|\tau)\mathbf{A}(\tau)\,\hat{E}(d\mathbf{Z}_\tau|\mathcal{K}_X{}^t)$$

Now, let $\mathcal{K}^t$ denote the smallest Hilbert space containing $\mathbf{X}_{t_0}$, $\mathbf{Y}_{t_0}$, and $\{\mathbf{Z}_\tau,\, \tau \leq t\}$. Because of (9.32) and (9.34), $\mathcal{K}_X{}^t$ is contained in $\mathcal{K}^t$. Because

$\mathbf{Z}_t$ is a process with orthogonal increments, and $\mathbf{X}_{t_0}$, $\mathbf{Y}_{t}$ are orthogonal to $\{\mathbf{Z}_t, t \geq t_0\}$,

$$\hat{E}(d\mathbf{Z}_\tau | \mathcal{H}^t) = 0 \qquad \tau \geq t$$

Hence,

$$\hat{E}(d\mathbf{Z}_\tau | \mathcal{H}_X^t) = \hat{E}[\hat{E}(d\mathbf{Z}_\tau | \mathcal{H}^t) | \mathcal{H}_X^t] = 0 \qquad \tau \geq t$$

It follows that

$$E(\mathbf{Y}_s | \mathcal{H}_X^t) = \Phi(s|t)\tilde{\mathbf{Y}}_t$$

which was to be derived.

To derive (9.37), we first note that every element in $\mathcal{H}_X^t$ can be written in the form of

$$\xi = \alpha \mathbf{X}_{t_0} + \int_{t_0}^t \beta(\tau) \, d\mathbf{X}_\tau$$

Since $\tilde{\mathbf{Y}}_t \in \mathcal{H}_X^t$ for each t, we can write

$$\tilde{\mathbf{Y}}_t = \alpha(t)\mathbf{X}_{t_0} + \int_{t_0}^t \mathbf{K}(t|\tau) \, d\mathbf{X}_\tau \tag{9.41}$$

Thus,

$$d\tilde{\mathbf{Y}}_t = \mathbf{K}(t|t) \, d\mathbf{X}_t + dt \left[\dot{\alpha}(t)\mathbf{X}_{t_0} + \int_{t_0}^t \frac{\partial}{\partial t} \mathbf{K}(t|\tau) \, d\mathbf{X}_\tau \right] \tag{9.42}$$

Since the bracketed terms in (9.42) are in $\mathcal{H}_X^t$, we can rewrite (9.42) as

$$d\tilde{\mathbf{Y}}_t = \mathbf{K}(t|t) \, d\mathbf{X}_t + \hat{E}[d\tilde{\mathbf{Y}}_t - \mathbf{K}(t|t) \, d\mathbf{X}_t | \mathcal{H}_X^t] \tag{9.43}$$

Now, from (9.34)

$$\hat{E}(d\mathbf{X}_t | \mathcal{H}_X^t) = \mathbf{H}(t)\tilde{\mathbf{Y}}_t \, dt \tag{9.44}$$

From (9.32), we have

$$\hat{E}(d\tilde{\mathbf{Y}}_t | \mathcal{H}_X^t) = \hat{E}(\hat{E}(\mathbf{Y}_{t+dt} | \mathcal{H}_X^{t+dt}) - \hat{E}(\mathbf{Y}_t | \mathcal{H}_X^t) | \mathcal{H}_X^t)$$
$$= \hat{E}(d\mathbf{Y}_t | \mathcal{H}_X^t) = \mathbf{F}(t)\tilde{\mathbf{Y}}_t \, dt \tag{9.45}$$

Using (9.44) and (9.45) in (9.42) yields

$$d\tilde{\mathbf{Y}}_t = \mathbf{F}(t)\tilde{\mathbf{Y}}_t \, dt + \mathbf{K}(t|t)[d\mathbf{X}_t - \mathbf{H}(t)\tilde{\mathbf{Y}}_t \, dt]$$

which is just (9.37) if we set $\mathbf{K}(t|t) = \mathbf{K}(t)$.

To derive (9.39), we combine (9.32), (9.34) and (9.37) and obtain

$$d\boldsymbol{\varepsilon}_t = [\mathbf{F}(t) - \mathbf{K}(t)\mathbf{H}(t)]\boldsymbol{\varepsilon}_t + [\mathbf{A}(t) - \mathbf{K}(t)\mathbf{B}(t)] \, d\mathbf{Z}_t \tag{9.46}$$

9. LINEAR PREDICTION AND FILTERING

Now,

$$\begin{aligned}
d\mathbf{\Sigma}(t) &= \mathbf{\Sigma}(t+dt) - \mathbf{\Sigma}(t) \\
&= E\boldsymbol{\varepsilon}_{t+dt}\boldsymbol{\varepsilon}_{t+dt}^+ - E\boldsymbol{\varepsilon}_t\boldsymbol{\varepsilon}_t^+ \\
&= E(\boldsymbol{\varepsilon}_{t+dt} - \boldsymbol{\varepsilon}_t)(\boldsymbol{\varepsilon}_{t+dt} - \boldsymbol{\varepsilon}_t)^+ + E(\boldsymbol{\varepsilon}_{t+dt} - \boldsymbol{\varepsilon}_t)\boldsymbol{\varepsilon}_t^+ + E\boldsymbol{\varepsilon}_t(\boldsymbol{\varepsilon}_{t+dt} - \boldsymbol{\varepsilon}_t)^+ \\
&= E\,d\boldsymbol{\varepsilon}_t\,d\boldsymbol{\varepsilon}_t^+ + E\,d\boldsymbol{\varepsilon}_t\boldsymbol{\varepsilon}_t^+ + E\boldsymbol{\varepsilon}_t\,d\boldsymbol{\varepsilon}_t^+
\end{aligned}$$

Using (9.46) and the fact that $d\mathbf{Z}_t$ is orthogonal to $\mathcal{K}^t$, we find

$$\begin{aligned}
d\mathbf{\Sigma}(t) &= [\mathbf{A}(t) - \mathbf{K}(t)\mathbf{B}(t)][\mathbf{A}(t) - \mathbf{K}(t)\mathbf{B}(t)]^+\,dt \\
&\quad + [\mathbf{F}(t) - \mathbf{K}(t)\mathbf{H}(t)]\mathbf{\Sigma}(t)\,dt + \mathbf{\Sigma}(t)[\mathbf{F}(t) - \mathbf{K}(t)\mathbf{H}(t)]^+\,dt
\end{aligned}$$

which is just (9.39).

Finally, to derive (9.40), we begin by writing $\boldsymbol{\varepsilon}_t$ as the solution of the differential equation (9.46) in the form

$$\boldsymbol{\varepsilon}_t = \boldsymbol{\psi}(t|t_0)\boldsymbol{\varepsilon}_{t_0} + \int_{t_0}^t \boldsymbol{\psi}(t|\tau)[\mathbf{A}(\tau) - \mathbf{K}(\tau)\mathbf{B}(\tau)]\,d\mathbf{Z}_\tau \qquad (9.47)$$

where $\boldsymbol{\psi}$ satisfies

$$\begin{aligned}
\frac{d}{dt}\boldsymbol{\psi}(t|t_0) &= [\mathbf{F}(t) - \mathbf{K}(t)\mathbf{H}(t)]\boldsymbol{\psi}(t|t_0) \qquad t > t_0 \\
\boldsymbol{\psi}(t|t_0) &= \mathbf{I}
\end{aligned} \qquad (9.48)$$

Since $\boldsymbol{\varepsilon}_t$ is orthogonal to $\mathcal{K}_X^t$, we have for $s \leq t$,

$$E\boldsymbol{\varepsilon}_t \mathbf{X}_s^+ = 0 = E\boldsymbol{\varepsilon}_t\left[\int_{t_0}^s \mathbf{H}(\tau)\mathbf{Y}_\tau\,d\tau + \int_{t_0}^s \mathbf{B}(\tau)\,d\mathbf{Z}_\tau\right]^+ \qquad (9.49)$$

Furthermore, for $\tau \leq t$, $\tilde{\mathbf{Y}}_t \in \mathcal{K}_X^t$, hence

$$E\boldsymbol{\varepsilon}_t\mathbf{Y}_\tau^+ = E\boldsymbol{\varepsilon}_t(\mathbf{Y}_\tau - \tilde{\mathbf{Y}}_\tau)^+ = E\boldsymbol{\varepsilon}_t\boldsymbol{\varepsilon}_\tau^+ \qquad (9.50)$$

Therefore, (9.49) becomes

$$0 = \int_{t_0}^s (E\boldsymbol{\varepsilon}_t\boldsymbol{\varepsilon}_\tau^+)\mathbf{H}^+(\tau)\,d\tau + \int_{t_0}^s \boldsymbol{\psi}(t|\tau)[\mathbf{A}(\tau) - \mathbf{K}(\tau)\mathbf{B}(\tau)]\mathbf{B}^+(\tau)\,d\tau \qquad s \leq t \quad (9.51)$$

whence for $s < t$,

$$(E\boldsymbol{\varepsilon}_t\boldsymbol{\varepsilon}_s^+)\mathbf{H}^+(s) + \boldsymbol{\psi}(t|s)\mathbf{A}(s)\mathbf{B}^+(s) = \boldsymbol{\psi}(t|s)\mathbf{K}(s)\mathbf{B}(s)\mathbf{B}^+(s) \qquad (9.52)$$

Letting $s \uparrow t$ and noting the continuity assumptions, we get

$$\mathbf{K}(t)\mathbf{B}(t)\mathbf{B}^+(t) = \mathbf{\Sigma}(t)\mathbf{H}^+(t) + \mathbf{A}(t)\mathbf{B}^+(t)$$

which is (9.40). If we use (9.40) in (9.39), it is not hard to show that (9.39) can be rewritten as

$$\dot{\mathbf{\Sigma}}(t) = \mathbf{A}(t)\mathbf{A}^+(t) - \mathbf{K}(t)\mathbf{B}(t)\mathbf{B}^+(t)\mathbf{K}^+(t) \\
+ \mathbf{F}(t)\mathbf{\Sigma}(t) + \mathbf{\Sigma}(t)\mathbf{F}^+(t) \qquad (9.53)$$

which is a useful alternative form to (9.39). ∎

For an example which illustrates these procedures and illustrates the difference between the Wiener filter and the Kalman-Bucy filter, consider the following problem:

We want to estimate Y_t using data $\{Y_\tau + N_\tau,\ t_0 < \tau \leq t\}$ where $\{Y_t, N_t,\ -\infty < t < \infty\}$ are two uncorrelated wide-sense stationary processes with spectral densities

$$S_Y(\nu) = \frac{1}{1 + (2\pi\nu)^2} \tag{9.54}$$

and

$$S_N(\nu) = \frac{1}{4 + (2\pi\nu)^2} \tag{9.55}$$

In order to use the Kalman-Bucy procedure, we first have to convert the problem into a standard form. From our earlier discussions concerning spectral factorization, we know that Y_t and N_t can be represented as

$$Y_t = \int_{-\infty}^{t} e^{-(t-\tau)}\, dZ_\tau \tag{9.56}$$

and

$$N_t = \frac{1}{2} \int_{-\infty}^{t} e^{-2(t-\tau)}\, dV_\tau \tag{9.57}$$

where Z_τ, V_τ are mutually uncorrelated processes with orthogonal increments. This means that Y_t and N_t satisfy

$$dY_t = -Y_t\, dt + dZ_t \tag{9.58}$$
$$dN_t = -2N_t\, dt + dV_t \tag{9.59}$$

Now, let

$$X_t = e^{2t}(Y_t + N_t) \tag{9.60}$$

Then, X_t satisfies

$$dX_t = e^{2t}(dY_t + 2Y_t\, dt + dN_t + 2N_t\, dt)$$
$$= e^{2t}Y_t\, dt + e^{2t}\, dZ_t + e^{2t}\, dV_t \tag{9.61}$$

We can now identify the quantities F, H, A, B of (9.32) and (9.34) as follows:

$$\begin{aligned}F(t) &= -1 \\ H(t) &= e^{2t} \\ \mathbf{A}(t) &= [1\ 0] \\ \mathbf{B}(t) &= [e^{2t}\ e^{2t}]\end{aligned} \tag{9.62}$$

Equation (9.40) becomes

$$2e^{4t}K(t) = e^{2t} + e^{2t}\Sigma(t) \tag{9.63}$$

9. LINEAR PREDICTION AND FILTERING

and (9.53) becomes

$$\dot{\Sigma}(t) = 1 - 2e^{4t}K^2(t) - 2\Sigma(t)$$
$$= \tfrac{1}{2} - 3\Sigma(t) - \tfrac{1}{2}\Sigma^2(t) \qquad (9.64)$$

Equation (9.64) can be transformed into a linear equation by the substitution $\Sigma(t) = 2\dot{\sigma}(t)/\sigma(t)$, giving us

$$\ddot{\sigma}(t) + 3\dot{\sigma}(t) - \tfrac{1}{4}\sigma(t) = 0 \qquad (9.65)$$

Solving (9.65) with initial condition $2\dot{\sigma}(t_0)/\sigma(t_0) = \Sigma(t_0)$, we find

$$\Sigma(t) = \frac{2\dot{\sigma}(t)}{\sigma(t)}$$
$$= \Sigma(t_0) \left\{ \frac{1 - [3/\sqrt{10} - 1/\sqrt{10}\,\Sigma(t_0)]\tanh(\sqrt{10}/2)(t-t_0)}{1 + [3/\sqrt{10} + \Sigma(t_0)/\sqrt{10}]\tanh(\sqrt{10}/2)(t-t_0)} \right\} \qquad (9.66)$$

The initial value $\Sigma(t_0)$ can be evaluated as follows: First, we note that the linear least-squares estimator of Y_{t_0} given X_{t_0} has the form αX_{t_0}, where α is determined by

$$E(Y_{t_0} - \alpha X_{t_0})\bar{X}_{t_0} = 0 \qquad (9.67)$$

This yields

$$\alpha = \frac{EY_{t_0}\bar{X}_{t_0}}{E|X_{t_0}|^2}$$
$$= e^{-2t_0}\frac{E|Y_{t_0}|^2}{E|Y_{t_0}|^2 + E|N_{t_0}|^2}$$
$$= \tfrac{1}{2}e^{-2t_0} \qquad (9.68)$$

Finally,

$$\Sigma(t_0) = E|Y_{t_0} - \alpha X_{t_0}|^2 = E|\tfrac{1}{2}Y_{t_0} - \tfrac{1}{2}N_{t_0}|^2$$
$$= \tfrac{1}{4}(E|Y_{t_0}|^2 + E|N_{t_0}|^2) = \tfrac{1}{4} \qquad (9.69)$$

This completes the solution for $\Sigma(t)$, and via (9.63), also completes the solution for $K(t)$, and hence the Kalman-Bucy filter.

If we let $t_0 \to -\infty$ in (9.66), we get

$$\Sigma(t) \xrightarrow[t_0 \to -\infty]{} (\sqrt{10} - 3) \qquad (9.70)$$

which yields $K(t) = (\sqrt{5/2} - 1)e^{-2t}$ and (9.37) becomes

$$d\tilde{Y}_t = -\sqrt{\tfrac{5}{2}}\,\tilde{Y}_t\,dt + (\sqrt{\tfrac{5}{2}} - 1)e^{-2t}\,dX_t \qquad (9.71)$$

This gives us

$$\tilde{Y}_t = \int_{-\infty}^{t} (\sqrt{\tfrac{5}{2}} - 1) \exp[-\sqrt{\tfrac{5}{2}}(t-\tau)]e^{-2\tau}\,dX_\tau$$
$$= \int_{-\infty}^{t} (\sqrt{\tfrac{5}{2}} - 1) \exp[-\sqrt{\tfrac{5}{2}}(t-\tau)]\,(d\xi_\tau + 2\xi_\tau\,d\tau) \qquad (9.72)$$

where we have set $\xi_\tau = e^{-2t}X_\tau = Y_\tau + N_\tau$. The final expression in (9.72) represents the output of a linear time-invariant filter with $Y_t + N_t$ as input and with a transfer function given by

$$\left(\sqrt{\tfrac{5}{2}} - 1\right)\frac{(2\pi i\nu) + 2}{(2\pi i\nu) + \sqrt{\tfrac{5}{2}}} = \frac{3}{\sqrt{5} + \sqrt{2}}\frac{2 + i2\pi\nu}{\sqrt{5} + \sqrt{2}\,i2\pi\nu}$$

which should be compared with (9.31a).

EXERCISES

1. Test whether each of the following functions is nonnegative definite.

(a) $R(t,s) = \begin{cases} 1 - |t-s| & |t-s| \leq 1 \\ 0 & |t-s| > 1 \end{cases}$

(b) $R(t,s) = e^{-|t-s|} \qquad -\infty < t, s < \infty$

(c) $R(t,s) = e^{|t-s|}$

(d) $R(t,s) = \begin{cases} 1 & |t-s| \leq 1 \\ 0 & |t-s| > 1 \end{cases}$

2. Let $\lambda_0, \lambda_1, \ldots$ be the eigenvalues of (4.12).

(a) Show that

$$R(t,t) - \sum_{n=0}^{N} \lambda_n |\varphi_n(t)|^2 = E\left| X_t - \sum_{n=0}^{N} \varphi_n(t) \int_a^b \bar{\varphi}_n(s) X_s\,ds \right|^2 \geq 0$$

(b) Show that

$$\sum_{n=0}^{N} \lambda_n \leq \int_a^b R(t,t)\,dt < \infty$$

Hence, $\sum_{n=0}^{N} \lambda_n$ must converge and $\lambda_n \xrightarrow[n\to\infty]{} 0$.

3. Suppose that a q.m. continuous and wide-sense stationary process $\{X_t, -\infty < t < \infty\}$ has a covariance function $R(\cdot)$ which is periodic with period T, that is,

$R(\tau + T) = R(\tau) \qquad$ for all τ in $(-\infty, \infty)$

EXERCISES

Define for $n = 0, \pm 1, \pm 2, \ldots$

$$Z_n = \frac{1}{T} \int_0^T X_t e^{-in(2\pi/T)t}$$

(a) Show that $\{Z_n\}$ are mutually orthogonal, that is,

$$EZ_m \bar{Z}_n = 0 \quad \text{whenever } m \neq n$$

(b) Show that for each t

$$X_t = \sum_{n=-\infty}^{\infty} Z_n e^{in(2\pi/T)t}$$

(c) Suppose that

$$R(\tau) = \sum_{n=0}^{\infty} \frac{1}{1+n^2} \cos n \left(\frac{2\pi}{T}\right) \tau$$

find the eigenvalues and a complete set of orthonormal eigenfunctions to

$$\int_0^T R(t-s)\varphi(s)\, ds = \lambda \varphi(t) \quad 0 \leq t \leq T$$

4. Let $R(t,s)$ be given by

$$R(t,s) = \begin{cases} 1 - |t-s| & 0 \leq |t-s| \leq 1 \\ 0 & \text{elsewhere} \end{cases}$$

Find the eigenvalues and a complete set of O-N eigenfunctions for

$$\int_0^{\frac{1}{2}} R(t,s)\varphi(s)\, ds = \lambda \varphi(t) \quad 0 \leq t \leq \tfrac{1}{2}$$

Hint: $\dfrac{\partial^2 R(t,s)}{\partial t^2} = -2\delta(t-s) \quad \text{for } 0 < t, s < \tfrac{1}{2}$

5. (a) Suppose that $R(t,s)$ satisfies

$$\sum_{k=0}^{n} a_k \frac{\partial^{2k}}{\partial t^{2k}} R(t,s) = \sum_{k=0}^{m} b_k \frac{\partial^{2k}}{\partial t^{2k}} \delta(t-s) \quad a < t, s < b$$

Show that the integral equation

$$\int_a^b R(t,s)\varphi(s)\, ds = \lambda \varphi(t) \quad a \leq t \leq b$$

can be reduced to the differential equation

$$\lambda \sum_{k=0}^{n} a_k \frac{\partial^{2k} \varphi(t)}{\partial t^{2k}} = \sum_{k=0}^{m} b_k \frac{\partial^{2k} \varphi(t)}{\partial t^{2k}} \quad a < t < b$$

together with appropriate boundary conditions.

(b) Verify that, formally at least, if $R(t,s)$ satisfies

$$R(t,s) = \int_{-\infty}^{\infty} \frac{\sum_{k=0}^{m} b_k (2\pi i \nu)^{2k}}{\sum_{k=0}^{n} a_k (2\pi i \nu)^{2k}} e^{i 2\pi \nu (t-s)} d\nu$$

then it also satisfies

$$\sum_{k=0}^{n} a_k \frac{\partial^{2k}}{\partial t^{2k}} R(t,s) = \sum_{k=0}^{m} b_k \frac{\partial^{2k}}{\partial t^{2k}} \delta(t-s)$$

(c) Reduce the equation

$$\int_a^b e^{-|t-s|} \varphi(s) \, ds = \lambda \varphi(t) \qquad a \le t \le b$$

to a differential equation and two linearly independent boundary conditions on $\varphi(a)$, $\varphi(b)$, $\dot\varphi(a)$, and $\dot\varphi(b)$. [Hint: Each of these four quantities can be expressed as a linear combination of $\int_a^b e^{-s} \varphi(s) \, ds$ and $\int_a^b e^{s} \varphi(s) \, ds$.]

6. Suppose that $\{W_t, t \ge 0\}$ is a standard Brownian motion and

$$X_t(\omega) = f(t) W_{\tau(t)}(\omega) \qquad 0 \le t \le T$$

where $f(t)$ is a continuous function and $\tau(t)$ is a continuous nondecreasing function with $\tau(0) = 0$. Using the results on the solutions of (4.31), find an expansion of this form

$$X_t(\omega) = \sum_{n=0}^{\infty} \alpha_n(t) Z_n(\omega)$$

where $\{Z_n, n = 0, 1, 2, \ldots\}$ are independent Gaussian random variables with $EZ_m Z_n = \delta_{mn}$. What conditions do we need, if any, in order that each Z_n belongs to $\mathcal{H}_X$?

7. Suppose that $\{X_t, -\infty < t < \infty\}$ has a covariance function

$R(\tau) = \frac{1}{8} e^{-|\tau|}(3 \cos \tau + \sin |\tau|)$

Find its spectral-density function $S(\nu)$, $-\infty < \nu < \infty$.

8. Suppose that $\{X_t, -\infty < t < \infty\}$ has the spectral density given in Exercise 7, and $\{Y_t, -\infty < t < \infty\}$ is such that:

(a) $Y_t \in \mathcal{H}_X$ for every t

(b) $EY_t \bar X_s = \frac{1}{8} e^{-|t-s|}[\cos(t-s) + \sin(t-s)]$. Find $EY_t \bar Y_s$.

9. Suppose that $\{X_t, -\infty < t < \infty\}$ is wide-sense stationary. Show that for a fixed constant W, $\{e^{i 2\pi W t} X_t, -\infty < t < \infty\}$ is again wide-sense stationary. Is $\{\cos 2\pi W t X_t, -\infty < t < \infty\}$ wide-sense stationary?

EXERCISES

10. Suppose that $X_t = \sum_{k=1}^{N} X_{kt}$, $-\infty < t < \infty$, is the sum of N wide-sense stationary and q.m. continuous processes $X_{1t}, X_{2t}, \ldots, X_{Nt}$. Show that we have a representation

$$X_t = \int_{-\infty}^{\infty} e^{i2\pi\nu t}\, d\hat{X}_\nu$$

Is the process $\{\hat{X}_\nu, -\infty < \nu < \infty\}$ always a process with orthogonal increments? (Note: $X_t = \cos 2\pi W t Z_t$, with Z_t stationary, is an example of such a process.)

11. For a process of the type given in Exercise 10, show that

$$\int_{-\infty}^{\infty} h(t-\tau) X_\tau\, d\tau = \int_{-\infty}^{\infty} \hat{h}(\nu) e^{i2\pi\nu t}\, d\hat{X}_\nu$$

12. Let $\{X_t, -\infty < t < \infty\}$ be a wide-sense stationary process with a spectral representation

$$X_t = \int_{-\infty}^{\infty} e^{i2\pi\nu t}\, d\hat{X}_\nu$$

Let

$$Y_t = \int_{-\infty}^{\infty} \eta(\nu) e^{i\varphi(\nu) t}\, d\hat{X}_\nu$$

where $\varphi(\nu)$, $-\infty < \nu < \infty$, is a real-valued function. Show that $\{Y_t, -\infty < t < \infty\}$ is wide-sense stationary.

13. Suppose that $\{X_t, -\infty < t < \infty\}$ is a real-valued wide-sense stationary process, and let $\tilde{X}_t$ denote its Hilbert transform

$$\tilde{X}_t = \int_{-\infty}^{\infty} (-i\, \mathrm{sgn}\, \nu) e^{i2\pi\nu t}\, d\hat{X}_\nu$$

Let $EX_t X_s = e^{-|t-s|}$.

(a) Show that $\tilde{X}_t$ is real valued

(b) Find $E\tilde{X}_t\tilde{X}_s$ and $E\tilde{X}_t X_s$

(c) Let $Z_t = \cos 2\pi\nu_0 t X_t + \sin 2\pi\nu_0 t \tilde{X}_t$. Show that Z_t is of the form

$$Z_t = \int_{-\infty}^{\infty} e^{i\varphi(\nu) t}\, d\hat{X}_\nu$$

with $\varphi(\nu) = 2\pi[\nu - \nu_0\, \mathrm{sgn}\, \nu]$. [Hence, it must be wide-sense stationary (see Exercise 12).]

14. Let $\{Z_t, -\infty < t < \infty\}$ be a process with orthogonal increments such that $EZ_t = 0$, $E|Z_t - Z_s|^2 = |t - s|$ and $Z_0 = 0$.

(a) Show that

$$EZ_t \bar{Z}_s = \tfrac{1}{2}(|t| + |s| - |t - s|)$$

(b) Let $\{\zeta_{nt}, -\infty < t < \infty\}$ be defined by

$$\zeta_{nt} = n(Z_{t+1/n} - Z_t) \qquad n = 1, 2, \ldots$$

Show that for each n, $\{\zeta_{nt}, -\infty < t < \infty\}$ is wide-sense stationary and find its spectral-density function $S_n(\nu)$, $-\infty < \nu < \infty$.

(c) Show that ζ_{nt} converges to a white noise in the sense of (8.2) and (8.3).

15. Let f be a differentiable function such that its derivative $\dot{f}$ is continuous on $[a,b]$. Show that

$$\int_a^b f(t)\, dZ_t + \int_a^b \dot{f}(t) Z_t\, dt = f(b) Z_b - f(a) Z_a$$

where $\{Z_t, -\infty < t < \infty\}$ is the process described in Exercise 14. (Hint: Make use of the sequence $\{\zeta_{nt}\}$ defined in Exercise 14 and show that

$$\int_a^b \dot{f}(t) \int_0^t \zeta_{ns}\, ds\, dt \xrightarrow[n \to \infty]{\text{q.m.}} \int_a^b \dot{f}(t) Z_t\, dt$$

The rest is easy.)

16. Use the results of Exercise 15 and show that the solution of the integral equation

$$Y_t = Y_a - \int_a^t Y_s\, ds + Z_t - Z_a$$

is given by $Y_t = Y_a e^{-(t-a)} + \int_a^t e^{-(t-s)}\, dZ_s$.

17. Suppose that $\{X_t, a \leq t \leq b\}$ is a q.m. continuous process such that

$$\frac{\partial^2}{\partial t\, \partial s} E X_t \bar{X}_s = \delta(t - s) + \rho(t,s) \qquad a < t < b$$

where ρ is a continuous nonnegative definite function on $[a,b]$.

(a) Show that $X(f) \equiv \int_a^b f(t)\, dX_t$ is well defined for all f satisfying $\int_a^b |f(t)|^2\, dt$ such that

(1) $X(\cdot)$ is linear, that is, $X(\alpha f + \beta g) = \alpha X(f) + \beta X(g)$
(2) If $a \leq c < d \leq b$ and $f = I_{cd}$ is the indicator function of $[c,d)$, then $X(f) = X_d - X_c$

(b) Find an expression for $EX(f)\bar{X}(g)$

(c) Show that every element Y in $\mathcal{K}_X$ can be represented as

$$Y = \int_a^b \eta(t)\, dX_t + k X_a$$

18. Let $\{W_t, t \geq 0\}$ be a standard Brownian motion process, and let A be a real-valued second-order random variable ($EA = 0$, $EA^2 = 1$) independent of the W process. Suppose that the process

$$X_t = At + W_t \qquad t \geq 0$$

is observed. Let $\tilde{A}_t$ denote the linear least-squares estimator of A given X_s, $0 \leq s \leq t$. Find an explicit expression for $\tilde{A}_t$ in the form of

$$\tilde{A}_t = \int_0^t h(t,s)\, dX_s$$

19. Let $X_t = Y \cos 2\pi W t + N_t$, where $\{N_t,\ -\infty < t < \infty\}$ is a second-order process with zero mean and $E N_t \bar{N}_s = e^{-\nu_0|t-s|}$, and Y is a random variable with $EY = 0$, $E|Y|^2 = 1$ and $E Y \bar{N}_t = 0$ for all t. Let $\tilde{Y}_t$ denote the linear least-squares estimator of Y given X_s, $0 \leq s \leq t$.

(a) Find $\tilde{Y}_t$ in the form of

$$\tilde{Y}_t = \int_0^t h(t,s)\, dX_s + X_0$$

(b) Find $\tilde{Y}_t$ as the solution of a Kalman filtering problem with observation process $Z_t = e^{\nu_0 t} X_t$. That is, show that $\tilde{Y}_t$ satisfies

$$\dot{\tilde{Y}}_t = \alpha(t)\tilde{Y}_t + K(t)[\dot{Z}_t - \beta(t)\tilde{Y}_t]$$

and find $\alpha(\cdot)$, $K(\cdot)$, and $\beta(\cdot)$.

20. Suppose that $\{X_t,\ -\infty < t < \infty\}$ has a spectral density given by

$$S_x(\nu) = \frac{1}{4 + (2\pi\nu)^4} \qquad -\infty < \nu < \infty$$

Find the predictor $\hat{E}(X_{t+\alpha}|\mathcal{K}_X{}^t)$, and express it in the form

$$\hat{E}(X_{t+\alpha}|\mathcal{K}_X{}^t) = \int_{-\infty}^{\infty} e^{i2\pi\nu t} H_\alpha(\nu)\, d\hat{X}_\nu$$

21. Let $\{X_t,\ -\infty < t < \infty\}$ be as in Exercise 20, and let $\{N_t,\ -\infty < t < \infty\}$ have spectral density $S_N(\nu) = 1/[1 + (2\pi\nu)^2]$. Assume that $E X_t \bar{N}_s = 0$ for all t and s and define

$$Y_t = X_t + N_t \qquad -\infty < t < \infty$$

Find $\hat{E}(X_t|\mathcal{K}_Y{}^t)$ and express it in the form

$$\hat{E}(X_t|\mathcal{K}_Y{}^t) = \int_{-\infty}^{\infty} e^{i2\pi\nu t} H(\nu)\, d\hat{Y}_\nu$$

22. Suppose that the quantities **A**, **B**, **F**, and **H** appearing in (9.32) and (9.34) are given as follows:

$$\mathbf{A}(t) = \begin{bmatrix} 1 \\ 0 \end{bmatrix}$$
$$\mathbf{B}(t) = 1$$
$$\mathbf{F}(t) = \begin{bmatrix} 0 & -1 \\ 1 & 0 \end{bmatrix}$$
$$\mathbf{H}(t) = [0 \quad 1]$$

Find the matrix $\mathbf{K}(t)$ appearing in (9.37) for $t \geq 0$, assuming $\mathbf{\Sigma}(0) = \mathbf{0}$.

23. Suppose that Y_t satisfies a differential equation

$$\ddot{Y}_t = Y_t + \eta_t$$

where $E\eta_t\bar{\eta}_s = \delta(t-s)$. Let X_t satisfy

$$\dot{X}_t = Y_t + \xi_t$$

with $E\eta_t\bar{\xi}_s = 0$ and $E\xi_t\bar{\xi}_s = \delta(t-s)$. Reexpress these equations in the form of (4.32) and (4.34) with

$$\mathbf{Y}_t = \begin{bmatrix} \dot{Y}_t \\ Y_t \end{bmatrix} \quad \text{and} \quad \zeta_t = \begin{bmatrix} \eta_t \\ \xi_t \end{bmatrix}$$

24. Suppose that in Exercise 23 the process ξ_t, instead of being white, satisfies $E\xi_t\bar{\xi}_s = e^{-|t-s|}$. Reexpress the two differential equations in the form of (4.32) and (4.34) with suitable choices for $\mathbf{X}_t$ and ζ_t. (Hint: Now $(d/dt)\xi_t + \xi_t$ is a white noise.)

4
Stochastic Integrals and Stochastic Differential Equations

1. INTRODUCTION

Roughly speaking, stochastic differential equations are differential equations driven by Gaussian white noise. Here, we are using the term "stochastic differential equations" in a restricted sense and not merely to denote differential equations with some probabilistic aspects. The importance of stochastic differential equations is largely due to the fact that the solution of such an equation is a sample-continuous Markov process, and conversely, a large and important class of sample-continuous Markov processes can be modeled by the solutions of stochastic differential equations. From the point of view of applications, this is a direct benefit of using white noise as a noise model, and this fact accounts for its popularity. After all, white noise is, at best, a tolerable abstraction and is never a completely faithful representation of a physical noise source. Its *raison d'être* is the simplicity of analysis that it brings about. We have seen this in connection with Kalman filtering, and we shall see it again in the Markovian nature of solutions to stochastic differential equations.

Stochastic differential equations are defined in terms of stochastic integrals, which in turn need to be defined. The generally accepted definition for a stochastic integral is the one due to Ito and is often referred to as the Ito integral. We shall call it simply the stochastic integral. In addition to its role in stochastic differential equations, the stochastic integral is of tremendous importance in its own right. The main source of its importance is due to the fact that an important class of martingales can be represented as stochastic integrals. We shall explore this aspect of stochastic integrals a little later.

It turns out that a calculus based on Ito's definition of stochastic integrals is not compatible with rules of ordinary calculus. Since the differential equations that we encounter in practice must usually be interpreted in terms of ordinary calculus, it raises the question whether stochastic differential equations can really be used to model differential equations driven by "approximately white" noise. This is an important question, since the uniqueness, existence, and Markov property of the solution to a stochastic differential equation are all based on the interpretation of the equation in terms of stochastic integrals. We shall pose this question in a more precise way and give an answer. The answer is roughly that a differential equation driven by "nearly white" Gaussian noise can indeed be modeled by a stochastic differential equation, but in general, a "correction term" has to be added to the equation.

The solution of a stochastic differential equation not only is a sample-continuous Markov process, but under quite general conditions, its transition-probability density function can be shown to satisfy a pair of partial differential equations called the backward and forward equations of Kolmogorov, or the diffusion equations. These equations were derived by Kolmogorov some time before Ito's work on stochastic differential equations, and they took on new light in view of Ito's results.

Both Kolmogorov and Ito were motivated in their work by the goal of discovering conditions under which "local" properties of a Markov process completely determine its probability law. This problem is brought into a sharper focus in the case of Markov processes with stationary transition probabilities. In that case, the results of semigroup theory can be used to give a more complete answer to the question, "When do local properties determine completely the probability law?" Theory of Markov semigroups is sufficiently different to require a separate treatment. It will be treated in the next chapter.

Finally, we should mention that by and large we shall restrict ourselves to scalar-valued processes in this chapter. For applications, extension to vector-valued processes is of great importance. This extension is not difficult in spirit, but is rather complicated in detail and involves cumbersome notations.

2. STOCHASTIC INTEGRALS

Let $(\Omega, \mathcal{A}, \mathcal{P})$ be a fixed probability space. Let $\{\mathcal{A}_t, -\infty < t < \infty\}$ be an increasing family of sub-σ algebras of $\mathcal{A}$, and let $\{W_t, -\infty < t < \infty\}$ be a Brownian motion process such that for each s, the aggregate $\{W_t - W_s, t \geq s\}$ is independent of $\mathcal{A}_s$ and W_t is $\mathcal{A}_t$ measurable for each t. It follows that for $s \geq 0$

$$E^{\mathcal{A}_t} W_{t+s} = W_t$$
$$E^{\mathcal{A}_t}(W_{t+s} - W_t)^2 = s \quad \text{a.s.} \qquad (2.1)$$

We recall that we refer to this situation by saying that $\{W_t, \mathcal{A}_t, -\infty < t < \infty\}$ is a Brownian motion. By a **stochastic integral** we mean a quantity of the form

$$I(\varphi) = \int_a^b \varphi(\omega, t) \, dW(\omega, t) \qquad (2.2)$$

Because W_t is neither differentiable nor of bounded variation, (2.2) does not have a well-defined interpretation as an integral in the ordinary sense. If the integrand φ does not depend on ω, then (2.2) can be treated as a second-order stochastic integral as introduced in Sec. 3.6. Interpreted in that way, only the property of a Brownian motion as a process with orthogonal increments is made use of, and not its other properties. If φ is random, i.e., it depends on ω, then (2.2) has to be defined anew. It turns out that its definition now depends in a crucial way on the martingale property (2.1) of W_t.

We make the following assumptions on the integrand φ:

The function φ is jointly measurable in (ω, t) (with respect to $\mathcal{A}$ in ω and the Lebesgue measure in t). For each t, φ_t is measurable with respect to $\mathcal{A}_t$ (2.3)

φ satisfies $\int_a^b E|\varphi_t|^2 \, dt < \infty$ (2.4)

The stochastic integral is now defined in the following way:

1. If there exist times $t_0, t_1, \ldots, t_n$ independent of ω, such that $a = t_0 < t_1 < \cdots < t_n = b$ and

$$\varphi(\omega, t) = \varphi_\nu(\omega) \quad \begin{array}{l} t_\nu \leq t < t_{\nu+1} \\ \nu = 0, \ldots, n-1 \end{array} \qquad (2.5)$$

and if φ satisfies (2.3) and (2.4), then we call φ an (ω, t)-**step function** and define the stochastic integral by

$$\int_a^b \varphi(\omega, t) \, dW(\omega, t) = \sum_{\nu=0}^{n-1} \varphi_\nu(\omega)[W(\omega, t_{\nu+1}) - W(\omega, t_\nu)] \qquad (2.6)$$

2. If φ satisfies (2.3) and (2.4), then we shall show that there exists a sequence of (ω,t)-step functions $\{\varphi_n(\omega,t)\}$ satisfying (2.3) and (2.4) such that

$$\|\varphi - \varphi_n\|^2 = \int_a^b E|\varphi(\cdot,t) - \varphi_n(\cdot,t)|^2 \, dt \xrightarrow[n\to\infty]{} 0 \tag{2.7}$$

It will then follow that $\int_a^b \varphi_n(\omega,t) \, dW(\omega,t)$ converges in q.m. as $n \to \infty$, and this q.m. limit is the same for any sequence of (ω,t)-step functions which satisfy (2.7). Therefore, we can define

$$\int_a^b \varphi(\omega,t) \, dW(\omega,t) = \lim_{n\to\infty} \text{in q.m.} \int_a^b \varphi_n(\omega,t) \, dW(\omega,t) \tag{2.8}$$

where $\{\varphi_n\}$ is any sequence of steps satisfying (2.7). We assume that both a and b are finite. If not, the stochastic integral is defined as the limit in q.m. as $a \to -\infty$, or $b \to \infty$, or both.

Proposition 2.1. Let $\{W_t, -\infty < t < \infty\}$ be a Brownian motion, and let $\varphi(\omega,t)$ satisfy (2.3) and (2.4). Then,
  (a) There exists a sequence of (ω,t) steps $\{\varphi_n\}$ satisfying (2.3) and (2.4) such that

$$\|\varphi - \varphi_n\|^2 = \int_a^b E|\varphi(\cdot,t) - \varphi_n(\cdot,t)|^2 \, dt \xrightarrow[n\to\infty]{} 0 \tag{2.9}$$

  (b) For each n, $I(\varphi_n) = \int_a^b \varphi_n(\omega,t) \, dW(\omega,t)$ is well defined by (2.6), and $\{I(\varphi_n)\}$ converges in q.m. as $n \to \infty$.
  (c) If $\{\varphi_n\}$ and $\{\varphi'_n\}$ are two sequences of (ω,t) steps satisfying (2.3) and (2.4) such that $\|\varphi - \varphi_n\|$ and $\|\varphi - \varphi'_n\|$ both go to zero as $n \to \infty$, then

$$\lim_{n\to\infty} \text{in q.m.} \ I(\varphi_n) = \lim_{n\to\infty} \text{in q.m.} \ I(\varphi'_n) \tag{2.10}$$

Proof:

(a) Suppose $E\varphi(\cdot,t)\bar\varphi(\cdot,s)$ is continuous on $[a,b] \times [a,b]$, that is, φ_t is q.m. continuous on $[a,b]$, then an approximating sequence of (ω,t) step functions $\{\varphi_n\}$ can be constructed by partitioning $[a,b]$, sampling $\varphi(\omega,t)$ at partition points $t_\nu^{(n)}$, defining $\varphi_n(\omega,t) = \varphi(\omega,t_\nu^{(n)})$, $t_\nu^{(n)} \le t < t_{\nu+1}^{(n)}$, and refining the partitions to zero $[\max_\nu (t_{\nu+1}^{(n)} - t_\nu^{(n)}) \xrightarrow[n\to\infty]{} 0]$. Since φ_t is q.m. continuous, $E|\varphi(\cdot,t) - \varphi_n(\cdot,t)|^2 \xrightarrow[n\to\infty]{} 0$ for every t in $[a,b]$. By the dominated convergence theorem, we have

$$\int_a^b E|\varphi(\cdot,t) - \varphi_n(\cdot,t)|^2 \, dt \xrightarrow[n\to\infty]{} 0$$

More generally, if φ merely satisfies (2.3) and (2.4) and is not necessarily q.m. continuous on $[a,b]$, we construct a sequence of approximating (ω,t)-

2. STOCHASTIC INTEGRALS

step functions in the following manner: First, let f_n be φ with its real and imaginary parts truncated to $\pm n$, then

$$\int_a^b E|\varphi(\cdot,t) - f_n(\cdot,t)|^2 \, dt \xrightarrow[n \to \infty]{} 0 \qquad (2.11)$$

Thus, we can always assume φ to be bounded. If φ is bounded, define

$$g_n(\omega, t) = \int_0^\infty e^{-\tau} \varphi\left(\omega, t - \frac{\tau}{n}\right) d\tau \qquad (2.12)$$

Then, $g_n(\cdot,t)$ is q.m. continuous on $[a,b]$ and satisfies (2.3) and (2.4). Now,

$$\int_a^b E|\varphi(\cdot,t) - g_n(\cdot,t)|^2 \, dt \leq \int_a^b \int_0^\infty e^{-\tau} E \left|\varphi(\cdot,t) - \varphi\left(\cdot, t - \frac{\tau}{n}\right)\right|^2 dt \, d\tau \qquad (2.13)$$

Since $\int_a^b \left|\varphi(\cdot,t) - \varphi\left(\cdot, t - \frac{\tau}{n}\right)\right|^2 dt \xrightarrow[n \to \infty]{} 0$ whenever φ is bounded and Lebesgue measurable, we have

$$\int_a^b E|\varphi(\cdot,t) - g_n(\cdot,t)|^2 \, dt \xrightarrow[n \to \infty]{} 0 \qquad (2.14)$$

It is now clear that we can construct a sequence of (ω,t)-step functions approximating φ by sampling g_n at the partition points of a sequence of partitions refining to zero. This completes the proof for (a).

(b) We assume that $\{\varphi_n\}$ is a sequence of (ω,t)-step functions such that $\|\varphi - \varphi_n\| \xrightarrow[n \to \infty]{} 0$ and define $I(\varphi_n)$ by (2.6) as

$$I(\varphi_n) = \sum_\nu \varphi_{n\nu}(\omega) [W(\omega, t_{\nu+1}^{(n)}) - W(\omega, t_\nu^{(n)})] \qquad (2.15)$$

Hence, if we write $\Delta_\nu^n W = W(t_{\nu+1}^{(n)}) - W(t_\nu^{(n)})$, we have

$$E|I(\varphi_n)|^2 = \sum_\nu \sum_\mu E(\varphi_{n\nu} \bar\varphi_{n\mu} \Delta_\nu^n W \Delta_\mu^n W)$$

If $\mu > \nu$, then $\Delta_\mu^n W$ is independent of $\mathcal{C}_{t_\mu}^{(n)}$ while $(\varphi_{n\nu} \bar\varphi_{n\mu} \Delta_\nu^n W)$ is $\mathcal{C}_{t_\mu}^{(n)}$ measurable. Hence

$$E(\varphi_{n\nu} \bar\varphi_{n\mu} \Delta_\nu^n W \Delta_\mu^n W) = 0 \quad \text{if } \mu \neq \nu$$

and

$$\begin{aligned} E|I(\varphi_n)|^2 &= \sum_\nu E[|\varphi_{n\nu}|^2 (\Delta_\nu^n W)^2] \\ &= \sum_\nu E|\varphi_{n\nu}|^2 E^{\mathcal{C}_{t_\nu}^{(n)}} (\Delta_\nu^n W)^2 \\ &= \sum_\nu E|\varphi_{n\nu}|^2 (t_{\nu+1}^{(n)} - t_\nu^{(n)}) \\ &= \int_a^b E|\varphi_n(\cdot,t)|^2 \, dt \end{aligned} \qquad (2.16)$$

Now, $I(\varphi_{m+n}) - I(\varphi_n) = I(\varphi_{m+n} - \varphi_n)$ and $\varphi_{m+n} - \varphi_n$ is again a step. Therefore,

$$\begin{aligned}E|I(\varphi_{m+n}) - I(\varphi_n)|^2 &= \int_a^b E|\varphi_{m+n}(\cdot,t) - \varphi_n(\cdot,t)|^2\,dt \\ &\leq 2\int_a^b E|\varphi_{m+n}(\cdot,t) - \varphi(\cdot,t)|^2\,dt \\ &\quad + 2\int_a^b E|\varphi_n(\cdot,t) - \varphi(\cdot,t)|^2\,dt \xrightarrow[n\to\infty]{} 0 \end{aligned} \quad (2.17)$$

Hence, $\{I(\varphi_n)\}$ is a mutually q.m. convergent sequence, and there exists a second-order random variable $I(\varphi)$ such that

$$E|I(\varphi_n) - I(\varphi)|^2 \xrightarrow[n\to\infty]{} 0$$

and (b) is proved.

(c) Suppose $\{\varphi_n\}$ and $\{\varphi'_n\}$ are (ω,t) steps such that $\|\varphi - \varphi_n\| \xrightarrow[n\to\infty]{} 0$ and $\|\varphi - \varphi'_n\| \xrightarrow[n\to\infty]{} 0$. Then,

$$\|\varphi_n - \varphi'_n\| \leq \sqrt{2}\,(\|\varphi_n - \varphi\|^2 + \|\varphi'_n - \varphi\|^2)^{\frac{1}{2}} \xrightarrow[n\to\infty]{} 0$$

Therefore, $E|I(\varphi_n) - I(\varphi'_n)|^2 \xrightarrow[n\to\infty]{} 0$ and

$$\lim_{n\to\infty} \text{in q.m. } I(\varphi_n) = \lim_{n\to\infty} \text{in q.m. } I(\varphi'_n)$$

This proves (c). ∎

Proposition 2.2. Let $\varphi(\omega,t)$ and $\psi(\omega,t)$ satisfy (2.3) and (2.4). Then

$$E\left[\int_a^b \varphi(\cdot,t)\,dW_t \cdot \overline{\int_a^b \psi(\cdot,t)\,dW_t}\right] = \int_a^b E\varphi_t\bar{\psi}_t\,dt \quad (2.18)$$

Proof: It is enough to prove (2.18) for $\varphi = \psi$, because

$$\begin{aligned}EI(\varphi)\overline{I(\psi)} &= \tfrac{1}{4}[E|I(\varphi) + I(\psi)|^2 - E|I(\varphi) - I(\psi)|^2] \\ &\quad + i\tfrac{1}{4}[E|I(-i\varphi) + I(\psi)|^2 - E|I(-i\varphi) - I(\psi)|^2] \\ &= \tfrac{1}{4}[E|I(\varphi + \psi)|^2 - E|I(\varphi - \psi)|^2] \\ &\quad + \frac{i}{4}[E|I(-i\varphi + \psi)|^2 - E|I(-i\varphi - \psi)|^2] \end{aligned} \quad (2.19)$$

which means that (2.18) follows from the seemingly more special case $E|I(\varphi)|^2 = \int_a^b E|\varphi_t|^2\,dt$. Now if φ is an (ω,t)-step function, we have already proved in (2.16) that

$$E|I(\varphi)|^2 = \int_a^b E|\varphi(\cdot,t)|^2\,dt$$

3. PROCESSES DEFINED BY STOCHASTIC INTEGRALS

If φ is not a step, let $\{\varphi_n\}$ be an approximating sequence of steps, then

$$E|I(\varphi)|^2 = E|I(\varphi - \varphi_n) + I(\varphi_n)|^2$$
$$= E|I(\varphi_n)|^2 + 2\operatorname{Re} EI(\varphi - \varphi_n)\overline{I(\varphi_n)} + E|I(\varphi - \varphi_n)|^2$$

Since $E|I(\varphi - \varphi_n)|^2 \xrightarrow[n\to\infty]{} 0$, we have

$$E|I(\varphi)|^2 = \lim_{n\to\infty} E|I(\varphi_n)|^2$$
$$= \lim_{n\to\infty} \int_a^b E|\varphi_n(\cdot,t)|^2\, dt = \int_a^b E|\varphi(\cdot,t)|^2\, dt \quad\blacksquare$$

3 PROCESSES DEFINED BY STOCHASTIC INTEGRALS

Proposition 3.1. Let $\{W_t, \mathfrak{A}_t\}$ be a Brownian motion and let φ satisfy (2.3) and (2.4). Define a process $\{X_t, a \leq t \leq b\}$ by

$$X(\omega,t) = \int_a^t \varphi(\omega,s)\, dW(\omega,s) \qquad (3.1)$$

Then, $\{X_t, \mathfrak{A}_t, a \leq t \leq b\}$ is a martingale, that is,

$$t > s \Rightarrow E^{\mathfrak{A}_s} X_t = X_s \qquad \text{a.s.} \qquad (3.2)$$

Remark: This important martingale property is intimately connected with the fact that a stochastic integral is inherently defined by forward difference approximations. We recall that if φ is an (ω,t)-step function, then

$$\int_a^b \varphi(\omega,t)\, dW(\omega,t) = \sum_\nu \varphi_\nu(\omega)[W(\omega,t_{\nu+1}) - W(\omega,t_\nu)]$$

where each summand consists of a term φ_ν measurable with respect to $\mathfrak{A}_{t_\nu}$ and a forward increment $\Delta_\nu W$ which is independent of $\mathfrak{A}_{t_\nu}$. For a general φ, the stochastic integral is defined by approximating φ with steps, and it involves forward difference approximation once again.

Proof: First, suppose that the integrand φ in (3.1) is an (ω,t)-step function. Let $t > s$, and let $t_1, t_2, \ldots, t_n$ be jump points of φ between s and t. We can write

$$X(\omega,t) - X(\omega,s) = \int_s^t \varphi(\omega,\tau) dW(\omega,\tau)$$
$$= \varphi_0(\omega)[W(\omega,t_1) - W(\omega,s)] + \varphi_1(\omega)$$
$$[W(\omega,t_2) - W(\omega,t_1)] + \cdots + \varphi_n(\omega)[W(\omega,t) - W(\omega,t_n)] \qquad (3.3)$$

where φ_0 is $\mathfrak{A}_s$ measurable and φ_k is $\mathfrak{A}_{t_k}$ measurable, $k = 1, \ldots, n$. By successively taking conditional expectation with respect to $\mathfrak{A}_{t_n}, \mathfrak{A}_{t_{n-1}},$

..., $\mathcal{A}_s$, we find

$$E^{\mathcal{A}_s}(X_t - X_s) = E^{\mathcal{A}_s} E^{\mathcal{A}_{t_1}} \cdots E^{\mathcal{A}_{t_n}}(X_t - X_s)$$
$$= 0 \quad \text{a.s.}$$

Since X_s is obviously $\mathcal{A}_s$ measurable, this proves the proposition for φ equal to a step. If φ is not a step, let $\{\varphi_n\}$ be step approximations to φ, and define

$$X_n(\omega,t) = \int_a^t \varphi_n(\omega,t)\, dW(\omega,\tau) \tag{3.4}$$

For each n, $\{X_n(\omega,t), \mathcal{A}_t, a \leq t \leq b\}$ is a martingale, and

$$E^{\mathcal{A}_s}(X_t - X_s) = E^{\mathcal{A}_s}[X(\cdot,t) - X_n(\cdot,t)] - E^{\mathcal{A}_s}[X(\cdot,s) - X_n(\cdot,s)]$$

Since $E|X(\cdot,t) - X_n(\cdot,t)|^2 \xrightarrow[n \to \infty]{} 0$, we have

$$E^{\mathcal{A}_s}(X_t - X_s) = E^{\mathcal{A}_s}[X(\cdot,t) - X_n(\cdot,t)] - E^{\mathcal{A}_s}[X(\cdot,s) - X_n(\cdot,s)]$$
$$\xrightarrow[n \to \infty]{q.m.} 0 \tag{3.5}$$

Hence, $E^{\mathcal{A}_s}(X_t - X_s) = 0$, a.s., and the proof is complete. ∎

A process $\{X_t, a \leq t \leq b\}$ as defined by (3.1) is obviously q.m. continuous. Thus, we can choose a version of $\{X_t, a \leq t \leq b\}$ which is separable and measurable. If we choose such a version and if we assume that the Brownian process $\{W_t, a \leq t \leq b\}$ in (3.1) is also separable, then $\{X_t, a \leq t \leq b\}$ is sample continuous with probability 1. When φ is an (ω,t)-step function, sample continuity is obvious since $\{X_t, a \leq t \leq b\}$ is then a separable Brownian motion pieced together in a continuous way. If φ is not a step, let $\{\varphi_n\}$ be a sequence of (ω,t)-step functions satisfying (2.3) and (2.4) such that

$$\|\varphi - \varphi_n\|^2 = \int_a^b E|\varphi(\cdot,s) - \varphi_n(\cdot,s)|^2\, ds \leq \frac{1}{n^4}$$

Such a sequence can always be obtained by choosing a subsequence of any sequence $\{\psi_m\}$ such that $\|\psi_m - \varphi\| \xrightarrow[m \to \infty]{} 0$.

If we set $X_{nt} = \int_a^t \varphi_{ns}\, dW_s$ and choose it to be separable, then for each n, $\{X_{nt}, a \leq t \leq b\}$ is sample continuous with probability 1. For each n, $\{X_{nt} - X_t, a \leq t \leq b\}$ is a separable second-order martingale. If we apply the version of Proposition 2.3.2 for complex-valued martingales, we get

$$\mathcal{P}\left(\sup_{a \leq t \leq b} |X_{nt} - X_t| \geq \frac{1}{n}\right) \leq n^2 E|X_{nb} - X_b|^2 = n^2\|\varphi_n - \varphi\|^2 \leq \frac{1}{n^2}$$

It follows that $\sum_n \mathcal{P}(\sup_{a \leq t \leq b} |X_{nt} - X_t| \geq 1/n) < \infty$ and the Borel-Cantelli

3. PROCESSES DEFINED BY STOCHASTIC INTEGRALS

lemma implies that

$$\Lambda = \limsup_{n} \left\{ \omega : \sup_{a \le t \le b} |X_{nt}(\omega) - X_t(\omega)| \ge \frac{1}{n} \right\}$$

is a null set, that is, for every $\omega \notin \Lambda$, $\sup_t |X_{nt}(\omega) - X_t(\omega)| \ge \frac{1}{n}$ for, at most, a finite number of n. Therefore, for $\omega \notin \Lambda$,

$$\lim_{n \to \infty} \sup_{a \le t \le b} |X_{nt}(\omega) - X_t(\omega)| = 0$$

and $X_t(\omega)$, $a \le t \le b$, being the uniform limit of a sequence of continuous functions, is itself continuous. This proves the sample continuity of $\{X_t, a \le t \le b\}$.

One immediate consequence of the martingale property is that a stochastic integral does not behave like an ordinary integral. Consider, for example, the stochastic integral $\int_0^t W_s \, dW_s$. If the integral is like an ordinary integral, surely it must be equal to $\frac{1}{2}(W_t^2 - W_0^2) = \frac{1}{2} W_t^2$. However, $\frac{1}{2} W_t^2$ is not a martingale, as is seen from the relationship

$$E^{as}(\tfrac{1}{2} W_t^2) = \tfrac{1}{2} W_s^2 + \tfrac{1}{2}(t - s)$$

Therefore, $\int_0^t W_s \, dW_s$ cannot be equal to $\frac{1}{2} W_t^2$. What $\int_0^t W_s \, dW_s$ is will be clarified by the so-called **Ito's differentiation rule,** which will be stated below.

To state the differentiation rule for stochastic integrals under its natural hypotheses, we need to generalize the definition of stochastic integrals to include integrands φ which satisfy (2.3), but instead of (2.4), the weaker condition

$$\int_a^b |\varphi(\omega,t)|^2 \, dt < \infty \qquad \text{almost surely} \tag{3.6}$$

This generalization is discussed in detail in Sec. 6 of this chapter (see, in particular, Proposition 6.1). For now, we merely note that if φ satisfies (2.3) and (3.6), then the stochastic integral $\int_a^b \varphi_t \, dW_t$ is defined by

$$\int_a^b \varphi(\omega,t) \, dW(\omega,t) = \lim_{n \to \infty} \text{in p.} \int_a^b \varphi^n(\omega,t) \, dW(\omega,t)$$

where φ^n is defined by

$$\varphi^n(\omega,t) = \begin{cases} \varphi(\omega,t) & \text{if } \int_a^t |\varphi(\omega,s)|^2 \, ds \le n \\ 0 & \text{otherwise} \end{cases}$$

Stochastic integrals appearing in the following proposition will be assumed to be defined in this way if the integrands satisfy (3.6) rather than (2.4).

Proposition 3.2. Let $X_1(\omega,t), X_2(\omega,t), \ldots, X_n(\omega,t)$ be processes defined in terms of a single Brownian motion $W(\omega,t)$ as follows:

$$X_k(\omega,t) = \int_a^t f_k(\omega,t')\, dt' + \int_a^t \varphi_k(\omega,t')\, dW(\omega,t') \qquad k = 1, \ldots, n \tag{3.7}$$

Let $Y(\omega,t) = \psi(X_{1t}, X_{2t}, \ldots, X_{nt}, t)$, where ψ is once continuously differentiable with respect to t and has continuous second partials with respect to the X's. Then, with probability 1,

$$Y(\omega,t) = Y(\omega,a) + \int_a^t \dot\psi(\mathbf{X}(\omega,t'), t')\, dt' + \sum_{k=1}^n \int_a^b \psi_k(\mathbf{X}(\omega,t'), t')\, dX_k(\omega,t') + \sum_{j=1}^n \sum_{k=1}^n \frac{1}{2} \int_a^b \psi_{jk}(\mathbf{X}(\omega,t'), t') \varphi_j(\omega,t') \varphi_k(\omega t')\, dt' \tag{3.8}$$

where $\dot\psi = \partial\psi/\partial t$, $\psi_k = \partial\psi/\partial x_k$, and $\psi_{jk} = \partial^2\psi/\partial x_j\,\partial x_k$.

Remark: The surprising thing about (3.8) is the last term. It comes about in roughly the following way. We recall that a Brownian motion W_t has the curious property $(dW_t)^2 \cong dt$. Therefore, $dX_j(t)\, dX_k(t) \cong \varphi_j \varphi_k\, dt$. Now,

$$dY_t = Y_{t+dt} - Y_t = \psi(\mathbf{X}_{t+dt}, t+dt) - \psi(\mathbf{X}_t, t)$$
$$= \dot\psi\, dt + \sum_k \psi_k\, dX_k(t) + \frac{1}{2} \sum_j \sum_k \psi_{jk}\, dX_j(t)\, dX_k(t) + \cdots \tag{3.9}$$

Both the first and the third term in (3.9) are of order dt, hence

$$dY_t = \dot\psi\, dt + \sum_k \psi_k\, dX_k(t) + \frac{1}{2} \sum_{j,k} \psi_{jk} \varphi_j \varphi_k\, dt + o(dt) \tag{3.10}$$

which is nothing but a symbolic way of writing (3.8). We note that $dX_k(t)$ can be replaced by $f_k\, dt + \varphi_k\, dW_t$ in both (3.8) and (3.10), permitting Y_t to be written in the form of

$$Y_t - Y_a = \int_a^t g(\omega,t')\, dt' + \int_a^t \gamma(\omega,t')\, dW(\omega,t')$$

If we apply (3.8) to $Y_t = \tfrac{1}{2}W_t^2$, we find immediately

$$dY_t = W_t\, dW_t + \tfrac{1}{2} dt$$

or

$$Y_t = \int_0^t W_s\, dW_s + \tfrac{1}{2} t$$

or

$$\int_0^t W_s\, dW_s = Y_t - \tfrac{1}{2}t = \tfrac{1}{2}W_t^2 - \tfrac{1}{2}t \tag{3.11}$$

which is indeed a martingale.

4. STOCHASTIC DIFFERENTIAL EQUATIONS

It might be useful to isolate two special cases of Proposition 3.2. First, suppose $Y_t = \psi(X_t,t)$ depends only on a single X process, and $dX_t = f_t\,dt + \varphi_t\,dW_t$. Then, (3.8) becomes

$$Y_t = Y_a + \int_a^t \dot\psi(X_s,s)\,ds + \int_a^t \psi'(X_s,s)\,dX_s$$
$$+ \frac{1}{2}\int_a^t \psi''(X_s,s)\varphi^2(\omega,s)\,ds \quad (3.12)$$

where prime denotes differentiation with respect to the first variable. For the second special case, consider the product $Y(\omega,t) = X_1(\omega,t)X_2(\omega,t)$, where X_1 and X_2 satisfy (3.7) with $k = 1, 2$. Then (3.8) becomes

$$Y_t = Y_a + \int_a^t X_2(\omega,t')\,dX_1(\omega,t') + \int_a^t X_1(\omega,t')\,dX_2(\omega,t')$$
$$+ \frac{1}{2}\int_a^t \varphi_1(\omega,t')\varphi_2(\omega,t')\,dt' \quad (3.13)$$

4. STOCHASTIC DIFFERENTIAL EQUATIONS

From the point of applications, a major motivation for studying stochastic differential equations is to give meaning to an equation of the form

$$\frac{d}{dt}X_t = m(X_t,t) + \sigma(X_t,t)\zeta_t \quad (4.1)$$

where ζ_t is a Gaussian white noise. At least formally, we know that $\int_0^t \zeta_s\,ds$ has all the attributes of a Brownian motion W_t. Hence, formally again, (4.1) appears to be equivalent to

$$X_t = X_a + \int_a^t m(X_s,s)\,ds + \int_a^t \sigma(X_s,s)\,dW_s \quad (4.2)$$

With stochastic integrals having been defined, (4.2) is capable of taking on a precise interpretation. Whether the interpretation is the one that we really want to give to (4.1) is something else again. We postpone examination of this question until Sec. 5. For the time being, we confine ourselves to a study of (4.2) as an equation in the unknown X_t and with the last integral interpreted as a stochastic integral.

By a **stochastic differential equation,** we mean an equation of the form

$$dX(\omega,t) = m(X(\omega,t),t)\,dt + \sigma(X(\omega,t),t)\,dW(\omega,t) \quad (4.3)$$

which is nothing more or less than a symbolic way of writing (4.2). A process $\{X_t,\, t \geq a\}$ is said to satisfy (4.3) with initial condition $X_a = X$ if (1) for each t

$$\int_a^t \sigma(X_s,s)\,dW_s$$

is capable of being interpreted as a stochastic integral and (2) for each t, X_t is almost surely equal to the random variable defined by

$$X + \int_a^t m(X_s,s)\, ds + \int_a^t \sigma(X_s,s)\, dW_s$$

Under the conditions that we shall assume, we can in fact assert a stronger result than a.s. equality of the two random variables, viz., q.m. difference between the two is zero.

We shall first state and prove an existence and uniqueness theorem following Ito.

Proposition 4.1. Let $\{W_t, \mathcal{Q}_t, a \leq t \leq T < \infty\}$ be a separable Brownian motion. Let X be a random variable measurable with respect to $\mathcal{Q}_a$ and satisfy $EX^2 < \infty$. Let $m(x,t)$ and $\sigma(x,t)$, $-\infty < x < \infty$, $a \leq t \leq T$, be Borel measurable functions in the pair (x,t). Let m and σ satisfy the following conditions:

$$|m(x,t) - m(y,t)| + |\sigma(x,t) - \sigma(y,t)| \leq K|x - y| \quad (4.4)$$
$$|m(x,t)| + |\sigma(x,t)| \leq K\sqrt{1 + x^2} \quad (4.5)$$

Under these hypotheses, there exists a separable and measurable process $\{X_t, a \leq t \leq T\}$ with the following properties:

P_1: For each t in $[a,T]$ X_t is $\mathcal{Q}_t$-measurable
P_2: $\int_a^T EX_t^2\, dt < \infty$
P_3: $\{X_t, a \leq t \leq T\}$ satisfies (4.2) with $X_a = X$
P_4: With probability 1, $\{X_t, a \leq t \leq T\}$ is sample continuous
P_5: $\{X_t, a \leq t \leq T\}$ is unique with probability 1
P_6: $\{X_t, a \leq t \leq T\}$ is a Markov process

Remark: Condition (4.4) is known as the **uniform Lipschitz condition.** Without loss of generality, the constants K in (4.4) and (4.5) can be assumed to be the same.

Proof: We shall give a proof by constructing a solution. Since we shall be dealing with a sequence of stochastic process, we shall write $X(\omega,t)$ or $X(\cdot,t)$ rather than X_t, because the subscript will be used to index terms in the sequence. First, define a sequence of processes $\{X_n(\cdot,t), a \leq t \leq T\}$ as follows:

$$X_0(\omega,t) = X(\omega)$$
$$X_{n+1}(\omega,t) = X(\omega) + \int_a^t m(X_n(\omega,s), s)\, ds$$
$$+ \int_a^t \sigma(X_n(\omega,s), s)\, dW(\omega,s) \quad (4.6)$$

We need to show that the last integral is well defined as a stochastic

4. STOCHASTIC DIFFERENTIAL EQUATIONS

integral for each n. That is, we need to show that for each n

$\sigma(X_n(\omega,t), t)$ is jointly measurable in (ω,t), and for each t
is $\mathcal{C}_t$ measurable (4.7)

and

$$\int_a^T E\sigma^2(X_n(\cdot,t), t)\, dt < \infty \qquad (4.8)$$

This can be done by induction. First, we verify (4.7) and (4.8) for $n = 0$. Since $X_0(\omega,t) = X(\omega)$, (4.7) is satisfied, because σ is a Borel measurable function, and $\sigma(X,t)$ is not only $\mathcal{C}_t$ measurable, it is $\mathcal{C}_a$ measurable. Using (4.5), we have

$$\sigma^2(X,t) \leq K^2(1 + X^2)$$

so that

$$\int_a^T E\sigma^2(X_0(\cdot,t), t)\, dt \leq K^2(1 + EX^2)(T - a) < \infty$$

and (4.8) is verified for $n = 0$. Now, assume that (4.7) and (4.8) are both satisfied for $n = 0, 1, 2, \ldots, k$. Then from (4.6),

$$X_{k+1}(\omega,t) = X + \int_a^t m(X_k(\omega,s), s)\, ds + \int_a^t \sigma(X_k(\omega,s), s)\, dW(\omega,s)$$

each of the three terms on the right is $\mathcal{C}_t$ measurable, because $\{X_k(\cdot,s), a \leq s \leq t\}$ is $\mathcal{C}_t$ measurable. Next, we note that for $a \leq t_0 \leq t \leq T$,

$$[X_{k+1}(\omega,t) - X_{k+1}(\omega,t_0)]^2 \leq 2\left\{\left[\int_{t_0}^t m(X_k(\omega,s), s)\, ds\right]^2 \right.$$
$$\left. + \left[\int_{t_0}^t \sigma(X_k(\omega,s), s)\, dW(\omega,s)\right]^2\right\} \qquad (4.9)$$

By using the Schwarz inequality on the second term and (2.18) on the last term, we get

$$E[X_{k+1}(\cdot,t) - X_{k+1}(\cdot,t_0)]^2 \leq 2\left[(t - t_0)\int_{t_0}^t Em^2(X_k(\cdot,s), s)\, ds \right.$$
$$\left. + \int_{t_0}^t E\sigma^2(X_k(\cdot,s), s)\, ds\right] \qquad (4.10)$$

Now, (4.5) can be applied again, and we get

$$E[X_{k+1}(\cdot,t) - X_{k+1}(\cdot,t_0)]^2$$
$$\leq 2\left\{K^2[1 + (t - t_0)]\int_{t_0}^t [1 + EX_k^2(\cdot,s)]\, ds\right\} \qquad (4.11)$$

Therefore, $\{X_{k+1}(\cdot,t), a \leq t \leq T\}$ is q.m. continuous and a measurable version can be chosen. Furthermore,

$$\int_a^T EX_{k+1}^2(\cdot,t)\, dt \leq 2\int_a^T E[X_{k+1}(\cdot,t) - X]^2\, dt + 2\int_a^T EX^2\, dt$$
$$\leq 4K^2[1 + (T - a)](T - a)\int_a^T [1 + EX_k^2(\cdot,s)]\, ds$$
$$+ 2(T - a)EX^2 < \infty \qquad (4.12)$$

The induction is complete, and we have verified (4.7) and (4.8) for every n. Therefore, the sequence of processes $\{X_n(\cdot,t), a \leq t \leq T, n = 0, 1, \ldots\}$ is well defined.

Next we prove that for each t, $\{X_n(\cdot,t), n = 0, 1, \ldots\}$ converges in quadratic mean. To do this, define

$$\begin{aligned} \Delta_0(\omega,t) &= X(\omega) \\ \Delta_n(\omega,t) &= X_n(\omega,t) - X_{n-1}(\omega,t) \quad n = 1, 2, \ldots \end{aligned} \quad (4.13)$$

Using (4.6), we get

$$\begin{aligned} \Delta_{n+1}(\omega,t) &= \int_a^t [m(X_n(\omega,s), s) - m(X_{n-1}(\omega,s), s)] \, ds \\ &+ \int_a^t [\sigma(X_n(\omega,s), s) - \sigma(X_{n-1}(\omega,s), s)] \, dW(\omega,s) \end{aligned} \quad (4.14)$$

If we make use of the inequality $(A + B)^2 \leq 2A^2 + 2B^2$, the Schwarz inequality, (2.18) on the stochastic integral, and the uniform Lipschitz condition, we find

$$E\Delta_{n+1}^2(\cdot,t) \leq 2K^2[1 + (T - a)] \int_a^t E\Delta_n^2(\cdot,s) \, ds \quad (4.15)$$

The inequality (4.15) can be iterated starting from $E\Delta_0^2(\cdot,t) = EX^2$, and we get

$$E\Delta_n^2(\cdot,t) \leq \{2K^2[1 + (T - a)]\}^n \frac{(t - a)^n}{n!} EX^2 \quad (4.16)$$

Now,

$$X_{n+m}(\omega,t) - X_n(\omega,t) = \sum_{k=1}^m \Delta_{n+k}(\omega,t)$$

and by the Cauchy-Schwarz inequality

$$[X_{n+m}(\omega,t) - X_n(\omega,t)]^2 \leq \left(\sum_{k=1}^m \frac{1}{2^{n+k}}\right)\left[\sum_{k=1}^m 2^{n+k}\Delta_{n+k}^2(\omega,t)\right] \quad (4.17)$$

Therefore, from (4.16) we get

$$\begin{aligned} \sup_{m \geq 0} E[X_{n+m}(\cdot,t) - X_n(\cdot,t)]^2 &\leq \sum_{k=1}^\infty 2^{n+k} E\Delta_{n+k}^2(\cdot,t) \\ &\leq EX^2 \sum_{k=n+1}^\infty \frac{\{4K^2[1 + (T - a)](t - a)\}^k}{k!} \end{aligned} \quad (4.18)$$

Since $\sum_{k=0}^\infty \alpha^k/k!$ converges to e^α for every finite α,

$$\sup_{m \geq 0} E[X_{n+m}(\cdot,t) - X_n(\cdot,t)]^2 \xrightarrow[n \to \infty]{} 0 \quad (4.19)$$

4. STOCHASTIC DIFFERENTIAL EQUATIONS

uniformly in t. Therefore, for every $t \in [a,T]$, $\{X_n(\cdot,t)\}$ is a q.m. convergent sequence. Let the q.m. limit be denoted by $X(\cdot,t)$.

Thus, we have obtained a process $\{X(\omega,t),\ a \leq t \leq T\}$ such that

$$\sup_{a \leq t \leq T} E[X_n(\cdot,t) - X(\cdot,t)]^2 \xrightarrow[n \to \infty]{} 0 \tag{4.20}$$

Because for each n, $\{X_n(\cdot,t),\ a \leq t \leq T\}$ is q.m. continuous, the limit process $\{X_t,\ a \leq t \leq T\}$ is also q.m. continuous, hence continuous in probability. It follows from Proposition 2.2.3 that a separable and measurable version can be chosen for $\{X_t,\ a \leq t \leq T\}$. We shall now show that $\{X_t,\ a \leq t \leq T\}$ so constructed satisfies $P_1 - P_5$.

First, for each t, $X_n(\cdot,t)$ is $\mathcal{A}_t$ measurable for every n. Therefore, $X(\cdot,t)$ is also $\mathcal{A}_t$ measurable for each t, and P_1 is proved. Next,

$$\int_a^T EX_t^2\, dt \leq 2\left\{\int_a^T E[X(\cdot,t) - X_n(\cdot,t)]^2\, dt + \int_a^T EX_n^2(\cdot,t)\, dt\right\} \tag{4.21}$$

From (4.18) we have for some constant α,

$$\sup_n \int_a^T EX_n^2(\cdot,t)\, dt \leq 2\left[\int_a^T e^{\alpha t}\, dt + (T - a)\right] EX^2 = A < \infty \tag{4.22}$$

Hence, using (4.20) on (4.21) we get

$$\int_a^T EX_t^2\, dt \leq 2A < \infty$$

which proves P_2. Together, P_1 and P_2 ensure that $\int_a^t \sigma(X_s,s)\, dW_s$ is well defined as a stochastic integral.

To prove that the process $\{X_t,\ a \leq t \leq T\}$ is indeed a solution to (4.2) with $X_a = X$, we define

$$D_t = X_t - X - \int_a^t m(X_s,s)\, ds - \int_a^t \sigma(X_s,s)\, dW_s$$

Using (4.6), we can rewrite D_t as

$$D_t = [X(\cdot,t) - X_{n+1}(\cdot,t)] - \int_a^t [m(X(\cdot,s), s) - m(X_n(\cdot,s), s)]\, ds$$
$$- \int_a^t [\sigma(X(\cdot,s), s) - \sigma(X_n(\cdot,s), s)]\, ds$$

It is now easy to show that each of the three terms on the right-hand side goes to zero in quadratic mean as $n \to \infty$. Therefore

$$ED_t^2 = 0$$

and for each $t \in [a,T]$,

$$X_t = X + \int_a^t m(X_s,s)\, ds + \int_a^t \sigma(X_s,s)\, dW_s \tag{4.23}$$

with probability 1. Further, both sides of (4.23) represent separable processes if we choose a separable version for the stochastic integral. Hence,

$$\mathcal{P}\left(X_t = X + \int_a^t m(X_s,s)\,ds + \int_a^t \sigma(X_s,s)\,dW_s,\ a \leq t \leq T\right) = 1 \quad (4.24)$$

and P_3 has been proved. Since the right-hand side of (4.23) represents a sample-continuous process, $\{X_t,\ a \leq t \leq T\}$ is also sample continuous, and P_4 is proved. To prove uniqueness, suppose that X_t and $\tilde{X}_t$ are both solutions of (4.2) with $X_a = \tilde{X}_a = X$. Then, we can write

$$(X_t - \tilde{X}_t) = \int_a^t [m(X_s,s) - m(\tilde{X}_s,s)]\,ds$$
$$+ \int_a^t [\sigma(X_s,s) - \sigma(\tilde{X}_s,s)]\,dW_s \quad (4.25)$$

Equation (4.25) has the same form as (4.14), and the same arguments yield

$$E(X_t - \tilde{X}_t)^2 \leq 2K^2[1 + (T - a)] \int_a^t E(X_s - \tilde{X}_s)^2\,ds \quad (4.26)$$

Inequality (4.26) has the form

$$\frac{d}{dt}f(t) \leq cf(t) \qquad t > a \quad (4.27)$$

with $f(t) \geq 0$ and $f(a) = 0$. Rewriting, we get

$$\frac{d}{dt}(e^{-ct}f(t)) \leq 0 \qquad t > a$$

Therefore, by integrating we get $e^{-ct}f(t) \leq f(a)e^{-ca}$, and it follows that

$$0 \leq f(t) \leq f(a)e^{c(t-a)} = 0 \qquad a \leq t \leq T$$

This proves that $f(t) = \int_a^t E(X_s - \tilde{X}_s)^2\,ds = 0,\ a \leq t \leq T$. From (4.26) we have

$$E(X_t - \tilde{X}_t)^2 = 0 \qquad a \leq t \leq T$$

Therefore, for each t in $[a,T]$,

$$\mathcal{P}(X_t \neq \tilde{X}_t) = 0$$

and by σ additivity,

$$\mathcal{P}(X_t \neq \tilde{X}_t \text{ at one or more rational points in } [a,T]) = 0$$

If X_t and $\tilde{X}_t$ are both chosen to be separable, then they are both sample continuous and

$$\mathcal{P}(X_t = \tilde{X}_t \text{ for all } t \in [a,T]) = 1$$

5. WHITE NOISE AND STOCHASTIC CALCULUS

This proves that with probability 1 there is only one sample-continuous solution to (4.2) with the same initial condition.

Finally, we prove that $\{X_t,\ a \leq t \leq T\}$ is a Markov process. Using (4.2) we write

$$X_t = X_s + \int_s^t m(X_\tau,\tau)\, d\tau + \int_s^t \sigma(X_\tau,\tau)\, dW_\tau \qquad a \leq s < t \leq T$$

which can be regarded as a stochastic integral equation on the interval $s \leq t \leq T$ with X_s as the initial condition. Thus, for each $t \in [s,T]$, X_t can be obtained as a function of X_s and $\{W_\tau - W_s,\ s \leq \tau \leq t\}$, that is, X_t is measurable with respect to the σ algebra generated by X_s and $\{W_\tau - W_s,\ s \leq \tau \leq t\}$. Since X_s is $\mathcal{C}_s$ measurable, and $\{W_\tau - W_s,\ s \leq \tau \leq t\}$ is independent of $\mathcal{C}_s$, X_t is conditionally independent of $\mathcal{C}_s$ given X_s. A fortiori, X_t is conditionally independent of $\{X_\tau,\ a \leq \tau \leq s\}$ given X_s, and this proves the Markov property. ∎

Summarizing the preceding results, we find that, under the conditions on X, m, and σ given in Proposition 4.1, the stochastic integral equation

$$X_t = X + \int_a^t m(X_s,s)\, ds + \int_a^t \sigma(X_s,s)\, dW_s \tag{4.28}$$

has a unique sample-continuous solution which is Markov. We emphasize again that, by definition, the last integral $\int_a^t \sigma(X_s,s)\, dW_s$ in the integral equation is to be interpreted as a stochastic integral. The question whether this stochastic integral equation adequately models a differential equation driven by Gaussian white noise

$$\begin{aligned} \frac{d}{dt} X_t &= m(X_t,t) + \sigma(X_t,t)\zeta_t \\ X_a &= X \end{aligned} \tag{4.29}$$

remains unanswered. Indeed, this question cannot be answered without more being said about what we want (4.29) to mean. As it stands, (4.29) is merely a string of symbols, nothing more. We shall take up this question in the next section.

Finally, we note that the existence of a solution to (4.2) is ensured even without the Lipschitz condition (4.4), but then the uniqueness is no longer guaranteed [Skorokhod, 1965, p. 59].

5. WHITE NOISE AND STOCHASTIC CALCULUS

In this section, we offer an interpretation of differential equations driven by white noise, and examine its relationship with stochastic differential equations. The equation that we would like to give a precise meaning to is the following:

$$\frac{d}{dt} X(\omega,t) = m(X(\omega,t), t) + \sigma(X(\omega,t), t)\zeta(\omega,t) \tag{5.1}$$

where ζ_t is a Gaussian white noise. Since white noise is an abstraction and not a physical process, what one really means by (5.1) in practice is probably an equation driven by a stationary Gaussian process with a spectral density that is flat over a very wide range of frequencies. If we take ζ_t to be such a process in (5.1), then there is no difficulty interpreting (5.1) as an ordinary differential equation for each sample function, provided that the spectral density of ζ_t eventually goes to zero sufficiently rapidly so that the sample functions are well behaved. While this is probably what we want (5.1) to mean, this is not how we want (5.1) to be handled mathematically. If we take ζ_t to be a process with well-behaved sample functions, we lose some of the simple statistical properties of X_t, the primary one being the Markov property. In practice, the interpretation of (5.1) that we really want is probably the following. Take a sequence of Gaussian processes $\{\zeta_n(\cdot,t)\}$ which "converges" in some suitable sense to a white Gaussian noise, and yet for each n $\zeta_n(\cdot,t)$ has well-behaved sample functions. Now, for each n the equation

$$\frac{d}{dt} X_n(\omega,t) = m(X_n(\omega,t), t) + \sigma(X_n(\omega,t), t)\zeta_n(\omega,t) \qquad a \leq t \leq T \tag{5.2}$$

together with the initial condition $X_n(\omega,a) = X(\omega)$ can be solved. We assume that m and σ are such that the solution exists and is unique for almost all sample functions. Thus, we obtain a sequence of processes $\{X_n(\cdot,t), a \leq t \leq T\}$. Suppose that as $n \to \infty$, $\{\zeta_n(\cdot,t)\}$ converges in a suitable sense to white noise, and the sequence $\{X_n(\cdot,t), a \leq t \leq T\}$ converges almost surely, or in quadratic mean, or even merely in probability, to a process $\{X(\cdot,t), a \leq t \leq T\}$. Then it is natural to say that X_t is the solution of

$$\dot{X}_t = m(X_t,t) + \sigma(X_t,t)\zeta_t$$

where ζ_t is a Gaussian white noise. This makes precise the interpretation of (5.1). We still have to determine whether (5.1) can be modeled by a stochastic differential equation as defined in the last section.

In order to make precise the notion of $\{\zeta_n(\cdot,t)\}$ converging to a white noise, we define

$$W_n(\omega,t) = \int_a^t \zeta_n(\omega,s)\, ds \tag{5.3}$$

5. WHITE NOISE AND STOCHASTIC CALCULUS

and rewrite (5.2) as an integral equation

$$X_n(\omega,t) = X_n(\omega,a) + \int_a^t m(X_n(\omega,s), s)\, ds + \int_a^t \sigma(X_n(\omega,s), s)\, dW_n(\omega,s) \quad (5.4)$$

Since a Gaussian white noise ζ_t is the formal derivative of a Brownian motion, we make precise the notion of $\{\zeta_n(\cdot,t)\}$ converging to a Gaussian white noise by requiring that

$$W_n(\cdot,t) \xrightarrow[n\to\infty]{\text{a.s.}} K[W(\cdot,t) - W(\cdot,a)] \quad (5.5)$$

where K is a constant and $\{W(\cdot,t),\ a \le t \le T\}$ is a Brownian motion process. Since the constant K can always be absorbed into σ in (5.4), we shall assume it to be 1. We want to resolve the following two questions: First, under what conditions will $\{X_n(\cdot,t), a \le t \le T\}$ converge? Secondly, if $\{X_n(\cdot,t),\ a \le t \le T\}$ converges, does the limit $\{X(\cdot,t),\ a \le t \le T\}$ satisfy a stochastic differential equation, and if so, what stochastic differential equation?

Before stating the precise results that can be proved concerning these questions, we shall give a preliminary and heuristic discussion of what we can expect. This is especially important since what can be proved precisely at present is a little complicated and undoubtedly falls far short of what is in fact true. To begin with, consider a sequence of processes $\{Y_n(\cdot,t)\}$ defined by

$$Y_n(\omega,t) = \int_a^t \varphi(W_n(\omega,t), t)\, dW_n(\omega,t) \quad (5.6)$$

where φ is a known function and $\{W_n(\cdot,t)\}$ converges to a Brownian motion, and we want to determine what $\{Y_n(\cdot,t)\}$ converges to. Suppose we define a function $\psi(x,t)$ by

$$\psi(x,t) = \int_0^x \varphi(z,t)\, dz \quad (5.7)$$

If we denote $(\partial/\partial t)\psi(x,t)$ by $\dot\psi(x,t)$, we find

$$d\psi(W_n(\omega,t), t) = \varphi(W_n(\omega,t), t)\, dW_n(\omega,t) + \dot\psi(W_n(\omega,t), t)\, dt \quad (5.8)$$

In other words, we have

$$\psi(W_n(\omega,t), t) - \psi(W_n(\omega,a), a) - \int_a^t \dot\psi(W_n(\omega,s), s)\, ds = Y_n(\omega,t) \quad (5.9)$$

Now, if ψ and $\dot\psi$ are reasonable functions, we would certainly expect that as $W_n(\omega,t) \xrightarrow[n\to\infty]{} W(\omega,t)$,

$$\psi(W_n(\omega,t), t) \xrightarrow[n\to\infty]{} \psi(W(\omega,t), t)$$
$$\dot\psi(W_n(\omega,t), t) \xrightarrow[n\to\infty]{} \dot\psi(W(\omega,t), t)$$

Therefore, if all this is true, then

$$Y_n(\omega,t) \xrightarrow[n\to\infty]{} Y(\omega,t) = \psi(W(\omega,t), t) - \psi(W(\omega,a), a)$$
$$- \int_a^t \dot{\psi}(W(\omega,s), s)\, ds \quad (5.10)$$

Now, by the Ito's differentiation rule (Proposition 3.2),

$$\psi(W(\omega,t), t) = \psi(W(\omega,a), a) + \int_a^t \dot{\psi}(W(\omega,s), s)\, ds$$
$$+ \int_a^t \psi'(W(\omega,s), s)\, dW(\omega,s) + \frac{1}{2} \int_a^t \psi''(W(\omega,s), s)\, ds \quad (5.11)$$

Noting $\psi'(x,t) = \varphi(x,t)$, we get

$$Y(\omega,t) = \int_a^t \varphi(W(\omega,s), s)\, dW(\omega,s) + \frac{1}{2} \int_a^t \varphi'(W(\omega,s), s)\, ds \quad (5.12)$$

Comparing (5.12) against (5.6), we get the interesting result

$$\int_a^t \varphi(W_n(\omega,s), s)\, dW_n(\omega,s) \xrightarrow[n\to\infty]{} \int_a^t \varphi(W(\omega,s), s)\, dW(\omega,s)$$
$$+ \frac{1}{2} \int_a^t \varphi'(W(\omega,s), s)\, ds \quad (5.13)$$

where the first term on the right-hand side in (5.13) is a stochastic integral. The reason for the extra term is the same as the reason for the extra term in the Ito's differentiation formula (3.8). As we discussed it at that time, roughly speaking, the extra term is due to the fact that $(dW)_t^2$ is approximately dt.

In light of (5.13), we should expect a similar development for (5.4) as $n \to \infty$, namely, there will be an extra term. To find out what this extra term is, we first rewrite (5.4) as

$$dX_n(\omega,t) = m(X_n(\omega,t), t)\, dt + \sigma(X_n(\omega,t), t)\, dW_n(\omega,t) \quad (5.14)$$

Now define

$$\psi(x,t) = \int_0^x \frac{1}{\sigma(z,t)}\, dz \quad (5.15)$$

so that

$$d\psi(X_n(\omega,t), t) = \dot{\psi}(X_n(\omega,t), t)\, dt + \psi'(X_n(\omega,t), t)\, dX_n(\omega,t)$$
$$= \dot{\psi}(X_n(\omega,t), t) + \frac{m(X_n(\omega,t), t)}{\sigma(X_n(\omega,t), t)}\, dt + dW_n(\omega,t) \quad (5.16)$$

or

$$\psi(X_n(\omega,t), t) - \psi(X_n(\omega,a), a) = \int_a^t \mu(X_n(\omega,s), s)\, ds$$
$$+ W_n(\omega,t) - W_n(\omega,a) \quad (5.17)$$

5. WHITE NOISE AND STOCHASTIC CALCULUS

where we have set $\mu = (m/\sigma) + \dot{\psi}$. Suppose $\{W_n(\cdot,t)\}$ converges to a Brownian motion $W(\cdot,t)$ and suppose that $\{X_n(\cdot,t)\}$ converges to a process $X(\cdot,t)$. Then, under reasonable conditions, we would expect

$$\psi(X(\omega,t), t) - \psi(X(\omega,a), a) = \int_a^t \mu(X(\omega,s), s)\, ds$$
$$+ W(\omega,t) - W(\omega,a) \quad (5.18)$$

If we assume that $X(\omega,t)$ can be written in the form of

$$X(\omega,t) = X(\omega,a) + \int_a^t f(\omega,s)\, ds + \int_a^t \varphi(\omega,s)\, dW(\omega,s) \quad (5.19)$$

then we can apply Ito's differentiation formula (3.8) to $\psi(X_t,t)$ and get

$$\psi(X(\omega,t), t) - \psi(X(\omega,a), a) = \int_a^t \psi'(X(\omega,s), s)f(\omega,s)\, ds$$
$$+ \int_a^t \dot{\psi}(X(\omega,s), s)\, ds + \int_a^t \psi'(X(\omega,s), s)\varphi(\omega,s)\, dW(\omega,s)$$
$$+ \frac{1}{2}\int_a^t \psi''(X(\omega,s), s)\varphi^2(\omega,s)\, ds \quad (5.20)$$

We can now equate (5.18) with (5.20) and get

$$\varphi(\omega,s)\psi'(X(\omega,s), s) = 1$$

Therefore, by noting that $\psi' = 1/\sigma$, we get

$$\varphi(\omega,t) = \sigma(X(\omega,t), t) \quad (5.21)$$

Further,

$$\dot{\psi} + \frac{1}{\sigma}f + \tfrac{1}{2}\sigma^2\left(\frac{1}{\sigma}\right)' = \mu = \frac{m}{\sigma} + \dot{\psi}$$

Hence,

$$f(\omega,t) = m(X(\omega,t), t) + \tfrac{1}{2}\sigma(X(\omega,t), t)\sigma'(X(\omega,t), t) \quad (5.22)$$

Putting (5.21) and (5.22) into (5.19), we get

$$X_t = X_a + \int_a^t [m(X_s,s) + \tfrac{1}{2}\sigma(X_s,s)\sigma'(X_s,s)]\, ds + \int_a^t \sigma(X_s,s)\, dW_s \quad (5.23)$$

What we have shown, at least formally, is that if we interpret a white-noise-driven equation

$$\frac{d}{dt}X_t = m(X_t,t) + \sigma(X_t,t)\zeta_t \quad (5.24)$$

by a sequence of equations like (5.2), then the white-noise-driven differential equation is equivalent to a stochastic differential equation given by

$$dX_t = m(X_t,t)\, dt + \tfrac{1}{2}\sigma(X_t,t)\sigma'(X_t,t)\, dt + \sigma(X_t,t)\, dW_t \quad (5.25)$$

Again, we note the presence of an extra term $\frac{1}{2}\sigma\sigma'$, which will be referred to as the **correction term**.

We shall now state some convergence results concerning (5.13) and (5.23) [Wong and Zakai, 1965a and b, 1966]. We need to define some types of approximations $\{W_n(\omega,t)\}$ to a Brownian motion $W(\omega,t)$ as follows:

A_1: For each t, $W_n(\cdot,t) \xrightarrow[n\to\infty]{\text{a.s.}} W(\cdot,t)$. For each n and almost all ω, $W_n(\omega,\cdot)$ is sample continuous and of bounded variation on $[a,T]$.

A_2: A_1 and also for almost all ω, $W_n(\omega,\cdot)$ uniformly bounded, i.e., for almost all ω,
$$\sup_n \sup_{t\in[a,b]} |W_n(\omega,t)| < \infty$$

A_3: A_2 and for each n and almost all ω, $W_n(\omega,t)$ has a continuous derivative $\dot{W}_n(\omega,t)$.

A_4: For each n, $W_n(\omega,t)$ is a polygonal approximation of $W(\omega,t)$ defined by
$$W_n(\omega,t) = W(\omega,t_j^{(n)}) + [W(\omega,t_{j+1}^{(n)}) - W(\omega,t_j^{(n)})] \frac{t - t_j^{(n)}}{t_{j+1}^{(n)} - t_j^{(n)}}$$
$$t_j^{(n)} \leq t \leq t_{j+1}^{(n)} \quad (5.26)$$
where $a = t_0^{(n)} < t_1^{(n)} < \cdots < t_n^{(n)} = T$ and
$$\max_j (t_{j+1}^{(n)} - t_j^{(n)}) \xrightarrow[n\to\infty]{} 0$$

Proposition 5.1. Let $\varphi(x,t)$ have continuous partial derivatives $\varphi'(x,t) = (\partial/\partial x)\varphi(x,t)$ and $(\partial/\partial t)\varphi(x,t)$ in $-\infty < x < \infty$, $a \leq t \leq b$. Let $\{W_n(\omega,t)\}$ satisfy A_2, then
$$\int_a^b \varphi(W_n(\omega,t), t)\, dW_n(\omega,t) \xrightarrow[n\to\infty]{\text{a.s.}} \int_a^b \varphi(W(\omega,t), t)\, dW(\omega,t)$$
$$+ \frac{1}{2}\int_a^b \varphi'(W(\omega,t), t)\, dt \quad (5.27)$$

Further, if $\varphi(x,t)$ does not depend on t, then the conclusion holds with A_1 replacing A_2.

Proposition 5.2. Let $m(x,t)$, $\sigma(x,t)$, $\sigma'(x,t) = (\partial/\partial x)\sigma(x,t)$, and $\dot{\sigma}(x,t) = (\partial/\partial t)\sigma(x,t)$ be continuous in $-\infty < x < \infty$, $a \leq t \leq b$. Let $m(x,t)$, $\sigma(x,t)$, and $\sigma(x,t)\sigma'(x,t)$ satisfy a uniform Lipschitz condition, i.e., if f denotes any of three quantities m, σ, $\sigma\sigma'$, then
$$|f(x,t) - f(y,t)| \leq K|x - y| \quad (5.28)$$

Let $\{X_n(\omega,t), t \geq a\}$ satisfy (5.2), and let $\{X(\omega,t), t \geq a\}$ satisfy the

5. WHITE NOISE AND STOCHASTIC CALCULUS

stochastic differential equation

$$dX(\omega,t) = m(X(\omega,t), t)\, dt + \sigma(X(\omega,t), t)\, dW(\omega,t)$$
$$+ \tfrac{1}{2}\sigma(X(\omega,t), t)\sigma'(X(\omega,t), t)\, dt \quad (5.29)$$

Let $X_n(\omega,a) = X(\omega) = X(\omega,a)$, where X is independent of the aggregate of differences $\{W_t - W_a, t \geq a\}$ and $EX^2 < \infty$.

(a) If in addition, $|\sigma(x,t)| \geq \beta > 0$ and $|\sigma(x,t)| < K\sigma^2(x,t)$, then with $\{W_n(\omega,t)\}$ satisfying A_3

$$X_n(\omega,t) \xrightarrow[n\to\infty]{\text{a.s.}} X(\omega,t) \qquad a \leq t \leq b \quad (5.30)$$

(b) If $\{W_n(\omega,t)\}$ satisfies A_4 and $EX^4 < \infty$, then

$$X_n(\omega,t) \xrightarrow[n\to\infty]{\text{q.m.}} X(\omega,t) \qquad a \leq t \leq b \quad (5.31)$$

It should be mentioned that a symmetrized definition for stochastic integrals has been proposed [Fisk, 1963; Stratonovich, 1966] for which rules of ordinary calculus apply. Rewritten in terms of the Fisk-Stratonovich's integral, neither (5.13) nor (5.23) would contain an extra term. However, this approach has the disadvantage that conditions which guarantee the convergence of Fisk-Stratonovich's integral are less natural and more difficult to verify than those of the stochastic integral. Furthermore, the martingale property of Ito's integral would be lost. As we shall see in Chap. 6, an important application of the stochastic integral is in the representation of likelihood ratios and filtering operations, and this application depends on the martingale property. While these representations, under suitable restrictions, can be reexpressed in terms of the Fisk-Stratonovich integral, the resulting formulas will be considerably more complicated.

Equations (5.27) and (5.29) can be interpreted as expressions relating a white-noise integral $\int_a^t \varphi(\omega,s)\zeta(\omega,s)\, ds$ to the stochastic integral $\int_a^t \varphi(\omega,s)\, dW(\omega,s)$ for the following two special cases:

1. $\varphi(\omega,s) = \psi(W(\omega,s), s)$
2. $\varphi(\omega,s) = \psi(X(\omega,s), s)$, and X_s is related to W_s via a stochastic differential equation.

In general, $\varphi(\omega,t)$ may depend in a much more complicated way on $\{W_s, a \leq s \leq t\}$. The question arises as to whether it is possible to relate the white-noise integral to the corresponding stochastic integral in the general situation. This question has been resolved [Wong and Zakai, 1969]. Roughly speaking, the white-noise integral is equal to the corresponding stochastic integral plus a correction term. If $\varphi(\omega,t)$ is viewed as a functional on $\{W(\omega,s), a \leq s \leq t\}$, then the correction term can be expressed in terms of the Frechét differential of this functional.

In applications, differential equations driven by white noise frequently appear in a vector form as follows:

$$\dot{\mathbf{X}}_t = \mathbf{m}(\mathbf{X}_t,t) + \pmb{\sigma}(\mathbf{X}_t,t)\pmb{\zeta}_t \tag{5.32}$$

where $\mathbf{X}_t$ and $\mathbf{m}$ are n vectors, $\pmb{\zeta}_t$ is a p vector of independent Gaussian white noise, and $\pmb{\sigma}$ is a matrix of appropriate dimensions. There is no difficulty in extending the definition of stochastic integral to the form

$$\int_a^b \pmb{\phi}(\omega,t)\, d\mathbf{W}(\omega,t)$$

where $\mathbf{W}_t$ is a vector of independent Brownian motions, and $\pmb{\phi}_t$ is a matrix provided that

$$\int_a^b \sum_{i,j} E|\varphi_{ij}(\cdot,t)|^2\, dt < \infty$$

In terms of the extended definition of stochastic integral, stochastic differential equations in the vector form can be treated. Intuitively, it is highly plausible that the white-noise equation (5.32) is equivalent to a stochastic differential equation

$$d\mathbf{X}_t = \mathbf{f}(\mathbf{X}_t,t)\, dt + \mathbf{g}(\mathbf{X}_t,t)\, d\mathbf{W}_t \tag{5.33}$$

The problem is to determine $\mathbf{f}$ and $\mathbf{g}$. It was conjectured by Wong and Zakai [1965a] that $\mathbf{g} = \pmb{\sigma}$, but

$$f_k(\mathbf{x},t) = m_k(\mathbf{x},t) + \frac{1}{2}\sum_{l,m} \frac{\partial \sigma_{km}(\mathbf{x},t)}{\partial x_l}\sigma_{lm}(\mathbf{x},t) \tag{5.34}$$

This has since been verified [McShane, 1969, 1970] under suitable conditions.

As the final topic in this section, we briefly consider problems arising in simulation. Suppose that we want to simulate a white-noise differential equation

$$\dot{X}_t = m(X_t,t) + \sigma(X_t,t)\zeta_t \tag{5.35}$$

Roughly speaking, there is a time constant or a bandwidth associated with the equation. While it is not clear how a bandwidth should be defined, it clearly should be related to the maximum rate of change of X_t in some way. The following definition may be useful:

$$B = \sup_{x,t}\left[\frac{|m(x,t)|}{1+|x|} + \frac{\sigma^2(x,t)}{1+|x|^2}\right]$$

Under assumption (4.5), this quantity is always finite. If z_t is a stationary Gaussian process with a spectral density that is constant over a bandwidth much greater than B, then it is intuitively clear that (5.35) can

6. GENERALIZATIONS OF THE STOCHASTIC INTEGRAL

be simulated by replacing ζ_t by z_t. Hence, an analog and continuous-time simulation of (5.35) can be achieved by implementing

$$\dot{X}_t = m(X_t,t) + \sigma(X_t,t)z_t \tag{5.36}$$

with a wide-band noise source z_t. Of course, this also simulates the stochastic differential equation

$$dX_t = m(X_t,t)\, dt + \sigma(X_t,t)\, dW_t + \tfrac{1}{2}\sigma\sigma'(X_t,t)\, dt \tag{5.37}$$

The situation is less clear in discrete-time simulation. All depends on the noise source. If one uses a random-number generator which produces a sequence of *independent* Gaussian random variables $z_1, z_2, \ldots$, then the difference equation

$$X_{k+1} = X_k + \Delta m(X_k,t_k) + \Delta \tfrac{1}{2}\sigma\sigma'(X_k,t_k) + \sigma(X_k,t_k)z_k \tag{5.38}$$

simulates (5.37) well, provided that we choose $Ez_k^2 = \Delta$ and $\Delta \ll 1/B$. Hence, (5.35) can be simulated by implementing (5.38). On the other hand, suppose that the noise source is a wide-band noise generator with bandwidth $B_0 \gg B$. If we sample this noise source at a rate to permit a faithful reproduction of this noise, we would have to sample at $2B_0$ or more. If we do this and produce a sequence $z_1, z_2, \ldots$, then the difference equation

$$X_{k+1} = X_k + \frac{1}{2B_0} m(X_k,t_k) + \sigma(X_k,t_k)z_k \tag{5.39}$$

is a good approximation to (5.35). The difference here is that $z_1, z_2, \ldots$, are no longer independent.

6. GENERALIZATIONS OF THE STOCHASTIC INTEGRAL

For a Brownian motion $\{W_t, \mathcal{Q}_t\}$, we have defined the stochastic integral

$$I(\varphi,\omega) = \int_a^b \varphi(\omega,t)\, dW(\omega,t) \tag{6.1}$$

for integrands satisfying (1) φ jointly measurable in (ω,t), (2) for each t φ_t is $\mathcal{Q}_t$ measurable, and (3) $\int_a^b E|\varphi_t|^2\, dt < \infty$. The stochastic integral (6.1) can be generalized in two important directions. First, it can be defined for integrands satisfying (1), (2), and instead of (3), the weaker condition

$$\int_a^b |\varphi(\omega,t)|^2\, dt < \infty \qquad \text{a.s.} \tag{6.2}$$

Secondly, the Brownian motion $\{W_t, \mathcal{Q}_t\}$ in (6.1) can be replaced by a class of martingales $\{X_t, \mathcal{Q}_t\}$. In this section, we shall consider both these generalizations and their applications.

Proposition 6.1. Let $\{W_t, \mathcal{A}_t\}$ be a Brownian motion and let $\varphi(\omega,t)$ satisfy:
(a) φ is jointly measurable in (ω,t).
(b) For each t, φ_t is $\mathcal{A}_t$ measurable,
(c) $\int_a^b |\varphi(\omega,t)|^2\, dt < \infty$ almost surely.

Let φ^m be defined by

$$\varphi^m(\omega,t) = \begin{cases} \varphi(\omega,t) & \text{if } \int_a^t |\varphi(\omega,t)|^2\, dt \leq m \\ 0 & \text{otherwise} \end{cases} \tag{6.3}$$

and let $I(\varphi^m)$ denote the stochastic integral

$$I(\varphi^m) = \int_a^b \varphi^m(\omega,t)\, dW(\omega,t)$$

Then, $\{I(\varphi^m),\ m = 1, \ldots\}$ converges in probability as $m \to \infty$, and we define

$$I(\varphi) = \int_a^b \varphi(\omega,t)\, dW(\omega,t) = \lim_{m \to \infty} \text{in p. } I(\varphi^m) \tag{6.4}$$

Proof: Let φ^m be defined by (6.3). For each m, φ^m satisfies (2.3) and (2.4) so that $I(\varphi^m)$ is well defined. Now, for any ω such that

$$\int_a^b |\varphi(\omega,t)|^2\, dt \leq \min(m,n)$$

we have from (6.3)

$$\sup_t |\varphi^m(\omega,t) - \varphi^n(\omega,t)| = 0$$

which in turn implies that $\int_a^b \varphi^m(\omega,t)\, dW(\omega,t) = \int_a^b \varphi^n(\omega,t)\, dW(\omega,t)$. It follows that for every $\epsilon > 0$,

$$\mathcal{P}\left(|I(\varphi^m) - I(\varphi^n)| \geq \epsilon\right) \leq \mathcal{P}\left(\int_a^b |\varphi_t|^2\, dt > \min(m,n)\right) \xrightarrow[m,n \to \infty]{} 0$$

which proves that $\{I(\varphi^n)\}$ converges in probability so that (6.4) is an adequate definition for $I(\varphi)$. ∎

Remarks.

(a) If $\{\varphi^n\}$ is a sequence of functions satisfying conditions (2.3) and (6.2), if $|\varphi_m(\omega,t)| \leq |\varphi(\omega,t)|$, and if

$$\varphi_m \xrightarrow[m \to \infty]{} \varphi \text{ in } \mathcal{P} \times \mathcal{L} \text{ measure} \tag{6.5}$$

then

$$I(\varphi_m) \xrightarrow[m \to \infty]{\text{in p.}} I(\varphi) \tag{6.6}$$

6. GENERALIZATIONS OF THE STOCHASTIC INTEGRAL

(b) Now the process

$$X_t = \int_a^t \varphi(\omega,s) \, dW(\omega,s) \tag{6.7}$$

is no longer necessarily a martingale. Of course, a sufficient condition for X_t to be a martingale is precisely

$$\int_a^b E|\varphi_s|^2 \, ds < \infty \tag{6.8}$$

However, this is not a necessary condition. If we define $\tau_n(\omega) = \min t : \int_a^t \varphi^2(\omega,s) \, ds \geq n$ and set $\tau_n(\omega) = \infty$ if $\int_a^b \varphi^2(\omega,s) \, ds < n$, then for each n, $X_{nt} = X_{\min(t,\tau_n)}$ is a martingale. By definition X_t is said to be a **local martingale** [see, e.g., Kunita and Watanabe, 1967].

Next, we shall consider generalizations of the stochastic integral by replacing the Brownian motion W_t by a more general process. As a first step in this direction, we shall replace W_t by a process Z_t satisfying the following properties. Throughout, $\{\mathcal{C}_t\}$ again denotes an increasing family of σ algebras.

$$\{Z_t, \mathcal{C}_t, a \leq t \leq b\} \text{ is a martingale and } EZ_t^2 < \infty \tag{6.9}$$
$$E(Z_t - Z_s)^2 = E^{\mathcal{C}_s}(Z_t - Z_s)^2 \quad \text{a.s.} \tag{6.10}$$

Let $F(t)$ be a nondecreasing function so that

$$E(Z_t - Z_s)^2 = F(t) - F(s) \quad t \geq s \tag{6.11}$$

Then, the stochastic integral

$$\int_a^b \varphi(\omega,s) \, dZ(\omega,s) \tag{6.12}$$

is well defined for any φ satisfying

φ is jointly measurable in (ω,t) and for each t, φ_t is $\mathcal{C}_t$ measurable
$$\tag{6.13}$$

$$\int_a^b |\varphi(\omega,s)|^2 \, dF(s) < \infty \quad \text{a.s.} \tag{6.14}$$

The procedure for defining (6.12) is exactly the same as before and will not be repeated.

The class of processes satisfying both (6.9) and (6.10) is still quite restricted. In particular, if Z_t is almost surely sample continuous, then $F(t)$ is necessarily continuous [for convenience, we set $F(0) = 0$] and Z_t can be expressed as

$$Z_t = W_{F(t)} \tag{6.15}$$

where W_t is a Brownian motion. Therefore, if we consider only sample-continuous Z_t, then the stochastic integral (6.12) is really the same as what we already defined. The next step in generalizing the stochastic integral is to get rid of the restriction (6.10). We begin with the following result.

Proposition 6.2. Let $\{Z_t, \mathcal{C}_t, a \leq t \leq b\}$ be a sample-continuous second-order martingale. Then there is a unique decomposition

$$Z_t^2 = Z_{1t} + Z_{2t} \qquad a \leq t \leq b \tag{6.16}$$

where $\{Z_{2t}, \mathcal{C}_t, a \leq t \leq b\}$ is a sample-continuous first-order martingale, and $\{Z_{1t}, a \leq t \leq b\}$ is sample continuous, nondecreasing, with $Z_{1a} = 0$.

Remark: This proposition is a special case of the well-known supermartingale-decomposition theorem of Meyer [1966, Chap. 7]. We note that if Z_t is a Brownian motion, then Z_{1t} is simply t.

Proposition 6.3. Let $\{Z_t, \mathcal{C}_t, a \leq t \leq b\}$ be a sample-continuous second-order martingale. Let $\{Z_{1t}, a \leq t \leq b\}$ be defined as in (6.16). Suppose that $\varphi(\omega,t)$, $\omega \in \Omega$, $t \in [a,b]$, is a jointly measurable function[1] such that for each t, φ_t is $\mathcal{C}_t$ measurable and

$$\int_a^b \varphi_t^2 \, dZ_{1t} < \infty \tag{6.17}$$

with probability 1. Then the stochastic integral

$$I(\varphi,\omega) = \int_a^b \varphi(\omega,t) \, dZ(\omega,t) \tag{6.18}$$

is well defined by the following two properties:
(a) If φ is an (ω,t)-step function, then

$$I(\varphi,\omega) = \sum_\nu \varphi_\nu(\omega)[Z(\omega,t_{\nu+1}) - Z(\omega,t_\nu)]$$

(b) If $\int_a^b |\varphi(\omega,t) - \varphi_k(\omega,t)| \, dZ_1(\omega,t) \xrightarrow[k\to\infty]{\text{a.s.}} 0$, then $I(\varphi_k,\omega) \xrightarrow[k\to\infty]{\text{in p.}} I(\varphi,\omega)$

Remark: It is clear that Z_{1t} now plays the role played by t in the original definition of stochastic integral.

If X_t is of the form

$$X_t = \int_a^t f(\omega,s) \, ds + \int_a^t \varphi(\omega,s) \, dZ(\omega,s) \tag{6.19}$$

where the last integral is defined as in (6.18), then a transformation rule similar to Proposition 3.2 holds once again. Let $\psi(x,t)$ be twice continuously

---
[1] Here measurability in t refers to Borel measurability.

6. GENERALIZATIONS OF THE STOCHASTIC INTEGRAL

differentiable in x and once in t. Then

$$\psi(X_t,t) = \psi(X_a,a) + \int_a^t \psi'(X_s,s)\, dX_s + \int_a^t \dot\psi(X_s,s)\, ds$$
$$+ \frac{1}{2} \int_a^t \psi''(X_s,s)\varphi_s^2\, dZ_{1s} \quad (6.20)$$

with probability 1 [Kunita and Watanabe, 1967].

Suppose that there exists a continuous and nondecreasing function $F(t)$, $a \leq t \leq b$, such that for almost all ω, $Z_1(\omega,t)$ as a function of t is absolutely continuous with respect to the Borel measure generated by F. That is, there exists an a.s. nonnegative function $z(\omega,t)$ such that

$$Z_1(\omega,t) = \int_a^t z(\omega,s)\, dF(s) \quad \text{a.s.} \quad (6.21)$$

If such an F exists it can always be taken to be EZ_{1t}, because (6.21) implies that

$$Z_1(\omega,t) = \int_a^t \frac{z(\omega,s)}{Ez_s}\, d(EZ_{1s}) \quad (6.22)$$

If Z_{1t} has the representation (6.21), then Z_t can be represented as a stochastic integral

$$Z(\omega,t) = \int_a^t \sqrt{z(\omega,s)}\, dW(\omega,F(s)) \quad (6.23)$$

where $\{W_s, 0 \leq s \leq F(b)\}$ is a Brownian motion. We note that (6.23) is a stochastic integral of the type given by (6.12). Now, with (6.23) we can rewrite any stochastic integral in terms of Z_t,

$$I(\varphi,\omega) = \int_a^b \varphi_s\, dZ_s = \int_a^t \varphi_s \sqrt{z_s}\, dW_{F(s)} \quad (6.24)$$

Once again, we return to the basic definition of a stochastic integral in terms of a Brownian motion.

As the final step in generalizing the stochastic integral, consider a sample-continuous process $\{X_t, a \leq t \leq b\}$ which has a decomposition

$$X_t = Y_t + Z_t \quad (6.25)$$

where Y_t is almost surely of bounded variation, and Z_t is a second-order sample-continuous martingale. Clearly, we can define

$$\int_a^b \varphi_t\, dX_t = \int_a^b \varphi_t\, dY_t + \int_a^b \varphi_t\, dZ_t \quad (6.26)$$

provided that the first integral exists almost surely as a Stieltjes integral and the second as a stochastic integral. A process that can be decomposed as in (6.25) was termed a quasi-martingale by Fisk [1965] who also gave necessary and sufficient conditions for the existence of such a decomposition. Unfortunately, these conditions are not always easily verified.

As an example of applications of the generalized definition of stochastic integral, we consider the following important representation theorem due to Doob [1953, pp. 287–291].

Proposition 6.4. Let $\{X_t, a \leq t \leq b\}$ be a sample-continuous second-order process. Let $m(x,t)$ and $\sigma(x,t)$ be Borel functions of (x,t) satisfying

$$|m(x,t)| \leq K \sqrt{1 + x^2} \tag{6.27}$$
$$0 \leq \sigma(x,t) \leq K \sqrt{1 + x^2} \tag{6.28}$$

Let $\mathcal{A}_t$ denote the smallest σ algebra such that X_s, $s \leq t$, are all measurable, and suppose that $\{X_t, a \leq t \leq b\}$ satisfies the following conditions:

(a) There exists $\{Z(\omega,t), a \leq t \leq b\}$ such that $Z_t \geq 0, EZ_t < \infty$, and

$$\sup_{t > s} E^{\mathcal{A}_s} X_t^2 \leq Z_s \tag{6.29}$$

(b) There exists a nondecreasing function f with $\lim_{h \downarrow 0} f(h) = 0$ such that whenever $a \leq t < t + h \leq b$, we have with probability 1,

$$\left| E^{\mathcal{A}_t}(X_{t+h} - X_t) - \int_t^{t+h} m(X_s, s) \, ds \right| \leq hf(h)(1 + X_t^2) \tag{6.30}$$

$$\left| E^{\mathcal{A}_t}(X_{t+h} - X_t)^2 - \int_t^{t+h} \sigma^2(X_s, s) \, ds \right| \leq hf(h)(1 + X_t^2) \tag{6.31}$$

Under these conditions, $\{X_t, a \leq t \leq b\}$ is a Markov process and satisfies a stochastic differential equation

$$X_t = X_a + \int_a^t m(X_s, s) \, ds + \int_a^t \sigma(X_s, s) \, dW_s \tag{6.32}$$

where $\{W_t, a \leq t \leq b\}$ is a Brownian motion.

Remark: We have made no assumption that m and σ satisfy a Lipschitz condition. Without such an assumption, we cannot be sure that there is a unique solution to (6.32). One possible consequence of this is that the finite-dimensional distributions of $\{X_t, a \leq t \leq b\}$ may not be completely determined by m, σ, and the distribution of X_a.

Proof: We shall give an outline of the proof. Let $\{Z_t, a \leq t \leq b\}$ be defined by

$$Z_t = X_t - X_a - \int_a^t m(X_s, s) \, ds \tag{6.33}$$

Because of (6.30), we can show that $\{Z_t, \mathcal{A}_t, a \leq t \leq b\}$ is a sample-continuous martingale. Because of (6.27), it is also second order. Furthermore, if we define

$$Y_t = Z_t^2 - \int_a^t \sigma^2(X_s, s) \, ds \tag{6.34}$$

7. DIFFUSION EQUATIONS

then because of (6.31), $\{Y_t, \mathcal{A}_t, a \leq t \leq b\}$ is also a sample-continuous martingale. Therefore, the process $\{Z_{1t}, a \leq t \leq b\}$ defined by the decomposition (6.16) is simply

$$Z_{1t} = \int_a^t \sigma^2(X_s, s)\, ds \tag{6.35}$$

Clearly, Z_{1t} has the form of (6.21) with

$$F(t) = t \tag{6.36}$$

and

$$z(\omega, t) = \sigma^2(X(\omega, t), t) \tag{6.37}$$

From (6.23) we get

$$Z_t = \int_a^t \sigma(X_s, s)\, dW_s \tag{6.38}$$

Equation (6.38) combined with (6.33) yields (6.32). ∎

7. DIFFUSION EQUATIONS

In this section, we shall try to show that the transition probabilities of a process satisfying a stochastic differential equation can be obtained by solving either of a pair of partial differential equations. These equations are called the backward and forward equations of Kolmogorov or, alternatively, diffusion equations. The forward equation is also sometimes called the Fokker-Planck equation. The situation, however, is not completely satisfactory. As we shall see, the original derivation of Kolmogorov involved assumptions that cannot be directly verified. Attempts in circumventing these assumptions involve other difficulties. We begin with a derivation of the diffusion equations following the lines of Kolmogorov [1931].

Let $\{X_t, a \leq t \leq b\}$ be a Markov process, and denote

$$P(x,t|x_0,t_0) = \mathcal{P}(X_t < x | X_{t_0} = x_0) \tag{7.1}$$

We call $P(x,t|x_0,t_0)$ the **transition function** of the process. If there is a function $p(x,t|x_0,t_0)$ so that

$$P(x,t|x_0,t_0) = \int_{-\infty}^{x} p(u,t|x_0,t_0)\, du \tag{7.2}$$

then we call $p(x,t|x_0,t_0)$ the **transition density function**. Since $\{X_t, a \leq t \leq b\}$ is a Markov process, $P(x,t|x_0,t_0)$ satisfies the Chapman-Kolmogorov equation

$$P(x,t|x_0,t_0) = \int_{-\infty}^{\infty} P(x,t|z,s)\, dP(z,s|x_0,t_0) \tag{7.3}$$

We now assume the crucial conditions on $\{X_t, a \leq t \leq b\}$ which make the derivation of the diffusion equations possible. These conditions are very similar to conditions (6.30) and (6.31) which made it possible to represent a process as the solution of a stochastic differential equation.

Define for a positive ϵ,

$$M_k(x,t; \epsilon, \Delta) = \int_{|y-x| \leq \epsilon} (y-x)^k \, dP(y, t+\Delta|x,t) \qquad k = 0, 1, 2$$

$$M_3(x,t; \epsilon, \Delta) = \int_{|y-x| \leq \epsilon} |y-x|^3 \, dP(y, t+\Delta|x,t)$$
(7.4)

We assume that the Markov process $\{X_t, a \leq t \leq b\}$ satisfies the following conditions:

$$\frac{1}{\Delta}[1 - M_0(x,t; \epsilon, \Delta)] \xrightarrow[\Delta \downarrow 0]{} 0 \qquad (7.5)$$

$$\frac{1}{\Delta} M_1(x,t; \epsilon, \Delta) \xrightarrow[\Delta \downarrow 0]{} m(x,t) \qquad (7.6)$$

$$\frac{1}{\Delta} M_2(x,t; \epsilon, \Delta) \xrightarrow[\Delta \downarrow 0]{} \sigma^2(x,t) \qquad (7.7)$$

$$\frac{1}{\Delta} M_3(x,t; \epsilon, \Delta) \xrightarrow[\Delta \downarrow 0]{} 0 \qquad (7.8)$$

It is clear that if $1 - M_0(x,t; \epsilon, \Delta) \xrightarrow[\Delta \downarrow 0]{} 0$, then by dominated convergence,

$$\mathcal{P}(|X_{t+\Delta} - X_t| > \epsilon) = \int_{-\infty}^{\infty} [1 - M_0(x,t; \epsilon, \Delta)] \, dP(x,t) \xrightarrow[\Delta \downarrow 0]{} 0$$

Therefore, (7.5) is considerably stronger than continuity in probability. In addition, suppose that the transition function $P(x,t|x_0,t_0)$ satisfies the condition:

For each (x,t), $P(x,t|x_0,t_0)$ is once differentiable in t_0 and three-times differentiable in x_0, and the derivatives are continuous and bounded in (x_0,t_0) (7.9)

Now we can derive the background equation as follows. Write the Chapman-Kolmogorov equation in the form

$$P(x,t|x_0,t_0) = \int_{-\infty}^{\infty} P(x,t|z, t_0 + \Delta) \, dP(z, t_0 + \Delta|x_0,t_0) \qquad (7.10)$$

Because of (7.9), we can write, by virtue of the Taylor's theorem,

$$P(x,t|z, t_0 + \Delta) = P(x,t|x_0, t_0 + \Delta) + \frac{\partial P(x,t|x_0, t_0 + \Delta)}{\partial x_0}(z - x_0)$$

$$+ \frac{1}{2} \frac{\partial^2 P(x,t|x_0, t_0 + \Delta)}{\partial x_0^2}(z - x_0)^2$$

$$+ \frac{1}{6} \frac{\partial^3 P(x,t|z, t_0 + \Delta)}{\partial z^3}\bigg|_{z=\theta} (z - x_0)^3 \qquad |\theta - x_0| \leq |z - x_0| \quad (7.11)$$

7. DIFFUSION EQUATIONS

Using (7.11) in (7.10) and using (7.4), we can write

$$P(x,t|x_0,t_0) = \int_{|z-x_0|>\epsilon} P(x,t|z, t_0 + \Delta)\, dP(z, t_0 + \Delta|x_0,t_0)$$
$$+ \sum_{k=0}^{2} \frac{1}{k!} M_k(x_0,t_0;\epsilon,\Delta) \frac{\partial^k}{\partial x_0^k} P(x,t|x_0, t_0 + \Delta) + \frac{1}{6} \int_{|z-x_0|\leq\epsilon} \frac{\partial^3 P(x,t|z, t_0 + \Delta)}{\partial z^3}\bigg|_{z=\theta} (z - x_0)^3\, dP(z, t_0 + \Delta|x_0,t_0) \quad (7.12)$$

This means that

$$\left| \left[\frac{P(x,t|x_0,t_0) - P(x,t|x_0, t_0 + \Delta)}{\Delta} \right] - \frac{1}{\Delta} M_1(x_0,t_0;\epsilon,\Delta) \frac{\partial P(x,t|x_0, t_0 + \Delta)}{\partial x_0} - \frac{1}{2}\frac{1}{\Delta} M_2(x_0,t_0;\epsilon,\Delta) \frac{\partial^2 P(x,t|x_0, t_0 + \Delta)}{\partial x_0^2} \right| \leq \left| \left[\frac{1 - M_0(x_0,t_0;\epsilon,\Delta)}{\Delta} \right] + \frac{M_3(x_0,t_0;\epsilon,\Delta)}{6\Delta} \sup_{|z-x_0|\leq\epsilon} \left| \frac{\partial^3 P(x,t|z, t_0 + \Delta)}{\partial z^3} \right| \right| \quad (7.13)$$

If we let $\Delta \downarrow 0$ and use conditions (7.5) through (7.8), (7.13) becomes

$$-\frac{\partial}{\partial t_0} P(x,t|x_0,t_0) = m(x_0,t_0) \frac{\partial}{\partial x_0} P(x,t|x_0,t_0) + \tfrac{1}{2}\sigma^2(x_0,t_0) \frac{\partial^2}{\partial x_0^2} P(x,t|x_0,t_0) \qquad a < t_0 < t < b \quad (7.14)$$

The "terminal" condition for (7.14) is

$$\lim_{t_0 \uparrow t} P(x,t|x_0,t_0) = \begin{cases} 1 & x > x_0 \\ 0 & x < x_0 \end{cases} \quad (7.15)$$

Equation (7.14) is the **backward equation** of diffusion. The name is due to the fact that it is an equation in the pair of initial variables (x_0,t_0) moving backward from t.

The forward equation can be derived in the following indirect way: Let $f(x)$, $-\infty < x < \infty$, be a Schwartz function of rapid descent. That is, f is infinitely differentiable, and for any k and m,

$$|x|^k |f^{(m)}(x)| \xrightarrow[|x|\to\infty]{} 0 \quad (7.16)$$

As we did in Chap. 3, the space of all such functions is denoted by $\mathcal{S}$ (cf. 3.5.6c). Define the function

$$\hat{f}(t|x_0,t_0) = E[f(X_t)|X_{t_0} = x_0]$$
$$= \int_{-\infty}^{\infty} f(x)\, dP(x,t|x_0,t_0) \quad (7.17)$$

Now, write

$$\tilde{f}(t + \Delta | x_0, t_0) = \int_{-\infty}^{\infty} f(x) \, dP(x, t + \Delta | x_0, t_0)$$
$$= \int_{-\infty}^{\infty} \left[\int_{-\infty}^{\infty} f(x) \, dP(x, t + \Delta | z, t) \right] dP(z, t | x_0, t_0) \quad (7.18)$$

Since f is infinitely differentiable and satisfies (7.16), we have

$$f(x) = \sum_{k=0}^{2} \frac{1}{k!} f^{(k)}(z)(x - z)^k + \tfrac{1}{6} f^{(3)}(\theta)(x - z)^3 \quad (7.19)$$

Repeating the arguments leading to (7.14), we find

$$\frac{\partial \tilde{f}(t | x_0, t)}{\partial t} = \int_{-\infty}^{\infty} \left[m(z, t) \frac{df(z)}{dz} + \tfrac{1}{2} \sigma^2(z, t) \frac{d^2 f(z)}{dz^2} \right] dP(z, t | x_0, t_0) \quad (7.20)$$

Now, if $P(x, t | x_0, t_0)$ satisfies

For each (x_0, t_0), $P(x, t | x_0, t_0)$ is four-times differentiable in x
and once in t, and the derivatives are continuous in (x, t) $\qquad$ (7.21)

and if $\sigma^2(x, t)$ is twice continuously differentiable in x, and $m(x, t)$ is once continuously differentiable in x, then we have from (7.20) and integrations by parts,

$$\int_{-\infty}^{\infty} f(x) \left\{ \frac{\partial}{\partial t} p(x, t | x_0, t_0) - \frac{1}{2} \frac{\partial^2}{\partial x^2} [\sigma^2(x, t) p(x, t | x_0, t_0)] \right.$$
$$\left. + \frac{\partial}{\partial x} [m(x, t) p(x, t | x_0, t_0)] \right\} dx = 0 \quad (7.22)$$

Since (7.22) holds for all $f \in \mathcal{S}$, the quantity in the brackets must be zero for almost all x, but being continuous, it must be zero for all x. Therefore,

$$\frac{\partial}{\partial t} p(x, t | x_0, t_0) = \frac{1}{2} \frac{\partial^2}{\partial x^2} [\sigma^2(x, t) p(x, t | x_0, t_0)]$$
$$- \frac{\partial}{\partial x} [m(x, t) p(x, t | x_0, t_0)] \qquad b > t > t_0 > a \quad (7.23)$$

Equation (7.23) is the **forward equation** of diffusion, and is also called the **Fokker-Planck equation**. The initial condition to be imposed is

$$\int_{-\infty}^{\infty} f(x) p(x, t | x_0, t_0) \, dx \xrightarrow[t \downarrow t_0]{} f(x_0) \qquad \forall f \in \mathcal{S} \quad (7.24)$$

that is, $p(x, t_0 | x_0, t_0) = \delta(x - x_0)$. A solution of (7.23), satisfying (7.24), will be called its **fundamental solution**. Our derivation of the two equations of diffusion are now complete.

7. DIFFUSION EQUATIONS

If we view the two diffusion equations as possible means for determining the transition probabilities of the solution of a stochastic differential equation, then the situation is still not entirely satisfactory. This is because the diffusion equations have been derived under differentiability assumptions (7.9) and (7.21). If we don't know $P(x,t|x_0,t_0)$, how do we know whether it is differentiable the required number of times? This difficulty is in part resolved by the following proposition.

Proposition 7.1. Let $m(x,t)$ and $\sigma(x,t)$ satisfy the following conditions on $-\infty < x < \infty$, $a \leq t \leq b$:

There exist positive constants σ_0 and K so that

$$|m(x,t)| \leq K\sqrt{1+x^2}$$
$$0 < \sigma_0 \leq \sigma(x,t) \leq K\sqrt{1+x^2} \qquad (7.25)$$

There exist positive constants γ and K so that

$$|m(x,t) - m(y,t)| \leq K|x-y|^\gamma$$
$$|\sigma(x,t) - \sigma(y,t)| \leq K|x-y|^\gamma \quad \text{(Hölder condition)} \qquad (7.26)$$

Then, the following conclusions are valid:
(a) The backward equation

$$\tfrac{1}{2}\sigma^2(x_0,t_0)\frac{\partial^2 P(x,t|x_0,t_0)}{\partial x_0^2} + m(x_0,t_0)\frac{\partial P(x,t|x_0,t_0)}{\partial x_0}$$
$$= -\frac{\hat{\partial}}{\partial t_0}P(x,t|x_0,t_0) \qquad t > t_0 \quad (7.27)$$

has a unique solution corresponding to condition (7.15). Further, for $t > t_0$, $P(x,t|x_0,t_0)$ is differentiable with respect to x so we have the transition density

$$p(x,t|x_0,t) = \frac{\partial}{\partial x}P(x,t|x_0,t_0) \qquad (7.28)$$

(b) There exists a sample-continuous Markov process $\{X_t, a \leq t \leq b\}$ with transition function $P(x,t|x_0,t_0)$.
(c) Conditions (7.5) to (7.8) are satisfied.
(d) If $m'(x,t), \sigma'(x,t), \sigma''(x,t)$ satisfy (7.25) and (7.26), then $p(x,t|x_0,t_0)$ is the unique fundamental solution of the forward equation.
(e) If γ can be taken to be 1 in (7.26), then $p(x,t|x_0,t_0)$ is the transition density of the unique solution to the stochastic integral equation

$$X_t = X_a + \int_a^t m(X_s,s)\,ds + \int_a^t \sigma(X_s,s)\,dW_s \qquad (7.29)$$

Example 1. Suppose that $\{X_t,\ t \geq 0\}$ satisfies a stochastic differential equation

$$dX_t = -X_t\,dt + \sqrt{2(1 + X_t^2)}\,dW_t \qquad (7.30)$$

Here, we have $m(x,t) = -x$ and $\sigma(x,t) = \sqrt{2(1 + x^2)}$. Therefore, the forward equation is given by

$$\frac{\partial^2}{\partial x^2}[(1 + x^2)p(x,t|x_0,t_0)] + \frac{\partial}{\partial x}[xp(x,t|x_0,t_0)] = \frac{\partial}{\partial t}p(x,t|x_0,t_0) \qquad (7.31)$$

Because m and σ do not depend on t, $p(x,t|x_0,t_0)$ will depend only on $t - t_0$ and not on t and t_0 separately. Furthermore, we can rewrite (7.30) as

$$X_t = e^{-(t-t_0)}X_{t_0} + \sqrt{2}\int_{t_0}^{t} e^{-(t-s)}\sqrt{1 + X_s^2}\,dW_s$$

and from this we expect that as $t \to \infty$, the conditional density $p(x,t|x_0,t_0)$ will approach the stationary density $p(x)$. We also expect that

$$\frac{\partial p(x,t|x_0,t_0)}{\partial t} \xrightarrow[t \to \infty]{} 0$$

Therefore,

$$\frac{d^2}{dx^2}[(1 + x^2)p(x)] + \frac{d}{dx}[xp(x)] = 0$$

or

$$\frac{d}{dx}[(1 + x^2)p(x)] + xp(x) = \text{constant}$$

Because $p'(x),\ p(x) \xrightarrow[|x| \to \infty]{} 0$, the constant must be 0, and by direct integration we get

$$p(x) = p(0)\frac{1}{(1 + x^2)^{\frac{3}{2}}} \qquad (7.32)$$

By requiring $\int_{-\infty}^{\infty} p(x)\,dx = 1$, we find $p(0) = \frac{1}{2}$.

The above procedure illustrates how (7.31) is very frequently used, namely, to find the stationary density when it exists. Actually, for this example, (7.31) can be solved completely. If we set

$$u(x,t;\ x_0,t_0) = \frac{\sinh^{-1} x - \sinh^{-1} x_0}{2\sqrt{t - t_0}} \qquad (7.33)$$

7. DIFFUSION EQUATIONS

then it can be shown that $p(x,t|x_0,t_0)$ is given by

$$p(x,t|x_0,t_0) = \frac{1}{2(1+x^2)^{\frac{3}{2}}} \left\{ \frac{1}{\sqrt{\pi(t-t_0)}} e^{-u^2} e^{-(t-t_0)} \right.$$
$$\left. + \frac{1}{\sqrt{\pi}} \int_{u-\sqrt{t-t_0}}^{u+\sqrt{t-t_0}} e^{-z^2} dz \right\} \quad (7.34)$$

Example 2. Suppose that $\{Y_t, t \geq 0\}$ and $\{Z_t, t \geq 0\}$ are two independent standard Brownian motion processes. Let

$$X_t = Y_t^2 + Z_t^2 \quad (7.35)$$

We shall use Proposition 6.4 and show that X_t is Markov and derive its stochastic differential equation. Let $\mathcal{B}_t$ denote the smallest σ algebra such that $Y_s, Z_s, s \leq t$ are all measurable. Let $\mathcal{A}_t$ denote the smallest σ algebra with respect to which $X_s, s \leq t$, are all measurable. Clearly, for each t, $\mathcal{B}_t \supset \mathcal{A}_t$. Now, for $h > 0$,

$$E^{\mathcal{A}_t}(X_{t+h} - X_t) = E^{\mathcal{A}_t} E^{\mathcal{B}_t}(X_{t+h} - X_t)$$
$$= E^{\mathcal{A}_t} E^{\mathcal{B}_t}\{(Y_{t+h}^2 - Y_t^2) + (Z_{t+h}^2 - Z_t^2)\}$$
$$= E^{\mathcal{A}_t}(h + h) = 2h$$

Similarly, we find that

$$E^{\mathcal{A}_t}(X_{t+h} - X_t)^2 = 8h^2 + 4hX_t$$

Clearly, the hypotheses of Proposition 6.4 are satisfied with

$$m(x,t) = 2 \quad (7.36)$$
$$\sigma^2(x,t) = 4x \quad (7.37)$$

Therefore, $\{X_t, t \geq 0\}$ satisfies the stochastic differential equation

$$dX_t = 2dt + \sqrt{X_t}\,dW_t \quad (7.38)$$

We note that σ does not satisfy a uniform Lipschitz condition. However, it turns out that (7.38) has a unique nonnegative solution for $X_0 = 0$ anyway [McKean, 1960, 1969].

According to Proposition 7.1 the conditional density for $\{X_t, t \geq 0\}$ must satisfy the Fokker-Planck equation

$$\frac{\partial^2}{\partial x^2}[xp(x,t|x_0,t_0)] - 2\frac{\partial}{\partial x}p(x,t|x_0,t_0) = \frac{\partial}{\partial t}p(x,t|x_0,t_0) \quad (7.39)$$

This equation has a unique fundamental solution given by

$$p(x,t|x_0,t_0) = \frac{1}{(t-t_0)} \exp{-\frac{1}{2}\left(\frac{x+x_0}{t-t_0}\right)} I_0\left(\frac{\sqrt{xx_0}}{t-t_0}\right) \quad (7.40)$$

which can be verified by direct computation. In this example $p(x,t|x_0,t_0)$ can also be obtained directly from the definition for X_t.

Finally, we shall state, but not derive, the diffusion equations in the multidimensional case. Let $\{\mathbf{W}_t,\ a \leq t \leq b\}$ be a vector, the components of which are independent Brownian motion processes. Let $\mathbf{X}$ be a vector random variable independent of $\{\mathbf{W}_t,\ a \leq t \leq b\}$. Under appropriate conditions on $\mathbf{m}$ and $\boldsymbol{\sigma}$, existence and uniqueness of the solution to the vector stochastic integral equation

$$\mathbf{X}_t = \mathbf{X} + \int_a^t \mathbf{m}(\mathbf{X}_s,s)\,ds + \int_a^t \boldsymbol{\sigma}(\mathbf{X}_s,s)\,d\mathbf{W}_s \tag{7.41}$$

can be established. The solution $\{\mathbf{X}_t,\ a \leq t \leq b\}$ is a vector Markov process in the sense that

$$\mathcal{P}(\mathbf{X}_t \in E|\mathbf{X}_\tau,\tau \leq s) = \mathcal{P}(\mathbf{X}_t \in E|\mathbf{X}_s) \tag{7.42}$$

whenever E is a Borel set and $s \leq t$. Furthermore, with probability 1,

$$\lim_{h \downarrow 0} \frac{1}{h} E[(\mathbf{X}_{t+h} - \mathbf{X}_t)|\mathbf{X}_t] = \mathbf{m}(\mathbf{X}_t,t) \tag{7.43}$$

$$\lim_{h \downarrow 0} \frac{1}{h} E[(\mathbf{X}_{t+h} - \mathbf{X}_t)(\mathbf{X}_{t+h} - \mathbf{X}_t)^T|\mathbf{X}_t] = \boldsymbol{\sigma}(\mathbf{X}_t,t)\boldsymbol{\sigma}^T(\mathbf{X}_t,t) = \boldsymbol{\beta}(\mathbf{X}_t,t) \tag{7.44}$$

Suppose that $\{\mathbf{X}_t,\ a \leq t \leq b\}$ is a vector Markov process satisfying (7.43) and (7.44). Then under some additional conditions similar to the scalar case, the conditional density $p(\mathbf{x},t|\mathbf{x}_0,t_0)$ can be shown to satisfy a pair of partial differential equations

$$\frac{1}{2}\sum_{i,j}\beta_{ij}(\mathbf{x}_0,t_0)\frac{\partial^2}{\partial x_{0i}\,\partial x_{0j}}p(\mathbf{x},t|\mathbf{x}_0,t_0) + \sum_i m_i(\mathbf{x}_0,t_0)\frac{\partial}{\partial x_{0i}}p(\mathbf{x},t|\mathbf{x}_0,t_0)$$
$$= -\frac{\partial}{\partial t_0}p(\mathbf{x},t|\mathbf{x}_0,t_0) \tag{7.45}$$

and

$$\frac{1}{2}\sum_{i,j}\frac{\partial^2}{\partial x_i\,\partial x_j}[\beta_{ij}(\mathbf{x},t)p(\mathbf{x},t|x_0,t_0)] - \sum_i \frac{\partial}{\partial x_i}[m_i(\mathbf{x},t)p(\mathbf{x},t|\mathbf{x}_0,t_0)]$$
$$= \frac{\partial}{\partial t}p(\mathbf{x},t|\mathbf{x}_0,t_0) \tag{7.46}$$

with initial condition $p(\mathbf{x},t_0|\mathbf{x}_0,t_0) = \delta(\mathbf{x} - \mathbf{x}_0)$. Under some additional conditions, $\{\mathbf{X}_t,\ a \leq t \leq b\}$ can also be shown to satisfy a stochastic differential equation of the form (7.41). Therefore, if $\mathbf{m}$ and $\boldsymbol{\sigma}$ are such that (7.41) has a unique solution, and (7.46) has a unique fundamental solution that is a density function, then the process $\{\mathbf{X}_t,\ a \leq t \leq b\}$ satisfying

EXERCISES

(7.41) must have a density function which satisfies (7.46). There is a major difference in the vector case, however. The best result that is known on the uniqueness of solution for (7.46) requires that **m** and **ó** be bounded, which is a strong assumption. In practice there is an abundance of examples to show that this boundedness condition is not essential. However, the situation remains not completely clear.

EXERCISES

1. Let φ satisfy (2.3) and (2.4). In addition, assume that its derivative $\dot\varphi$ is continuous on $[a,b]$. Show that

$$\int_a^b \dot\varphi(\omega,t) W(\omega,t)\, dt + \int_a^b \varphi(\omega,t)\, dW(\omega,t) = \varphi(\omega,b) W(\omega,b) - \varphi(\omega,a) W(\omega,a)$$

almost surely. (Hint: Let $\{W_n\}$ be a sequence of step approximations to the Brownian motion W and consider

$$\varphi_b W_b - \varphi_a W_a - \int_a^b \dot\varphi_t W_{nt}\, dt)$$

2. Suppose that Z_t satisfies

$$Z_t = 1 + \int_0^t Z_s \varphi_s\, dW_s$$

Show that Z_t is given by

$$Z_t = \exp\left(\int_0^t \varphi_s\, dW_s - \frac{1}{2}\int_0^t \varphi_s^2\, ds\right)$$

(Hint: Consider $\ln Z_t$, and use Ito's differentiation formula.)

3. Let X_t satisfy the stochastic differential equation

$$dX_t = X_t[m(t)\, dt + \sigma(t)\, dW_t] \qquad t > 0$$

where m and σ are nonrandom, i.e., independent of ω. Show that X_t, $t \geq 0$ has the form

$$X_t = X_0 \exp\left[\int_0^t \sigma(s)\, dW_s + \int_0^t f(s)\, ds\right]$$

and find f.

4. Let X_t satisfy a stochastic differential equation

$$dX_t = m(X_t,t)\, dt + \sigma(X_t,t)\, dW_t \qquad t > 0$$

where m and σ satisfy conditions given in Proposition 4.1. Show that

$$EX_t^2 \leq 3\left[EX_0^2 + K(1+t)\int_0^t (1 + EX_s^2)\, ds\right]$$

From this inequality, show that if $EX_0^2 < \infty$ then there exist finite constants A and α such that

$$EX_t^2 \leq A e^{\alpha t^2} \qquad 0 \leq t < \infty$$

5. Let $\varphi(x)$, $-\infty < x < \infty$, be a real-valued Borel function with bounded-continuous second derivative φ'', and let $\{W_t, t \geq 0\}$ be a standard Brownian motion. Show that

$$X_t = \varphi(W_t) - \frac{1}{2}\int_0^t \varphi''(W_s)\, ds$$

is a martingale. Thus, if $\varphi'' \geq 0$ ($\varphi'' \leq 0$) then $\varphi(W_t)$ is a submartingale (respectively, supermartingale).

6. Let φ be as in Exercise 5, and set

$$\mathbf{X}_t = \begin{bmatrix} X_{1t} \\ X_{2t} \end{bmatrix} = \begin{bmatrix} W_t \\ \varphi(W_t) \end{bmatrix}$$

Show that $\mathbf{X}_t$ is the solution of a vector stochastic differential equation

$$d\mathbf{X}_t = \mathbf{f}(\mathbf{X}_t)\, dt + \mathbf{g}(\mathbf{X}_t)\, dW_t$$

where

$$\mathbf{f}(\mathbf{x}) = \begin{bmatrix} 0 \\ \frac{1}{2}\varphi''(x_1) \end{bmatrix} \quad \text{and} \quad \mathbf{g}(\mathbf{x}) = \begin{bmatrix} 1 \\ \varphi'(x_1) \end{bmatrix}$$

Verify, both directly and by using (5.34), that $\mathbf{X}_t$ can also be considered to be the solution of a white-noise equation

$$\dot{\mathbf{X}}_t = \begin{bmatrix} 1 \\ \varphi'(X_{1t}) \end{bmatrix} \zeta_t$$

7. Let W_{kt}, $k = 1, \ldots, n$ be independent standard Brownian motion processes. Define

$$X_t = \sqrt{\sum_{k=1}^{n} W_{kt}^2}$$

Show that X_t satisfies a stochastic differential equation, and find this equation. Is X_t Markov?

8. Consider a white-noise equation

$$\dot{X}_t = \mu_t(b + \sigma\zeta_t) \qquad t > 0$$

where X_t is interpreted as the state at t, μ_t is the control at t, ζ_t is a standard Gaussian white noise, and b, σ are constants.

(a) For open-loop control, μ_t can depend only on the initial state X_0 and t. Find the open-loop control $\mu(X_0, t)$ which minimizes EX_T^2.

(b) For linear closed-loop control, we permit μ_t to take on the form $\mu_t = \alpha(t)X_t$. Find the function $\alpha(t)$ which minimizes EX_T^2.

9. Consider the forward equation (7.23) for the case where m and σ^2 are functions only of x and not t, that is,

$$\frac{\partial}{\partial t} p(x,t|x_0,t_0) = \frac{1}{2}\frac{\partial^2}{\partial x^2}[\sigma^2(x)p(x,t|x_0,t_0)] - \frac{\partial}{\partial x}[m(x)p(x,t|x_0,t_0)]$$

Let w be any positive solution of the equation

$$\frac{1}{2}\frac{d}{dx}[\sigma^2(x)w(x)] = m(x)w(x)$$

Show that $p(x,t|x_0,t_0)$ can be written in the form

$$p(x,t|x_0,t_0) = w(x)\int_\Lambda e^{-\lambda(t-t_0)}\varphi_\lambda(x)\psi_\lambda(x_0)\,\mu(d\lambda)$$

where Λ is a Borel subset of the real line, μ a Borel measure, and φ_λ, ψ_λ are both solutions of the Sturm-Liouville equation

$$\frac{1}{2}\frac{d}{dx}\left[\sigma^2(x)w(x)\frac{df(x)}{dx}\right] + \lambda w(x)f(x) = 0$$

10. For $\sigma^2(x) = 1$, $m(x) = 0$, we have the Brownian motion case and

$$p(x,t|x_0,t_0) = \frac{1}{\sqrt{2\pi(t-t_0)}}\exp\left[-\frac{1}{2}\frac{(x-x_0)^2}{(t-t_0)}\right]$$

Show that $p(x,t|x_0,t_0)$ has the form prescribed by Exercise 9. In particular,

$$p(x,t|x_0,t_0) = \frac{1}{2\pi}\int_{-\infty}^{\infty} e^{-\frac{1}{2}\nu^2(t-t_0)}e^{i\nu(x-x_0)}\,d\nu$$

11. Suppose that $\sigma^2(x,t) = \beta(t)$ and $m(x,t) = \alpha(t)x$. Show that the fundamental solution of the Fokker-Planck equation has the form

$$p(x,t|x_0,t_0) = \frac{1}{\sqrt{2\pi a^2(t,t_0)}}\exp\left\{-\frac{1}{2a^2(t,t_0)}[x - b(t,t_0)x_0]^2\right\}$$

How are $a^2(t,t_0)$ and $b(t,t_0)$ obtained?

12. Suppose that $\sigma^2(x,t) = 1$ and $m(x,t) = -\,\text{sgn}\,x$ where $\text{sgn}\,x$ is $+1$ or -1 according as $x \geq 0$ or $x < 0$.

(a) Find the limiting density $p(x) = \lim_{t\to\infty} p(x,t|x_0,t_0)$.

(b) Find $p(x,t|x_0,t_0)$.

5
One-Dimensional Diffusions

1. INTRODUCTION

This chapter is an introduction to the semigroup treatment of Markov processes with stationary transition functions. Modern theory of Markov processes is primarily a semigroup theory. Even though much of this theory has not found its way into applications in physical problems, the elucidation that is made possible with the semigroup approach makes it indispensable in any treatment of Markov processes.

Consider a Markov process $\{X(\omega,t),\ t \in [0,\infty)\}$ defined on a probability space $(\Omega,\mathcal{C},\mathcal{P})$. We assume that the transition function

$$\mathcal{P}(X_t < b | X_s = a) = P(b,t|a,s) \qquad t > s$$

depends only on $t - s$ and not on t and s separately. We call such transition functions **stationary transition functions.** A process with a stationary transition function need not be a stationary process. For example, Brownian motion has a stationary transition function, but is not a stationary process.

It is rather important that the set of values that $X(\omega,t)$ can assume be

1. INTRODUCTION

explicitly stated. We assume that this set is an *interval S* which we shall call the **state space**. The interval S can be finite or infinite, closed or open at either end. Let $\mathcal{R}$ denote the σ algebra of Borel sets in S. That is, $\mathcal{R}$ is the smallest σ algebra containing all subintervals of S. For $E \in \mathcal{R}$, we denote

$$P_a(E,t) = \mathcal{P}(X_{t+s} \in E | X_s = a) \tag{1.1}$$

and call $\{P_a(E,t), a \in S, E \in \mathcal{R}, t \in [0, \infty)\}$ the transition function. Since $\{X_t, 0 \leq t \leq \infty\}$ is Markov, $P_a(E,t)$ satisfies the Chapman-Kolmogorov equation, viz.,

$$P_a(A, t+s) = \int_S P_b(E,t) P_a(db, s) \tag{1.2}$$

If a transition density $p_a(b,t)$ exists for $t > 0$, that is, if $P_a(E,t)$ can be written as

$$P_a(E,t) = \int_E p_a(b,t)\, db \quad \begin{array}{c} E \in \mathcal{R} \\ t > 0 \end{array} \tag{1.3}$$

then $p_a(b,t)$ satisfies

$$p_a(b, t+s) = \int_S p_a(x,s) p_x(b,t)\, dx \tag{1.4}$$

Equations (1.2) and (1.4) reveal the basic theme of this chapter. If $P_a(\cdot,t)$ is known for $0 \leq t \leq \epsilon$, then $P_a(\cdot,t)$ for any t can be determined by iterating (1.2), no matter how small ϵ is. Therefore, the knowledge of $P_a(\cdot,t)$ for $t \approx 0$ suffices to determine $P_a(\cdot,t)$ for all t. It turns out that under suitable continuity conditions and with proper interpretation, $P_a(\cdot,0)$ and $(\partial/\partial t)P_a(\cdot,t)\big|_{t=0}$ completely determine $P_a(\cdot,t)$ for all t. Furthermore, since $P_a(\cdot,0)$ is always the same, viz.,

$$P_a(A,0) = \begin{cases} 1 & \text{if } a \in A \\ 0 & \text{if } a \notin A \end{cases}$$

the behavior of Markov processes with stationary transition functions depends entirely on $(\partial/\partial t)P_a(\cdot,t)\big|_{t=0}$. The goal is to deduce everything about the behavior of a Markov process from $(\partial/\partial t)P_a(\cdot,t)\big|_{t=0}$. We note that this is very much the same motivation underlying the derivation of the diffusion equations (backward and forward equations of Kolmogorov). The only difference is that in the stationary case, the setting is much more general, and the answers are more clear cut. Of course, when we say "everything about the behavior of the process" we mean everything up to an arbitrary distribution of X_0, since we only make use of the transition functions. In this sense, we do not distinguish between processes differing only in the distribution of X_0. For example, we shall define in this chapter a Brownian motion as any process $\{X_t, t \geq 0\}$ such that $\{X_t - X_0, t \geq 0\}$ is a Brownian motion in the earlier sense. That is, here a Brownian motion is any Markov process with transition density

$$p_a(b,t) = \frac{1}{\sqrt{2\pi t}} e^{-(1/2t)(b-a)^2} \tag{1.5}$$

Incidentally, this broadening of the definition of Brownian motion also avoids the awkwardness in interpreting $P_a(E,t)$ as

$$P_a(E,t) = \mathcal{P}(X_t \in E | X_0 = a)$$

when our earlier definition specified $X_0 = 0$, a.s.

2. THE MARKOV SEMIGROUP

Let $\{X_t, 0 \leq t < \infty\}$ be a Markov process with a stationary transition function P_a and continuous in probability. We shall adopt the notations

$$E_a Z = E(Z | X_0 = a) \tag{2.1}$$

and

$$\mathcal{P}_a(B) = \mathcal{P}(B | X_0 = a) \tag{2.2}$$

Because of the continuity in probability, for *every open set* E containing a,

$$\lim_{t \downarrow 0} \mathcal{P}_a(X_t \in E) = \lim_{t \downarrow 0} P_a(E,t) = 1 \tag{2.3}$$

Let **B** denote the space of all complex-valued bounded functions defined on the state space S and measurable with respect to the Borel σ algebra $\mathcal{R}$. With the usual norm

$$\|f\| = \sup_{x \in S} |f(x)| \tag{2.4}$$

the space **B** is a Banach space. For each $t \in [0, \infty)$, we define an operator H_t by

$$(H_t f)(a) = E_a f(X_t) \tag{2.5}$$

Since

$$|E_a f(X_t)| = \left| \int_S f(x) P_a(dx, t) \right| \leq \sup_{x \in S} |f(x)| \int_S P_a(dx, t) = \|f\|$$

H_t maps **B** into **B** and is a contraction, that is,

$$\|H_t f\| \leq \|f\| \tag{2.6}$$

2. THE MARKOV SEMIGROUP

Because of (2.6), for each t, H_t is continuous, that is,

$$\|f_n - f\| \xrightarrow[n \to \infty]{} 0 \Rightarrow \|H_t f_n - H_t f\| \xrightarrow[n \to \infty]{} 0 \tag{2.7}$$

A sequence $\{f_n\}$ in **B** is said to **converge strongly** to f if $\|f_n - f\| \xrightarrow[n \to \infty]{} 0$, and we shall use the notation

$$s \lim_{n \to \infty} f_n = f \tag{2.8}$$

The one-parameter family of operators $\{H_t, 0 \leq t < \infty\}$ that we have defined has some additional important properties. First and foremost, is the **semigroup property**. From the Chapman-Kolmogorov equation (1.2), we can write

$$(H_{t+s}f)(a) = E_a f(X_{t+s}) = \int_S f(x) P_a(dx, t+s)$$
$$= \int_S f(x) \int_S P_b(dx,t) P_a(db,s)$$
$$= \int_S (H_t f)(b) P_a(db,s) = (H_s H_t f)(a)$$

Therefore,

$$H_{t+s} = H_s H_t = H_t H_s \tag{2.9}$$

In addition, because of (2.3) we also have

$$(H_t f)(x) \xrightarrow[t \downarrow 0]{} f(x) \tag{2.10}$$

at every continuity point of f. Roughly speaking, (2.9) and (2.10) imply that H_t must be of the form

$$H_t = e^{tA}$$

where A is necessarily given by $A = (d/dt) H_t \big|_{t=0}$. Thus, the first step is to define $(d/dt) H_t \big|_{t=0}$.

Let $\mathfrak{D}_A$ denote the set of all functions f in **B** such that the limit

$$s \lim_{t \downarrow 0} \frac{1}{t} (H_t f - f)$$

exists. This limit defines a linear operator A mapping $\mathfrak{D}_A$ into **B**, that is,

$$Af = s \lim_{t \downarrow 0} \frac{1}{t} (H_t f - f) \tag{2.11}$$

The operator A is called the **generator** of the semigroup and of the Markov process. If $\mathfrak{D}_A = \mathbf{B}$ and if A is a bounded operator (i.e., there exists finite K such that $\|Af\| \leq K\|f\|$ for all $f \in \mathbf{B}$), then we truly would have

$$H_t = e^{tA} \equiv \sum_{n=0}^{\infty} \frac{t^n}{n!} A^n$$

However, this happens only in a few relatively uninteresting cases, for example, when S contains only a finite number of points. In the general case $\mathfrak{D}_A \neq \mathbf{B}$, A is unbounded and e^{tA} is not well defined. The main goal of this section is to show that if a Markov process is continuous in probability and has a stationary transition function, then its transition function is uniquely determined by its generator.

We note that if f is such that $H_t f \in \mathfrak{D}_A$, then

$$\frac{d}{dt} H_t f = A H_t f \tag{2.12}$$

and this is a version of the backward equation. To see that this is the case, let I_E denote the indicator function of E and suppose that $H_t I_E \in \mathfrak{D}_A$ for $t > 0$. Then

$$(H_t I_E)(a) = P_a(E, t)$$

and

$$\frac{\partial}{\partial t} P_a(E, t) = (AP.(E, t))(a) \tag{2.13}$$

As we shall see a little later, for a Brownian motion $H_t I_E \in \mathfrak{D}_A$ for $t > 0$ and

$$(Af)(a) = \frac{1}{2} \frac{d^2 f(a)}{da^2}$$

Therefore, for a Brownian motion,

$$\frac{\partial}{\partial t} P_a(E, t) = \frac{1}{2} \frac{\partial^2}{\partial a^2} P_a(E, t) \qquad t > 0$$

which is just the backward equation for that process.

The procedure of constructing the semigroup $\{H_t, 0 \leq t < \infty\}$, or equivalently, determining the transition function, from its generator A involves in an essential way the resolvent. First, define $\mathbf{B}_0$ as the set of all functions f in $\mathbf{B}$ such that

$$\|H_t f - f\| \xrightarrow[t \downarrow 0]{} 0 \tag{2.14}$$

It is clear that $\mathbf{B}_0 \supset \mathfrak{D}_A$. It turns out that $\mathfrak{D}_A$ is dense in $\mathbf{B}_0$. That is, every $f \in \mathbf{B}_0$ is the strong limit of a sequence from $\mathfrak{D}_A$. To show this, take any $f \in \mathbf{B}_0$ and set

$$f_n = n \int_0^{1/n} H_s f \, ds \tag{2.15}$$

Then, for $n = 1, 2, \ldots,$

$$\underset{t \downarrow 0}{\text{s lim}} \, (H_t f_n - f_n) = n(H_{1/n} f - f)$$

2. THE MARKOV SEMIGROUP

so that $f_n \in \mathcal{D}_A$ for each n. Furthermore,

$$\|f_n - f\| = \left\| n \int_0^{1/n} (H_t f - f)\, dt \right\| \leq \sup_{0 \leq t \leq 1/n} \|H_t f - f\| \xrightarrow[n \to \infty]{} 0$$

Therefore, $\mathcal{D}_A$ is dense in $\mathbf{B}_0$. For $f \in \mathbf{B}_0$ and $\lambda > 0$. We define

$$(R_\lambda f)(a) = \int_0^\infty e^{-\lambda t}(H_t f)(a)\, dt \qquad (2.16)$$

The family of operators $\{R_\lambda,\ 0 < \lambda < \infty\}$ is called the **resolvent** of the semigroup $\{H_t,\ 0 \leq t < \infty\}$. The importance of R_λ is that it is simply related to the generator A.

Proposition 2.1. For every $g \in \mathbf{B}_0$ and every $\lambda > 0$, $R_\lambda g \in \mathcal{D}_A$. Furthermore, $f = R_\lambda g$ is the unique solution to the equation

$$\lambda f - Af = g \qquad f \in \mathcal{D}_A \qquad (2.17)$$

Proof: We shall sketch a proof with some details omitted. First, we verify that $R_\lambda g \in \mathcal{D}_A$ by computing

$$\frac{1}{t}(H_t R_\lambda g - R_\lambda g) = \frac{1}{t} \int_0^\infty e^{-\lambda s}(H_{t+s} g - H_s g)\, ds$$

$$= \frac{1}{t}\left(e^{\lambda t} \int_t^\infty e^{-\lambda s} H_s g\, ds - \int_0^\infty e^{-\lambda s} H_s g\, ds\right)$$

$$= -e^{\lambda t} \frac{1}{t} \int_0^t e^{-\lambda s} H_s g\, ds + \frac{1}{t}(e^{\lambda t} - 1) R_\lambda g \xrightarrow[t \downarrow 0]{s} -g$$

$$+ \lambda R_\lambda g$$

where $\xrightarrow{s}$ denotes strong convergence. Therefore,

$$AR_\lambda g = -g + \lambda R_\lambda g$$

and

$$\lambda R_\lambda g - AR_\lambda g = g$$

Therefore, $f = R_\lambda g$ is a solution to (2.17). Next, we prove that it is the only solution.

Suppose that f_1 and f_2 are two solutions to (2.17). Set

$$\varphi = f_1 - f_2$$

Then $\varphi \in \mathcal{D}_A$ and

$$\lambda \varphi - A\varphi = 0$$

Therefore,

$$\frac{d}{dt}(e^{-\lambda t} H_t \varphi) = -H_t(\lambda \varphi - A\varphi) = 0$$

and
$$e^{-\lambda t} H_t \varphi = H_0 \varphi = \varphi$$
It follows that
$$0 \le \|\varphi\| = e^{-\lambda t}\|H_t \varphi\| \le e^{-\lambda t}\|\varphi\| \xrightarrow[t \to \infty]{} 0$$
so that $\|\varphi\| = 0$, which proves uniqueness. ∎

Proposition 2.1 shows that the mapping $R_\lambda: \mathbf{B}_0 \to \mathfrak{D}_A$ is one to one and onto. If we denote the identity operator by I, then we have
$$R_\lambda = (\lambda I - A)^{-1} \tag{2.18}$$
and
$$A = \lambda I - R_\lambda^{-1} \tag{2.19}$$
By using R_λ, we can now construct the semigroup $\{H_t,\ 0 \le t < \infty\}$ from A.

Proposition 2.2. Let $\{X_t,\ 0 \le t \le \infty\}$ be a Markov process with a stationary transition function $P_a(E,t)$, and let $\{X_t,\ 0 \le t < \infty\}$ be continuous in probability. Then its transition function, equivalently, its semigroup $\{H_t,\ 0 \le t < \infty\}$, is uniquely determined by its generator A.

Proof: First, we note that
$$\lambda R_\lambda f = \lambda \int_0^\infty e^{-\lambda t} H_t f\, dt = \int_0^\infty e^{-t} H_{t/\lambda} f\, dt$$
so that $\|\lambda R_\lambda f\| \le \|f\|$ and
$$\|\lambda R_\lambda f - f\| = \left\| \int_0^\infty e^{-t}(H_{t/\lambda} f - f)\, dt \right\|$$
which goes to zero as $\lambda \to \infty$ by dominated convergence. Therefore,
$$\lambda R_\lambda \xrightarrow[\lambda \to \infty]{} I \tag{2.20}$$
Next, define A_λ by
$$A_\lambda = \lambda A R_\lambda \tag{2.21}$$
From (2.19) we have
$$\|A_\lambda f\| = \lambda \|A R_\lambda f\| = \lambda \|\lambda R_\lambda f - f\| \le \lambda \|\lambda R_\lambda f\| + \lambda \|f\| \le 2\lambda \|f\| \tag{2.22}$$
so that A_λ is a bounded operator. We can define
$$e^{tA_\lambda} = \sum_{n=0}^\infty \frac{t^n}{n!} A_\lambda^n$$

2. THE MARKOV SEMIGROUP

Given A, we now determine $\{H_t, 0 \leq t < \infty\}$ as follows: For $f \in \mathfrak{D}_A$ we can show that

$$H_t f = s \lim_{\lambda \to \infty} e^{tA_\lambda} f \tag{2.23}$$

For $f \in \mathbf{B}_0$, let $\{f_n\}$ be a sequence from $\mathfrak{D}_A$ converging strongly to f. Then

$$H_t f = s \lim_{n \to \infty} H_t f_n$$

For any bounded-continuous f, we set

$$f_n = n \int_0^{1/n} H_s f \, ds \tag{2.24}$$

Then, $f_n \in \mathbf{B}_0$ for each n, and

$$(H_t f_n)(x) \xrightarrow[n \to \infty]{} (H_t f)(x) \tag{2.25}$$

for each $x \in S$ and each $t \geq 0$. Finally, we note that

$$f(u,x) = e^{iux}$$

is a bounded-continuous function in x for each real u. Therefore, the characteristic function

$$E_a e^{iuX_t} = F_a(u,t)$$

is uniquely determined by the generator A. This in turn implies that the transition function $P_a(E,t)$ is uniquely determined by A. What need to be shown are (2.23) and (2.25).

To prove (2.23), we note that

$$\frac{d}{dt} H_t f = A H_t f$$

and

$$\frac{d}{dt} e^{tA_\lambda} f = A_\lambda e^{tA_\lambda} f$$

Therefore,

$$\begin{aligned} \|H_t f - e^{tA_\lambda} f\| &= \left\| \int_0^t (A H_s f - A_\lambda e^{sA_\lambda} f) \, ds \right\| \\ &\leq \left\| \int_0^t (A - A_\lambda) H_s f \, ds \right\| + \left\| A_\lambda \int_0^t (H_s - e^{sA_\lambda}) f \, ds \right\| \\ &\leq \int_0^t \|(A - A_\lambda) H_s f\| \, ds + \int_0^t \|A_\lambda (H_s - e^{sA_\lambda}) f\| \, ds \end{aligned}$$

If we set $\varphi_t = \int_0^t \|H_s f - e^{sA_\lambda} f\| \, ds$ and make use of (2.22), we find

$$\dot{\varphi}_t \leq 2\lambda \varphi_t + \int_0^t \|(A - A_\lambda) H_s f\| \, ds \tag{2.26}$$

or

$$\frac{d}{dt}(e^{-2\lambda t}\varphi_t) \leq e^{-2\lambda t}\int_0^t \|(A - A_\lambda)H_s f\|\, ds$$

By direct integration, we get

$$e^{-2\lambda t}\varphi_t \leq \int_0^t e^{-2\lambda \tau}\int_0^\tau \|(A - A_\lambda)H_s f\|\, ds\, d\tau = -\frac{1}{2\lambda}e^{-2\lambda t}$$

$$\int_0^t \|(A - A_\lambda)H_s f\|\, ds + \frac{1}{2\lambda}\int_0^t e^{-2\lambda s}\|(A - A_\lambda)H_s f\|\, ds \quad (2.27)$$

Combining (2.26) and (2.27), we find

$$0 \leq \|H_t f - e^{tA_\lambda}f\| = \dot{\varphi}_t \leq \int_0^t e^{-2\lambda(t-s)}\|(A - A_\lambda)H_s f\|\, ds \xrightarrow[\lambda \to \infty]{} 0$$

which proves (2.23).

To prove (2.25), we write

$$(H_t f_n - H_t f)(a) = E_a f_n(X_t) - E_a f(X_t)$$
$$= \int_S P_a(dx,t)[f_n(x) - f(x)]$$
$$= \int_S P_a(dx,t)\, n \int_0^{1/n}[E_x f(X_s) - f(x)]\, ds$$

Because $\{X_t,\ 0 \leq t < \infty\}$ is continuous in probability and f is bounded continuous,

$$|(H_t f_n)(a) - (H_t f)(a)| \xrightarrow[n \to \infty]{} 0$$

by dominated convergence. ∎

Example. As an example, we shall derive the generator for a standard Brownian motion. We recall that a Brownian motion has a transition-density function given by

$$p_a(x,t) = \frac{1}{\sqrt{2\pi t}}e^{-(1/2t)(x-a)^2} \quad (2.28)$$

Let f be any function in **B** with a bounded-continuous second derivative f'', and let $\mathcal{C}^2$ denote the set of all such functions. Then,

$$\frac{1}{t}[E_a f(X_t) - f(a)] = \frac{1}{t}\int_{-\infty}^\infty \frac{1}{\sqrt{2\pi t}}e^{-(1/2t)(x-a)^2}[f(x) - f(a)]\, dx$$

By using the Taylor's theorem at a, and by making a change in the variable of integration, we get

$$\frac{1}{t}[E_a f(X_t) - f(a)] = \tfrac{1}{2}f''(a) + \frac{1}{2}\int_{-\infty}^\infty \frac{1}{\sqrt{2\pi}}[f''(\theta) - f''(a)]e^{-\frac{1}{2}z^2}\, dz$$

2. THE MARKOV SEMIGROUP

where θ lies between a and $a + \sqrt{t}\, z$. Because f'' is bounded continuous, we get

$$(Af)(a) = \tfrac{1}{2} f''(a) \tag{2.29}$$

for all $f \in \mathcal{C}^2$, and we have also shown that $\mathfrak{D}_A \supset \mathcal{C}^2$. What is more difficult is to determine $\mathbf{B}_0$ and $\mathfrak{D}_A$.

If we take the Laplace transform of (2.28), we get

$$\int_0^\infty e^{-\lambda t} p_a(x,t)\, dt = \frac{\exp(-\sqrt{2\lambda}\,|x-a|)}{\sqrt{2\lambda}} \qquad 0 < \lambda < \infty \tag{2.30}$$

Hence,

$$(R_\lambda f)(a) = \int_{-\infty}^\infty \frac{\exp(-\sqrt{2\lambda}\,|x-a|)}{\sqrt{2\lambda}} f(x)\, dx \tag{2.31}$$

For any bounded f, the right-hand side of (2.31) is a bounded-continuous function. If we denote the set of all bounded-continuous functions by $\mathcal{C}$, then we have

$$\mathfrak{D}_A = R_\lambda \mathbf{B}_0 \subset \mathcal{C}$$

Since $\mathbf{B}_0$ is the closure of $\mathfrak{D}_A$ under uniform convergence, we must also have

$$\mathbf{B}_0 \subset \mathcal{C}$$

On the other hand, for any bounded-continuous f

$$|E_a f(X_t) - f(a)| \xrightarrow[t\downarrow 0]{} 0$$

so that $\mathbf{B}_0 \supset \mathcal{C}$. Hence, $\mathbf{B}_0 = \mathcal{C}$. Now the only thing left to do is to find $\mathfrak{D}_A$.

For $f \in \mathcal{C} = \mathbf{B}_0$, let $g = R_\lambda f$. Then

$$g(a) = \frac{1}{\sqrt{2\lambda}} \left[\int_{-\infty}^\infty \exp(-\sqrt{2\lambda}\,|x-a|) f(x)\, dx \right]$$

$$g'(a) = -\int_a^\infty \exp[-\sqrt{2\lambda}\,(x-a)] f(x)\, dx$$
$$\qquad\qquad - \int_{-\infty}^a \exp[+\sqrt{2\lambda}\,(x-a)] f(x)\, dx$$

$$g''(a) = -2f(a) + 2\lambda g(a)$$

Therefore, $g'' \in \mathcal{C}$, that is, $g \in \mathcal{C}^2$. We have shown that $\mathfrak{D}_A \subset \mathcal{C}^2$. But earlier we showed that $\mathfrak{D}_A \supset \mathcal{C}^2$. Hence, $\mathfrak{D}_A = \mathcal{C}^2$. To summarize, for a Brownian motion we have

$$\mathbf{B}_0 = \mathcal{C} \qquad \mathfrak{D}_A = \mathcal{C}^2 \tag{2.32}$$

and

$$Af = \tfrac{1}{2}f'' \tag{2.33}$$

Further, for every $t > 0$, $p_a(x,t)$ is $\mathcal{C}^2$ in a so that

$$\frac{\partial}{\partial t} p_a(x,t) = \frac{1}{2} \frac{\partial^2}{\partial a^2} p_a(x,t)$$

which is the familiar backward equation for Brownian motion.

3. STRONG MARKOV PROCESSES

Let $\{X_t,\ 0 \leq t < \infty\}$ be a separable Markov process, continuous in probability, and with a stationary transition function $P_a(E,t)$. Let $\mathcal{Q}_t$ denote the smallest σ algebra with respect to which $\{X_\tau,\ \tau \leq t\}$ are all measurable. Then the Markov property can be stated as follows:

$$\mathcal{P}(X_{t+s} \in E | \mathcal{Q}_t) = P_{X_t}(E,s) \quad \text{a.s.} \tag{3.1}$$

for all $s \geq 0$. Often, we state (3.1) verbally as "future and present given the past and present depends only on the present." In (3.1) the present is a time t which is fixed. Roughly speaking, a strong Markov process satisfies (3.1) even when the present is a suitably restricted random time, i.e., it varies from sample function to sample function.

We assume from now on that all processes are sample right continuous with probability 1. Indeed, we shall shortly specialize to sample-continuous processes. With $\{\mathcal{Q}_t,\ 0 \leq t < \infty\}$ defined as before, a non-negative random variable τ is said to be a **Markov time** if for every $t > 0$,

$$\{\omega : \tau(\omega) < t\} \in \mathcal{Q}_t \tag{3.2}$$

This means that if we observe a sample function $X_s(\omega_0)$ on the interval $0 \leq s \leq t$, we can always determine where $\tau(\omega_0) < t$ or $\tau(\omega_0) \geq t$. However, we cannot always determine whether $\tau(\omega_0) = t$ or not. It is clear that deterministic times are always Markov times. Another important class of Markov times are the **first passage times** (for level a) defined by

$$\tau_a(\omega) = \min \{t : X_t(\omega) = a\} \tag{3.3}$$

To show that τ_a is a Markov time, we write

$$\{\omega : \tau_a(\omega) < t\} = \bigcup_{n=1}^{\infty} \left\{\omega : \tau_a(\omega) \leq t - \frac{1}{n}\right\}$$

$$= \bigcup_{n=1}^{\infty} \left\{\omega : X_s(\omega) = a \text{ for some } s \text{ in } \left[0, t - \frac{1}{n}\right]\right\}$$

3. STRONG MARKOV PROCESSES

Since for each n, the set $\{\omega : X_s(\omega) = a \text{ for some } s \text{ in } [0, t - 1/n]\}$ is in $\mathfrak{A}_t$,

$$\{\omega : \tau_a(\omega) < t\} \in \mathfrak{A}_t$$

Now, let S be the state space of the Markov process, and let a be in the interior of S. We define

$$\tau_{a+} = \lim_{b \downarrow a} \tau_b \tag{3.4}$$

$$\tau_{a-} = \lim_{b \uparrow a} \tau_b \tag{3.5}$$

and these are also Markov times. We should again note that if τ is a Markov time, the set $\{\omega : \tau(\omega) = t\}$ need not be in $\mathfrak{A}_t$. For example, neither $\{\tau_{a+} = t\}$ nor $\{\tau_{a-} = t\}$ is necessarily in $\mathfrak{A}_t$.

Let τ be a Markov time. We define the σ algebra $\mathfrak{A}_{\tau+}$ as follows:

$$E \in \mathfrak{A}_{\tau+} \quad \text{if and only if } E \cap \{\omega : \tau(\omega) < t\} \in \mathfrak{A}_t$$

It is obvious that τ is $\mathfrak{A}_{\tau+}$ measurable. If $\tau = t_0$ is a deterministic time, then

$$\mathfrak{A}_{t_0+} = \bigcap_{n=1}^{\infty} \mathfrak{A}_{t_0 + \frac{1}{n}}$$

Thus, we see that if τ represents the present, then $\mathfrak{A}_{\tau+}$ is a little bit more than the past and present.

Definition. $\{X_t, 0 \leq t < \infty\}$ is said to be a **strong Markov** process if for every Markov time τ,

$$\mathcal{P}(X_{\tau+s} \in E | \mathfrak{A}_{\tau+}) = P_{X_\tau}(E, s) \tag{3.6}$$

Every strong Markov process is Markov in the ordinary sense. This is because if (3.6) is satisfied, then

$$\mathcal{P}(X_{t+s} \in E | \mathfrak{A}_t) = E^{\mathfrak{A}_t} \mathcal{P}(X_{t+s} \in E | \mathfrak{A}_{t+})$$
$$= E^{\mathfrak{A}_t} P_{X_t}(E, s) = P_{X_t}(E, s) \quad \text{a.s.}$$

which is just the ordinary Markov property.

For an example of a Markov process which is not a strong Markov process, consider the following:

$$X_t(\omega) = \max(0, t - T(\omega)) \qquad 0 \leq t < \infty, \tag{3.7}$$

where $T(\omega)$ is a nonnegative random variable with

$$\mathcal{P}(T < t) = 1 - e^{-t} \tag{3.8}$$

This process is obviously Markov in the ordinary sense because:

1. Given $X_t = a > 0$, $X_{t+s} = a + s$ with probability 1

2. Given $X_t = 0$, X_s must be zero for $s \leq t$ so that it provides no further information

Now, the random variable T is a Markov time for this process because
$$\{\omega: T(\omega) \geq t\} = \{\omega: X_t(\omega) = 0\} \in \mathcal{A}_t \tag{3.9}$$
Given T, $X_{T+s} = s$ with probability 1. Therefore,
$$\mathcal{P}(X_{T+s} < x | \mathcal{A}_{T+}) = \begin{cases} 1 & x > s \\ 0 & x \leq s \end{cases} \tag{3.10}$$
On the other hand,
$$P_{X_T}(x,s) = P_0(x,s) = \mathcal{P}(X_{t+s} < x | X_t = 0)$$
$$= e^{-t} + \int_0^{\min(t,s)} e^{-(t-y)} \, dy \tag{3.11}$$
Obviously, (3.10) and (3.11) are not the same, so $\{X_t, 0 \leq t < \infty\}$ cannot be a strong Markov process.

There are two extremely useful criteria for determining whether a Markov process is also a strong Markov process:

1. If (3.6) is satisfied for the following classes of Markov times, then the process is strongly Markov:

$$\begin{aligned} \tau &= \tau_a & a &\in S \\ \tau &= \tau_{a+} & a &\in \text{int}(S) \\ \tau &= \tau_{a-} & a &\in \text{int}(S) \end{aligned}$$

2. If for every $t \geq 0$, the operator H_t maps bounded-continuous functions into bounded-continuous functions, then the process is a strong Markov process.

Processes satisfying (2) are called **Feller processes**. For example, a Brownian motion is a Feller process, hence, it is a strong Markov process.

For strong Markov processes with continuous-sample functions, the generator A is a local operator, that is, Af at a depends only on f in a neighborhood of a. This fact can be deduced from the following proposition.

Proposition 3.1. Let $\{X_t, 0 \leq t < \infty\}$ be a strong Markov process. Let $\tau(\omega)$ be a Markov time.

(a) Let $f \in \mathbf{B}_0$ and define
$$u_\lambda(a) = (R_\lambda f)(a) = E_a \int_0^\infty e^{-\lambda t} f(X_t) \, dt \tag{3.12}$$
Then
$$u_\lambda(a) = E_a \int_0^\tau e^{-\lambda t} f(X_t) \, dt + E_a[e^{-\lambda \tau} u_\lambda(X_\tau)] \tag{3.13}$$

4. CHARACTERISTIC OPERATORS

(b) Let $g \in \mathfrak{D}_A$, and let $E\tau < \infty$, then

$$E_a \int_0^\tau (Ag)(X_t)\, dt = E_a g(X_\tau) - g(a) \tag{3.14}$$

Remark: Equation (3.14) is generally known as **Dynkin's formula,** even though both are due to Dynkin [1965, pp. 132–133].

Proof: To prove (3.13), we only need to show

$$E_a \int_\tau^\infty e^{-\lambda t} f(X_t)\, dt = E_a e^{-\lambda \tau} u_\lambda(X_\tau)$$

which can be done as follows:

$$\begin{aligned}
E_a \int_\tau^\infty e^{-\lambda t} f(X_t)\, dt &= E_a \int_0^\infty e^{-\lambda(t-\tau)} f(X_{t+\tau})\, dt \\
&= E_a \int_0^\infty e^{-\lambda(t+\tau)} E[f(X_{t+\tau})|\mathfrak{A}_{\tau+}]\, dt \\
&= E_a \int_0^\infty e^{-\lambda(t+\tau)} E_{X_\tau}[f(X_t)]\, dt \\
&= E_a e^{-\lambda \tau} u_\lambda(X_\tau)
\end{aligned}$$

To prove (3.14), set $f_\lambda = (\lambda - A)g$, so that

$$g = (\lambda - A)^{-1} f_\lambda = R_\lambda f_\lambda$$

From (3.13), we have

$$g(a) = E_a \int_0^\tau e^{-\lambda t}[(\lambda - A)g](X_t)\, dt + E_a e^{-\lambda \tau} g(X_\tau)$$

which becomes (3.14) as $\lambda \to 0$, provided that $E_a \tau < \infty$. ∎

Equation (3.14) reveals the local character of the generator A when the process is sample continuous. To see this, let $a \in \text{int}(S)$, and let

$$\tau = \min(\tau_{a+\epsilon}, \tau_{a-\epsilon}) \tag{3.15}$$

Then, starting from a at $t = 0$, $X_t \in (a - \epsilon, a + \epsilon)$, $0 \le t < \tau$. From (3.14) we have

$$(Ag)(a) = \lim_{\epsilon \downarrow 0} \frac{E_a g(X_\tau) - g(a)}{E_a \tau} \tag{3.16}$$

where τ is given by (3.16). It is easy to see that the right-hand side of (3.16) depends only on g in a neighborhood of a. It turns out that under some additional assumptions we can show that A is always a differential operator. This will be taken up in the next section.

4. CHARACTERISTIC OPERATORS

For the remainder of this chapter we shall restrict ourselves to processes $\{X_t,\ 0 \le t < \infty\}$ satisfying the following conditions:

Every point in the state space S is reachable from every point in int (S) with a nonzero probability, that is,

$$\mathcal{P}_a(\tau_b < \infty) > 0 \quad \text{for every } a \in \text{int } (S) \text{ and every } b \in S \quad (4.1a)$$

$\{X_t,\ 0 \leq t < \infty\}$ is a strong Markov process with a stationary transition function. Starting from every point in S, every sample function is continuous with probability 1 \hfill (4.1b)

We observe that if (4.1a) is not satisfied, then S can be decomposed into sets, such that starting from a point in a set, X_t remains in that set for all t. In that case, the process can be decomposed into separate processes. Together, (4.1a) and (4.1b) imply that the process is a Feller process. We recall that this means the following: If f is bounded and continuous ($f \in C$), then $H_t f \in C$. In this case it is sufficient to consider the semigroup $\{H_t,\ 0 \leq t < \infty\}$ as acting on C rather than B. This is a great convenience, and we assume that this is done for the remainder of this chapter.

Let τ_a be the first passage time at the point a, defined as before by

$$\tau_a = \min\ (t\colon X_t = a) \quad (4.2)$$

Let (a,b) be an open interval such that its closure $[a,b]$ belongs to the state space S. Define the **first exit time** from (a,b),

$$\tau_{ab} = \min\ (t_a, t_b) \equiv t_a \wedge t_b \quad (4.3)$$

We have the following result on τ_{ab}.

Proposition 4.1. Under assumption (4.1) we have

$$\sup_{a \leq x \leq b} E_x \tau_{ab} < \infty \quad (4.4)$$

Proof: Under assumption (4.1), $\mathcal{P}_a(\tau_b < \infty) > 0$, so that for some $t < \infty$,

$$\mathcal{P}_a(\tau_b > t) = \alpha(t) < 1 \quad (4.5)$$

Now, for $a \leq x \leq b$,

$$\mathcal{P}_x(\tau_{ab} > t) \leq \mathcal{P}_x(\tau_b > t) \leq \mathcal{P}_a(\tau_b > t) = \alpha(t) < 1 \quad (4.6)$$

Next, by the Markov property, we have

$$\begin{aligned}\mathcal{P}_x(\tau_{ab} > t + s) &= \int_a^b \mathcal{P}_x(\tau_{ab} > s,\ X_s \subset dz) \mathcal{P}_z(\tau_{ab} > t) \\ &\leq \alpha(t) \int_a^b \mathcal{P}_x(\tau_{ab} > s,\ X_s \in dz) \\ &= \alpha(t) \mathcal{P}_x(\tau_{ab} > s)\end{aligned} \quad (4.7)$$

so that for some $t < \infty$,

$$\mathcal{P}_x(\tau_{ab} > nt) \leq \alpha^n(t) \quad (4.8)$$

4. CHARACTERISTIC OPERATORS

Now, we write

$$\begin{aligned}
E_x \tau_{ab} &= \int_0^\infty s \mathcal{P}_x(\tau_{ab} \in ds) \\
&= \sum_{n=0}^\infty \int_{nt}^{(n+1)t} s \mathcal{P}_x(\tau_{ab} \in ds) \\
&\leq \sum_{n=0}^\infty (n+1) t [\mathcal{P}_x(\tau_{ab} > nt) - \mathcal{P}_x(\tau_{ab} > (n+1)t)] \\
&= t \sum_{n=0}^\infty \mathcal{P}_x(\tau_{ab} > nt) \leq t \sum_{n=0}^\infty \alpha^n(t) = \frac{t}{1 - \alpha(t)}
\end{aligned} \quad (4.9)$$

Since there exists some $t < \infty$ such that $\alpha(t) < 1$, we have proved the proposition. ∎

Now, let $[a,b] \subset S$ and $x \in (a,b)$. Dynkin's formula (3.14) yields

$$E_x \int_0^{\tau_{ab}} (Ag)(X_t) \, dt = E_x g(X_{\tau_{ab}}) - g(x) \quad (4.10)$$

Since we are considering $\{H_t, 0 \leq t < \infty\}$ as acting on C, we can assume $g \in C$. It follows that $Ag \in C$, and by shrinking (a,b) down to x, we get

$$(Ag)(x) = \lim_{(a,b) \downarrow \{x\}} \frac{E_x g(X_{\tau_{ab}}) - g(x)}{E_x \tau_{ab}} \quad (4.11)$$

Equation (4.11) expresses the local nature of A in the interior of S, since $(Ag)(x)$ depends only on the values of g in a neighborhood of x. However, (4.11) does not completely specify A. For example, suppose that $S = [0, \infty)$ and for every $x \in (0, \infty)$,

$$(Ag)(x) = \frac{1}{2} \frac{d^2 g(x)}{dx^2} \quad (4.12)$$

Both of the following two processes satisfy (4.12):

1. Absorbing Brownian motion

$$\mathcal{P}_a(X_t = 0) = \frac{1}{\sqrt{2\pi}} \int_{-a/\sqrt{t}}^{a/\sqrt{t}} e^{-\frac{1}{2}z^2} \, dz$$

$$\mathcal{P}_a(0 < X_t < b) = \int_0^b \frac{1}{\sqrt{2\pi t}} [e^{-(1/2t)(z-a)^2} - e^{-(1/2t)(z+a)^2}] \, dz$$

2. Reflecting Brownian motion

$$\mathcal{P}_a(0 \leq X_t < b) = \int_0^b \frac{1}{\sqrt{2\pi t}} [e^{-(1/2t)(z-a)^2} + e^{-(1/2t)(z+a)^2}] \, dz$$

Obviously, these two processes have different generators.

We can now use (4.11) to define an extension to A as follows. Let J denote an interval which is the intersection of an open interval with S. We call such a J a **relatively open interval** (in S). Let $\mathfrak{D}_{\mathring{A}}'$ denote the set of all functions such that

$$\lim_{J \downarrow \{x\}} \frac{E_x g(X_{\tau_J}) - g(x)}{E_x \tau_J} = (\mathring{A} g)(x) \tag{4.13}$$

exists at every $x \in S$. The difference between A and $\mathring{A}$ is that A is defined by a uniform limit, and $\mathring{A}$ by a pointwise limit. We adopt the convention that if $E_x \tau_J = \infty$ for every relatively open J containing x, then we set $(\mathring{A} g)(x) = 0$. Of course, this can only happen if x is a closed endpoint of S, because if $x \in \text{int}\,(S)$, then (4.4) applies. We shall denote by $\mathfrak{D}_{\mathring{A}}$ the set of all functions in C such that the limit in (4.13) exists at every $x \in S$.

Let p_{ab}^+ and $m_{ab}(x)$ be defined by

$$p_{ab}^+(x) = \mathcal{P}_x(\tau_b < \tau_a) \tag{4.14}$$

and

$$m_{ab} = E_x \tau_{ab} \tag{4.15}$$

For $x \in \text{int}\,(S)$, we can write (4.13) as

$$(\mathring{A} g)(x) = \lim_{\substack{b \downarrow x \\ a \uparrow x}} \frac{p_{ab}^+(x) g(b) + [1 - p_{ab}^+(x)] g(a) - g(x)}{m_{ab}(x)} \tag{4.16}$$

It is clear that knowing the two functions p_{ab}^+ and m_{ab} for every $[a,b] \subset S$ completely determines $(\mathring{A} g)(x)$ at every interior point of S. It will turn out that they also elucidate the behavior of $\mathring{A} g$ at any closed endpoints. The converse is also true. That is, knowing $\mathring{A} g$ in int (S) completely determines p_{ab}^+ and m_{ab} for every $[a,b] \subset S$, as the following proposition shows.

Proposition 4.2. For every $[a,b] \subset S$, the functions p_{ab}^+ and m_{ab} are, respectively, the unique continuous solutions to the equations

$$\begin{aligned}(\mathring{A} p_{ab}^+)(x) &= 0 & a < x < b \\ (\mathring{A} m_{ab})(x) &= -1 & a < x < b\end{aligned} \tag{4.17} \tag{4.18}$$

subject to the boundary conditions

$$\begin{aligned} p_{ab}^+(b) &= 1 = 1 - p_{ab}^+(a) \\ m_{ab}(a) &= 0 = m_{ab}(b) \end{aligned} \tag{4.19} \tag{4.20}$$

4. CHARACTERISTIC OPERATORS

Proof: We can easily verify (4.17) as follows:

$$(\mathring{A} p_{ab}^+)(x) = \lim_{\substack{d \downarrow x \\ c \uparrow x}} \frac{p_{cd}^+(x) p_{ab}^+(d) + [1 - p_{cd}^+(x)] p_{ab}^+(c) - p_{ab}^+(x)}{m_{cd}(x)} \quad (4.21)$$

If $x \in (c,d) \subset (a,b)$, then to exit from (a,b) starting at x the process must first exit from (c,d). Using the strong Markov property, we get

$$p_{ab}^+(x) = p_{cd}^+(x) p_{ab}^+(d) + [1 - p_{cd}^+(x)] p_{ab}^+(c) \quad (4.22)$$

whenever $x \in (c,d) \subset (a,b)$. Using (4.22) in (4.21) results in (4.17). Equation (4.18) can be verified in a similar way.

To prove continuity, let $y > x$, and set $c = a$, $d = y$ in (4.22). Then we get

$$p_{ab}^+(x) = p_{ay}^+(x) p_{ab}^+(y)$$

But $p_{ay}^+(x) \xrightarrow[x \uparrow y]{} 1$, so that $p_{ab}^+(x) \xrightarrow[x \uparrow y]{} p_{ab}^+(y)$. Similarly, we can show $p_{ab}^+(x) \xrightarrow[x \downarrow y]{} p_{ab}^+(y)$, so that p_{ab}^+ is continuous. Continuity of m_{ab} can be proved in a similar way. Uniqueness is much more difficult to prove. It depends on the fact that $\mathring{A}$ satisfies a "minimum principle" [Dynkin, 1965, Chap. 1, p. 145]. ∎

If S is a closed interval, say $[0,1]$, let

$$u(x) = p_{01}^+(x) \quad (4.23)$$

Then, for $0 \leq a < b \leq 1$, we have from (4.22)

$$p_{ab}^+(x) = \frac{u(x) - u(a)}{u(b) - u(a)} \quad (4.24)$$

It turns out that (4.24) holds generally whether S is a closed interval or not. The function $u(x)$, unique only up to an affine transformation $u \to \alpha u + \beta$, is called the **scale function**. The scale function is always continuous and nondecreasing, but bounded only if S is closed. It is clear from (4.17) that

$$(\mathring{A} u)(x) = 0 \quad (4.25)$$

for all $x \in \text{int}(S)$. A process is said to be in its **natural scale** if $u(x)$ can be taken to be equal to x. For example, a Brownian motion is in its natural scale. If a process X_t has a scale function $u(x)$, then $u(X_t)$ is a process in its natural scale, so that it is no restriction to assume a process to be in its natural scale.

It also turns out that there exists a Borel measure μ, defined on the

Borel sets in int (S), such that

$$m_{ab}(x) = \int_{(a,b)} \frac{[u(x \wedge y) - u(a)][u(b) - u(x \vee y)]}{u(b) - u(a)} \mu(dy) \qquad (4.26)$$

where $x \wedge y = \min(x,y)$ and $x \vee y = \max(x,y)$. In terms of u and μ, $\mathring{A}$ can be written as a generalized differential operator in int (S) as

$$A = \frac{d}{d\mu}\frac{d}{du} \qquad (4.27)$$

Actually, we are primarily interested in diffusion processes for which both μ and u are absolutely continuous. In that case, the restriction of $\mathring{A}$ to twice-differentiable functions take on the form

$$(\mathring{A}g)(x) = \frac{1}{\mu'(x)}\frac{d}{dx}\left(\frac{1}{u'(x)}\frac{dg(x)}{dx}\right)$$

We won't give a precise definition to (4.27). Instead, we take up the subject of diffusion processes in the next section.

5. DIFFUSION PROCESSES

There is some disagreement in recent literature as to the definition of a diffusion process. Some authors call any process satisfying condition (4.1) a diffusion, while other restrict the name to a smaller class of processes. We shall adopt the more restrictive definition and define a **diffusion process** as any process satisfying (4.1) and for which the limits

$$\lim_{\substack{a \uparrow x \\ b \downarrow x}} \frac{E_x X_{\tau_{ab}} - x}{E_x \tau_{ab}} = m(x) \qquad (5.1)$$

and

$$\lim_{\substack{a \uparrow x \\ b \downarrow x}} \frac{E_x(X_{\tau_{ab}} - x)^2}{E_x \tau_{ab}} = \sigma^2(x) \qquad (5.2)$$

exist at every $x \in$ int (S). We shall always assume that m and σ^2 satisfy a Hölder condition (cf. 4.7) and

$$\sigma^2(x) > 0 \qquad x \in \text{int } (S) \qquad (5.3)$$

Proposition 5.1. Let $\{X_t, 0 \leq t < \infty\}$ be a diffusion process. Let $g \in C^2 \cap \mathfrak{D}_{\mathring{A}}$. Then for every $x \in$ int (S);

$$(\mathring{A}g)(x) = \tfrac{1}{2}\sigma^2(x)\frac{d^2g(x)}{dx^2} + m(x)\frac{dg(x)}{dx} \qquad (5.4)$$

5. DIFFUSION PROCESSES

Proof: By assumption g has a bounded-continuous second derivative so that by Taylor's theorem we can write

$$g(X_{\tau_{ab}}) = g(x) + g'(x)(X_{\tau_{ab}} - x) + \tfrac{1}{2}g''(\theta)(X_{\tau_{ab}} - x)^2 \quad x, \theta \in (a,b)$$

Since $\sup\limits_{\theta \in (a,b)} |g''(\theta) - g''(x)| \xrightarrow[(a,b) \downarrow \{x\}]{} 0$, (5.4) follows immediately from the definition of $\mathring{A}$. ∎

Because of Proposition 4.2, we know that if the equation

$$\tfrac{1}{2}\sigma^2(x)\frac{d^2g(x)}{dx^2} + m(x)\frac{dg(x)}{dx} = 0 \quad a < x < b \tag{5.5}$$

has a continuous solution satisfying the boundary conditions $g(a) = 0 = 1 - g(b)$, then it must be $p_{ab}^+(x)$. Such a solution is easily constructed, and we find

$$p_{ab}^+(x) = \frac{\int_a^x \exp\left[-\int_a^y \frac{2m(z)}{\sigma^2(z)}\,dz\right]dy}{\int_a^b \exp\left[-\int_a^y \frac{2m(z)}{\sigma^2(z)}\,dz\right]dy} \tag{5.6}$$

It is clear that we can take the scale function to be

$$u(x) = \int_c^x \exp\left[-\int_c^y \frac{2m(z)}{\sigma^2(z)}\,dz\right]dy \tag{5.7}$$

where c is any point in int (S).

The function m_{ab} can be found by solving

$$\begin{aligned}\tfrac{1}{2}\sigma^2(x)\frac{d^2m_{ab}(x)}{dx^2} + m(x)\frac{dm_{ab}(x)}{dx} &= -1 \quad a < x < b \\ m_{ab}(a) &= 0 = m_{ab}(b)\end{aligned} \tag{5.8}$$

To construct m_{ab}, define the Green's function

$$G_{ab}(x,y) = \frac{[u(x \wedge y) - u(a)][u(b) - u(x \vee y)]}{u(b) - u(a)} \quad x, y \in (a,b) \tag{5.9}$$

Because $(\mathring{A}u)(x) = 0$, $x \in$ int (S),

$$\tfrac{1}{2}\sigma^2(x)\frac{\partial^2 G_{ab}(x,y)}{\partial x^2} + m(x)\frac{\partial G_{ab}(x,y)}{\partial x} = 0$$

for $x \in (a,y)$ and $x \in (y,b)$. Furthermore, $G_{ab}(a,y) = G_{ab}(b,y) = 0$, and

$$\left.\frac{\partial G_{ab}(x,y)}{\partial x}\right|_{x=y^+} - \left.\frac{\partial G_{ab}(x,y)}{\partial x}\right|_{x=y^-} = -u'(y)$$

Therefore,

$$\tfrac{1}{2}\sigma^2(x) \frac{\partial^2 G_{ab}(x,y)}{\partial x^2} + m(x) \frac{\partial G_{ab}(x,y)}{\partial x} = -\tfrac{1}{2}\sigma^2(y)u'(y)\,\delta(x-y) \quad (5.10)$$

and

$$m_{ab}(x) = \int_a^b \frac{2}{\sigma^2(y)u'(y)}\, G_{ab}(x,y)\, dy \quad (5.11)$$

Given m and σ^2 in int (S), $p_{ab}{}^+$ and m_{ab} are determined, and thus A is completely determined in int (S). However, we know that if S is closed at one or both endpoints, knowing $\overset{\circ}{A}$ in the interior of S may not be enough to determine the semigroup $\{H_t, 0 \leq t < \infty\}$ (equivalently, the transition function) uniquely. To clarify the situation, we need to study the possible behavior at the boundaries. The first result in this direction is the following.

Proposition 5.2. Let a be the left endpoint of S. Then $a \in S$ if and only if $u(a) > -\infty$, and for some $c > a$,

$$\int_a^c \frac{u(y) - u(a)}{\sigma^2(y)u'(y)}\, dy < \infty \quad (5.12)$$

The right endpoint b of S belongs to S if and only if $u(b) < \infty$, and for some $c < b$,

$$\int_c^b \frac{u(b) - u(y)}{\sigma^2(y)u'(y)}\, dy < \infty \quad (5.13)$$

Proof: We shall only prove the first half, the proof for the second half being nearly identical.

First, suppose that $a \in S$. Let α be any point such that $[a,\alpha] \subset S$. Then from (4.4) we have

$$\sup_{a < x < \alpha} m_{a\alpha}(x) < \infty$$

From (5.11) we find that $\sup\limits_{a < x < \alpha} m_{a\alpha}(x)$ occurs at a point c satisfying

$$\int_a^c \frac{u(y) - u(a)}{\sigma^2(y)u'(y)}\, dy = \int_c^\alpha \frac{u(\alpha) - u(y)}{\sigma^2(y)u'(y)}\, dy$$

and $m_{a\alpha}(c)$ is given by

$$m_{ab}(c) = 2 \int_a^c \frac{u(y) - u(a)}{\sigma^2(y)u'(y)}\, dy$$

Therefore, $a \in S$ implies (5.12).

5. DIFFUSION PROCESSES

Conversely, assume that (5.12) holds, but suppose that $a \notin S$. Then, for any point $c \in S$,

$$E_x \tau_c = \lim_{z \downarrow a} m_{zc}(x) \qquad a < x < c$$

$$= \lim_{z \downarrow a} \int_z^c G_{zc}(x,y) \frac{2}{\sigma^2(y) u'(y)} \, dy$$

$$= 2 \left[\frac{u(c) - u(x)}{u(c) - u(u)} \right] \int_a^x \frac{u(y) - u(a)}{\sigma^2(y) u'(y)} \, dy$$

$$+ 2 \left[\frac{u(x) - u(a)}{u(c) - u(a)} \right] \int_x^c \frac{u(c) - u(y)}{\sigma^2(y) u'(y)} \, dy$$

If (5.12) holds, the last expression goes to zero as $x \downarrow a$. But $E_x \tau_c$ must be nondecreasing as $x \downarrow a$, because it takes longer and longer to get from x to c. Hence, there is a contradiction and (5.12) implies that $a \in S$. ∎

Proposition 5.2 shows that the knowledge of m and σ^2 in the interior of S is sufficient to determine whether S is closed at one or the other of the endpoints. As an example, suppose that int $(S) = (0, \infty)$, and $m(x) = 0$, $\sigma^2(x) = 1$, $0 < x < \infty$. In this case, we can take $u(x) = x$. Since for every finite c,

$$\int_0^c \frac{u(y) - u(0)}{\sigma^2(y) u'(y)} \, dy = \int_0^c y \, dy < \infty$$

S must be closed at the left endpoint 0, that is, $S = [0, \infty)$. If S is closed at an endpoint, then A at that point may not be determined by the knowledge of σ^2 and m in int (S). The situation is as follows: Suppose that a is a closed left endpoint of S and that for every $c > a$, $E_a \tau_c = \infty$. Then $(\mathring{A} g)(a) = 0$. However, if for some $c > a$, $E_a \tau_c < \infty$, then $(\mathring{A} g)(a)$ is not determined by m and σ^2 in int (S). We now state a criterion which separates these two cases.

Proposition 5.3. Let a be a closed left endpoint of S. If for some $c \in$ int (S),

$$\int_a^c \frac{1}{\sigma^2(y) u'(y)} \, dy < \infty \tag{5.14}$$

then for $x \in [a,c]$,

$$E_a \tau_x < \infty \tag{5.15}$$

On the other hand, if for every $c \in$ int (S),

$$\int_a^c \frac{1}{\sigma^2(y) u'(y)} \, dy = \infty \tag{5.16}$$

then $E_a \tau_c = \infty$ for every $c \in$ int (S).

Remarks:

(a) The point a is called a **regular boundary** if (5.14) is satisfied for some $c \in \text{int}(S)$, otherwise it is called an **exit boundary**.

(b) The criterion of Proposition 5.3 can be modified for a closed right endpoint in an obvious way, and we won't repeat it.

Let a be a regular left endpoint of S. Define

$$K_a = \lim_{c \downarrow a} \frac{E_a \tau_c}{c - a} \tag{5.17}$$

We note that K_a cannot be determined from m and σ^2. Indeed, each choice of K_a gives us a process with a different transition function. If b is a regular right endpoint of S, then we set

$$K_b = \lim_{c \downarrow b} \frac{E_b \tau_c}{b - c} \tag{5.18}$$

If a is a regular left endpoint, then

$$(\mathring{A}g)(a) = \frac{1}{K_a} g'(a^+) \tag{5.19}$$

and if b is a regular right endpoint, then

$$(\mathring{A}g)(b) = \frac{1}{K_b} g'(b^-) \tag{5.20}$$

To summarize the situation, we see that $\mathring{A}$ is completely summarized by the knowledge of m and σ^2 in $\text{int}(S)$ and the values of K_a and K_b at regular boundaries, if there are any. The following proposition shows that the transition function is also uniquely determined.

Proposition 5.4. Let $g \in C$ (that is, g is bounded continuous). Let f_λ be defined by

$$f_\lambda(x) = \int_0^\infty e^{-\lambda t} E_x g(X_t)\, dt \tag{5.21}$$

Then for every $\lambda > 0$, $f_\lambda \in \mathcal{D}_{\mathring{A}} \cap C^2$ and is the unique continuous solution to

$$(\mathring{A} f_\lambda)(x) = \lambda f_\lambda(x) - g(x) \qquad x \in S \tag{5.22}$$

Remark: We note that because $f_\lambda \in \mathcal{D}_{\mathring{A}} \cap C^2$, (5.22) takes on a differential form in the interior of S, viz.,

$$\tfrac{1}{2}\sigma^2(x) \frac{d^2 f_\lambda(x)}{dx^2} + m(x) \frac{df_\lambda(x)}{dx} = \lambda f_\lambda(x) - g(x) \qquad x \in \text{int}(S) \tag{5.23}$$

5. DIFFUSION PROCESSES

On a *closed boundary* of S, (5.22) takes on the form given in the following table:

Table 5.1

Nature of boundary	Closed left endpoint a	Closed right endpoint b
Exit	$\lambda f_\lambda(a^+) = g(a)$	$\lambda f_\lambda(b^-) = g(b)$
Regular	$\dfrac{1}{K_a} f'_\lambda(a^+) = \lambda f_\lambda(a^+) - g(a)$	$\dfrac{1}{K_b} f'_\lambda(b^-) = \lambda f_\lambda(b^-) - g(b)$

If $K_a = 0$, the boundary a is called a **reflecting boundary,** and we interpret the boundary condition to mean $f'_\lambda(a^+) = 0$. Similar comments apply if $K_b = 0$. On the other hand, if $K_a = \infty$ (or $K_b = \infty$), the boundary is called an **absorbing** boundary. We can now understand the two processes both satisfying (4.12). For the absorbing Brownian motion, the boundary at $x = 0$ is absorbing, and for the reflecting Brownian motion, the boundary at $x = 0$ is reflecting.

Let $F(x,u;\ t) = E_x e^{iuX_t}$ be the characteristic function. Because $g(x) = e^{iux}$ is bounded continuous, we can use (5.22) to obtain the Laplace transform of F. Hence, the characteristic function (hence, the transition function) can be uniquely determined by solving the differential equation (5.23) with the boundary conditions given in Table 5.1 for closed endpoints and by inverting a Laplace transform. Actually, the transition function can be found more directly.

As an example, let $S = [0, \infty)$ and set

$$\sigma^2(x) = 1 \qquad \mu(x) = 0 \qquad 0 < x < \infty \tag{5.24}$$

It is clear that $x = 0$ is a regular boundary. Thus, for a bounded-continuous g, $f_\lambda(x) = \int_0^\infty e^{-\lambda t} E_x g(X_t)\, dt$ is the unique bounded-continuous solution to

$$\tfrac{1}{2} f''_\lambda(x) - \lambda f_\lambda(x) = -g(x) \qquad 0 < x < \infty \tag{5.25}$$

subject to the boundary condition

$$\frac{1}{2K_0} f'_\lambda(0^+) = \lambda f_\lambda(0^+) - g(0) \tag{5.26}$$

We interpret (5.26) to mean $f'_\lambda(0^+) = 0$ if $K_0 = 0$, and $f_\lambda(0^+) = (1/\lambda)g(0)$ if $K_0 = \infty$. The solution for $f_\lambda(x)$ is rather easy. The best approach is to seek a solution of the form

$$f_\lambda(x) = \int_0^\infty F_\lambda(x,y) g(y)\, dy + v_\lambda(x) g(0) \tag{5.27}$$

where $F_\lambda(x,y)$ and $v_\lambda(x)$ are determined by the following differential equations and boundary conditions, together with the condition that both $F_\lambda(x,y)$ and $v_\lambda(x)$ are to be bounded in x:

$$\frac{1}{2}\frac{\partial^2}{\partial x^2}F_\lambda(x,y) - \lambda F_\lambda(x,y) = 0 \quad \begin{array}{l} 0 < y < x < \infty \\ 0 < x < y < \infty \end{array}$$

$$F_\lambda(y^+,y) = F_\lambda(y^-,y)$$

$$\frac{\partial}{\partial x}F_\lambda(x,y)\bigg|_{x=y^+} - \frac{\partial}{\partial x}F_\lambda(x,y)\bigg|_{x=y^-} = -2 \quad (5.28)$$

$$\frac{1}{2K_0}\frac{\partial F_\lambda(x,y)}{\partial x}\bigg|_{x=0^+} = \lambda F_\lambda(0^+,y)$$

$$\frac{1}{2}\frac{d^2}{dx^2}v_\lambda(x) - \lambda v_\lambda(x) = 0 \quad 0 < x < \infty$$

$$\frac{1}{2K_0}v'_\lambda(0^+) = \lambda v_\lambda(0^+) - 1 \quad (5.29)$$

It is obvious that the solutions must have the form

$$F_\lambda(x,y) = \begin{cases} A(y)\exp(-\sqrt{2\lambda}\,x) & x \geq y \\ B(y)\exp(-\sqrt{2\lambda}\,x) + C(y)\exp(\sqrt{2\lambda}\,x) & \\ & 0 \leq x \leq y \end{cases} \quad (5.30)$$

$$v_\lambda(x) = D\exp(-\sqrt{2\lambda}\,x) \quad 0 \leq x < \infty \quad (5.31)$$

where $A(y)$, $B(y)$, $C(y)$ and D are determined by the subsidiary conditions in (5.28) and (5.29).

Let $g(u,x) = e^{-ux}$. Then, $f_\lambda(u,x)$ is given by

$$f_\lambda(u,x) = \int_0^\infty e^{-\lambda t}(E_x e^{-uX_t})\,dt$$

$$= \int_0^\infty e^{-\lambda t}\int_0^\infty e^{-ub}P_x(db,t)\,dt \quad (5.32)$$

which is seen to be the double Laplace transform of the transition function $P_x(E,t)$. It is easy to see from (5.28) and (5.29) that neither v_λ nor F_λ depends on u, and we can write

$$f_\lambda(u,x) = v_\lambda(x) + \int_0^\infty F_\lambda(x,y)e^{-uy}\,dy \quad (5.33)$$

It is clear now what the probabilistic interpretations of $F_\lambda(x,y)$ and $v_\lambda(x)$ are. If we write

$$P_x(E,t) = I_E(0)\mathcal{P}_x(X_t = 0) + \int_E p_x(b,t)\,db \quad (5.34)$$

5. DIFFUSION PROCESSES

where $I_E(x) = 0, 1$ according as $x \notin E$ or $x \in E$, then

$$v_\lambda(x) = \int_0^\infty e^{-\lambda t} \mathcal{P}_x(X_t = 0)\, dt \tag{5.35}$$

$$F_\lambda(x,y) = \int_0^\infty e^{-\lambda t} p_x(y,t)\, dt \tag{5.36}$$

It is easy to find v_λ and F_λ, but to invert the Laplace transforms to get $\mathcal{P}_x(X_t = 0)$ and $p_x(y,t)$ is tedious.

It is fairly clear how the above example generalizes. We shall sketch an outline of the situation below. First, the equation

$$\left[\tfrac{1}{2}\sigma^2(x)\frac{d^2}{dx^2} + \mu(x)\frac{d}{dx}\right] f_\lambda(x) = \lambda f_\lambda(x) \qquad x \in \text{int } (S) \tag{5.37}$$

for $\lambda > 0$ has a pair of solutions $f_\lambda = \varphi_\lambda, \theta_\lambda$ with the following properties:

$$\begin{array}{l}\varphi_\lambda(x) > 0 \qquad \theta_\lambda(x) > 0 \qquad x \in \text{int } (S) \\ \varphi_\lambda(x) \text{ nondecreasing, } \theta_\lambda(x) \text{ nonincreasing} \\ \varphi_\lambda(x) \text{ bounded if and only if the right endpoint of } S \text{ is closed} \\ \theta_\lambda(x) \text{ bounded if and only if the left endpoint of } S \text{ is closed}\end{array} \tag{5.38}$$

The two functions φ_λ and θ_λ are linearly independent, and the Wronskian Δ is given by

$$\Delta(x) = \theta_\lambda(x)\varphi_\lambda'(x) - \varphi_\lambda(x)\theta_\lambda'(x) = u'(x) \tag{5.39}$$

where $u(x)$ is the scale function. Therefore, every solution of (5.37) is a linear combination of φ_λ and θ_λ. We now seek a bounded solution of

$$\tfrac{1}{2}\sigma^2(x)\frac{d^2}{dx^2} f_\lambda(x) + \mu(x)\frac{df_\lambda(x)}{dx} = \lambda f_\lambda(x) - g(x)$$

$$x \in \text{int } (S) \quad (5.40)$$

subject to the boundary conditions at closed endpoints given in Table 5.1. Imitating the procedure in the example, we seek a solution in the form of

$$f_\lambda(x) = \int_a^b F_\lambda(x,y) g(y)\, dy + g(a) v_\lambda(x) + g(b) u_\lambda(x) \tag{5.41}$$

where a and b are endpoints of S. The functions F, u_λ, v_λ are of the form

$$F_\lambda(x,y) = \begin{cases} A(y)\varphi_\lambda(x) + B(y)\theta_\lambda(x) & x > y \\ C(y)\varphi_\lambda(x) + D(y)\theta_\lambda(x) & x < y \end{cases} \tag{5.42}$$

$$v_\lambda(x) = \alpha\varphi_\lambda(x) + \beta\theta_\lambda(x) \tag{5.43}$$

$$u_\lambda(x) = \gamma\varphi_\lambda(x) + \delta\theta_\lambda(x) \tag{5.44}$$

The unknowns in (5.42) to (5.44) are determined by requiring

$$F_\lambda(y^+,y) = F_\lambda(y^-,y) \tag{5.45}$$

$$\frac{\partial}{\partial x} F_\lambda(x,y)\Big|_{x=y^+} - \frac{\partial}{\partial x} F_\lambda(x,y)\Big|_{x=y^-} = -\frac{2}{\sigma^2(y)} \tag{5.46}$$

and by imposing the following boundary conditions:

$F_\lambda(x,y)$, $v_\lambda(x)$, $u_\lambda(x)$ stay bounded as x approaches an open endpoint (5.47a)

At a closed endpoint, $F_\lambda(x,y)$ (as a function of x) satisfies this boundary conditions of Table 5.1 for $g \equiv 0$ (5.47b)

If the left endpoint is closed, $v_\lambda(x)(u_\lambda(x))$ satisfies the boundary conditions of Table 5.1 corresponding to $g \equiv 1$ ($g \equiv 0$). If the right endpoint is closed, $v_\lambda(x)(u_\lambda(x))$ satisfies the boundary conditions of Table 5.1 corresponding to $g \equiv 0$ ($g \equiv 1$) (5.47c)

The probabilistic interpretation of u_λ, v_λ, and F_λ is given by

$$F_\lambda(x,y)\,dy = \int_0^\infty e^{-\lambda t}\mathcal{P}_x(X_t \in dy)\,dt \qquad a < x < y < b \tag{5.48a}$$

$$u_\lambda(x) = \int_0^\infty e^{-\lambda t}\mathcal{P}_x(X_t = b)\,dt \tag{5.48b}$$

$$v_\lambda(x) = \int_0^\infty e^{-\lambda t}\mathcal{P}_x(X_t = a)\,dt \tag{5.48c}$$

As to be expected, the simplest case results when both endpoints are open. In that case, both u_λ and v_λ are identically zero, and $F_\lambda(x,y)$ is given by

$$F_\lambda(x,y) = \begin{cases} \dfrac{2}{\sigma^2(y)u'(y)}\,\varphi_\lambda(y)\theta_\lambda(x) & b \geq x \geq y > a \\[6pt] \dfrac{2}{\sigma^2(y)u'(y)}\,\theta_\lambda(y)\varphi_\lambda(x) & a \leq x \leq y < b \end{cases} \tag{5.49}$$

From (5.49) and the fact that $F_\lambda(x,y)$ is the Laplace transform of the transition density $p_x(y,t)$, we can finally deduce that $p_x(y,t)$ must satisfy both the forward and the backward equations of diffusion. Why it is so hard to prove that $p_x(y,t)$ satisfy the diffusion equations (without assuming the differentiability conditions) is not well understood.

EXERCISES

1. Let $\{X_t, -\infty < t < \infty\}$ be a random telegraph process which we define as a stationary Markov process such that

   (a) $X_t = +1$ or -1 with $\mathcal{P}(X_t = 1) = \frac{1}{2} = \mathcal{P}(X_t = -1)$

   (b) $\mathcal{P}(X_t = X_s) = \frac{1}{2}(1 + e^{-|t-s|}) = 1 - \mathcal{P}(X_t = -X_s)$

   Find the generator A, and show that $H = e^{tA}$.

EXERCISES

2. Let $\{X_t, 0 \leq t < \infty\}$ have a state space $S = [0, \infty)$, and let

$$\mathcal{P}(X_t < x | X_s = a)$$
$$= \int_0^x \frac{1}{\sqrt{2\pi(t-s)}} \left\{ \exp\left[-\frac{1}{2}\frac{(\xi-a)^2}{t-s}\right] - \exp\left[-\frac{1}{2}\frac{(\xi+a)^2}{t-s}\right]\right\} d\xi$$
$$+ \sqrt{\frac{2}{\pi}} \int_{a/\sqrt{t-s}}^\infty e^{-\frac{1}{2}z^2} dz \qquad \begin{array}{l} t > s \\ x > 0 \\ a > 0 \end{array}$$

For $a = 0$ we have $\mathcal{P}(X_t = 0 | X_s = 0) = 1$. Find the generator A.

3. Let $\{X, t \geq 0\}$ be a Brownian motion starting from 0, that is, $X_0 = 0$ with probability 1. Let τ_c be defined by

$$\tau_c = \min_{t>0} \{t, X_t = c\} \qquad c > 0$$

(a) Introduce a function f defined as

$$f(x) = \begin{cases} 1 & x > c \\ 0 & x \leq c \end{cases}$$

and use (3.13) to find $Ee^{-\lambda \tau_c}$.

(b) Use the same method and prove the reflection principle of D. Andre,

$$\mathcal{P}(\tau_c \leq t) = 2\mathcal{P}(X_t \geq c)$$

4. Let $\{X_t, -\infty < t < \infty\}$ be a zero mean Gaussian process with $EX_t X_s = e^{-|t-s|}$, namely, X is an Ornstein-Uhlenbeck process. Use (3.13) to find

$$\mathcal{P}(X_t \geq 0, 0 \leq t \leq 1)$$

5. Let W_t and V_t be standard Brownian motion processes with $W_0 = V_0 = 0$. Let $X_t = W_t^2 + V_t^2$. Show that $X_t, t \geq 0$, is a diffusion process, and find the functions σ^2 and m.

6. For the process in Exercise 5 find a suitable scale function $u(x)$, and determine whether the state space S is closed at its left endpoint 0. If S is closed at 0, determine whether it is a regular boundary or an exit boundary. If it is a regular boundary, determine $(\mathring{A}g)(0)$.

7. Show that, except for notational changes, (5.22) can be viewed as a generalization of (2.12), that is, if $H_t g \in \mathcal{D}_A$ for each t, then it follows from (2.12) that

$$A f_\lambda = \lambda f_\lambda - g$$

8. Let $X_t, t \geq 0$, be a diffusion process so that in int (S) $\mathring{A} = \frac{1}{2}\sigma^2(x) d^2/dx^2 + m(x) d/dx$. Let $h(x,t)$ be defined by

$$h(x,t) = E_x f(X_t) \exp\left[-\int_0^t k(X_s) \, ds\right]$$

Give a heuristic derivation of the equation

$$\frac{\partial}{\partial t} h(x,t) = \frac{1}{2}\sigma^2(x) \frac{\partial^2 h(x,t)}{\partial x^2} + m(x) \frac{\partial h(x,t)}{\partial x} - k(x) h(x,t) \qquad x \in \text{int } (S)$$

9. A precise and general statement of the results in Exercise 8 is known as Kac's theorem [Kac, 1951] stated as follows: Let k and f be in $\mathbf{B}_0$, and let k be nonnegative and f be continuous. Let $u_\lambda(x)$ be defined by

$$u_\lambda(x) = \int_0^\infty h(x,t)e^{-\lambda t}\,dt = E_x \int_0^\infty e^{-\lambda t} f(X_t) \exp\left[-\int_0^t k(X_s)\,ds\right] dt$$

Then u_λ is the unique solution of

$$\lambda u_\lambda - f = Au_\lambda - ku_\lambda$$

(which is roughly equivalent to $\partial h/\partial t = Ah - kh$).

Use Kac's theorem to prove that if τ is the amount of time that a Brownian motion is positive during [0,1], then

$$\mathcal{P}(\tau \leq t) = \frac{2}{\pi}\sin^{-1}\sqrt{t}$$

This is the celebrated arc-sine law of Lévy. [Hint: Set $k(x) = \beta((1 + \operatorname{sgn} x)/2)$ and $f(x) = 1$ in Kac's theorem.]

10. Let X_t, $t \geq 0$, be a standard Brownian motion. Use the result of Exercise 8 to find the distribution of the quadratic functional

$$\int_0^1 X_t^2\,dt$$

11. Suppose that int $(S) = (0, \infty)$. For the following pairs of σ^2 and m, determine:

   (a) Whether S is closed at 0.

   (b) If closed, is 0 a regular boundary?

$\sigma^2(x)$	$m(x)$
1	-1
1	$-x$
x	$-x$
x^2	$1 - \tfrac{3}{2}x$

6
Martingale Calculus

1. MARTINGALES

Let $(\mathcal{A}, \Omega, \mathcal{P})$ be a complete probability space, let T be a subset of $\mathbb{R}$ and let $\mathcal{A}_{\cdot} = (\mathcal{A}_t : t \in T)$ be a family of sub-σ-algebras of $\mathcal{A}$ which is increasing: $\mathcal{A}_s \subset \mathcal{A}_t$ for $s < t$. A stochastic process $X = (X_t : t \in T)$ is said to be adapted to $\mathcal{A}_{\cdot}$ if for each t in T, X_t is $\mathcal{A}_t$ measurable. For example, X is adapted to $\mathcal{A}_{\cdot}^X$, where for each t, $\mathcal{A}_t^X$ is the σ-algebra generated by the family of random variables $\{X_s : s \leq t\}$. We call $\mathcal{A}_{\cdot}^X$ the family of σ-algebras generated by the process X.

A complex-valued stochastic process X is called a martingale with respect to $\mathcal{A}_{\cdot}$ if it is adapted to $\mathcal{A}_{\cdot}$ and if $t \geq s$ implies that

$$E[X_t | \mathcal{A}_s] = X_s \quad \text{a.s.} \tag{1.1}$$

The process is said to be a submartingale (supermartingale) if it is real-valued and equality in (1.1) is replaced by $\geq$ (respectively, by $\leq$).

For example, if w is a Wiener process which is adapted to $\mathcal{A}_{\cdot}$, and if for each t, $\{w_s - w_t, s \geq t\}$ is independent of $\mathcal{A}_t$, then w is a martingale with respect to $\mathcal{A}_{\cdot}$ (cf. Sec. 2.3). We call w an $\mathcal{A}_{\cdot}$ Wiener martingale.

For another example, let B be a random variable with $E|B|$ finite. Then the random process X defined by

$$X_t = E[B|\mathcal{A}_t] \qquad t \in T$$

is a martingale with respect to $\mathcal{A}_\cdot$. Indeed, X is adapted to $\mathcal{A}_\cdot$ by the definition of conditional expectations and Eq. (1.1) is a consequence of the smoothing property of conditional expectation.

A key to the tractability of martingales is that they have versions with well-behaved sample paths. The sample paths can be chosen to be especially nice under some simple conditions on $\mathcal{A}_\cdot$.

Definition. We say that $\mathcal{A}_\cdot = (\mathcal{A}_t : t \geq 0)$ satisfies the **usual conditions** if

1. $\mathcal{A}_0$ contains all sets A in $\mathcal{A}$ with $\mathcal{P}(A) = 0$, and
2. (Right continuity) $\mathcal{A}_t = \mathcal{A}_{t+}$ for each t, where $\mathcal{A}_{t+} = \bigcap_{s>t} \mathcal{A}_s$.

Proposition 1.1. Suppose that $X = (X_t : t \geq 0)$ is an $\mathcal{A}_\cdot$ martingale (or submartingale or supermartingale) such that EX_t is right continuous in t, and suppose that $\mathcal{A}_\cdot$ satisfies the usual conditions. Then there exists a modification of X such that for all $t \geq 0$,

$$X_t(\omega) = \lim_{\substack{s \to t \\ s > t}} X_s(\omega)$$

and (1.2)

$$X_{t-}(\omega) = \lim_{\substack{s \to t \\ s < t}} X_s(\omega)$$

exists.

Remark: Proposition 1.1 and the closely related result, Proposition 1.2, are due to Doob. Proofs can be found in [Doob, 1953], [Doob, 1984], [Meyer, 1966] and in [Dellacherie and Meyer, 1982]. In typical proofs, Proposition 1.2 in the case $T = \mathbb{Z}_+$ is deduced first and then Propositions 1.1 and 1.2 are deduced.

A process X which satisfies the conditions (1.2) for all t for ω in a set (not depending on t) of probability one is said to have continuous-on-the-right with limits-on-the-left (corlol) sample paths, or is simply said to be a corlol process. A corlol process is clearly separable with any countable dense subset of $\mathbb{R}_+$ serving as a separating set.

Given two numbers a and b we use $a \wedge b$ to denote the smaller and $a \vee b$ to denote the larger of the two numbers. Given two σ-algebras $\mathcal{A}$ and

1. MARTINGALES

$\mathcal{B}$ of subsets of a given set we use $\mathcal{A} \vee \mathcal{B}$ to denote the smallest σ-algebra containing both $\mathcal{A}$ and $\mathcal{B}$.

Proposition 1.2 (Martingale Convergence Theorem). Suppose that $(X_t, \mathcal{A}_t : t \in T)$ with $T = \mathbb{R}_+$ or $T = \mathbb{Z}_+$ is a (separable, if $T = \mathbb{R}_+$) supermartingale such that $E(X_t \wedge 0)$ is bounded below by a constant not depending on t. Then there exists an a.s. finite random variable X_∞ such that

$$X_t \xrightarrow[t \to \infty]{\text{a.s.}} X_\infty$$

Remark: Clearly X_∞ is $\mathcal{A}_\infty$ measurable, where $A_\infty = \bigvee_t \mathcal{A}_t$, that is, $\mathcal{A}_\infty$ is the smallest σ-algebra containing the algebra $\bigcup_t \mathcal{A}_t$. A similar convergence theorem is true when $T = \mathbb{R}$ or $T = \mathbb{Z}$ and t tends to minus infinity.

An arbitrary family $\mathcal{L}$ of random variables is called **uniformly integrable** if

$$\lim_{c \to +\infty} \sup_{Y \in \mathcal{L}} E\left[|Y| I_{\{|Y| \geq c\}}\right] = 0$$

The following is a proposition from classical measure theory, the first half of which is a generalization of the dominated convergence theorem.

Proposition 1.3.

(a) Let (X_n) be a sequence of random variables, with $E|X_n|$ finite for each n, converging a.s. to a random variable X. Then ($E|X|$ is finite and X_n converges to X in 1-mean) if and only if the family of random variables $\{X_n\}$ is uniformly integrable.

(b) Let X be a random variable on $(\Omega, \mathcal{A})$ with $E|X| < +\infty$, and let $\{\mathcal{A}_\gamma : \gamma \in \Gamma\}$ be an arbitrary family of sub-σ-algebras of $\mathcal{A}$ with arbitrary parameter set Γ. Then

$$\mathcal{L} = \left\{ E\left[X | \mathcal{A}_\gamma\right] : \gamma \in \Gamma \right\}$$

is a uniformly integrable family of random variables. (See [Meyer 1966, pp. 18, 85] for a proof.)

A martingale (or submartingale or supermartingale) $(X_t : t \in T)$ is called uniformly integrable if the family of random variables $\{X_t : t \in T\}$ is uniformly integrable. We will deduce form Propositions 1.2 and 1.3 the following proposition:

Proposition 1.4. Let $T = \mathbb{R}_+$ or $T = \mathbb{Z}_+$.
(a) If $(X_t: t \in T)$ is a uniformly integrable separable martingale, then

$$\lim_{t \to \infty} X_t = X_\infty \quad \text{a.s. and in 1-mean}$$

where X_∞ is an $\mathcal{A}_\infty$ measurable random variable with $E|X_\infty|$ finite, and

$$X_t = E[X_\infty | \mathcal{A}_t] \quad \text{a.s. for } t \in T \tag{1.3}$$

(b) A random process $(X_t: t \in T)$ is a uniformly integrable martingale if and only if

$$X_t = E[B | \mathcal{A}_t] \quad \text{a.s. for } t \in T \tag{1.4}$$

for some random variable B with $E|B|$ finite.
(c) Any B satisfying (1.4) is related to X_∞ by

$$X_\infty = E[B | \mathcal{A}_\infty] \quad \text{a.s.} \tag{1.5}$$

Proof: Convergence a.s. in part (a) is guaranteed by Proposition 1.2. Convergence in 1-mean then is a consequence of the uniform integrability of X. Now for s and t with $s \geq t$,

$$E|E[X_s|\mathcal{A}_t] - E[X_\infty|\mathcal{A}_t]| \leq E[E[|X_s - X_\infty||\mathcal{A}_t]]$$
$$= E|X_s - X_\infty|$$

which tends to zero as s tends to infinity by part (a). Since $E[X_s|\mathcal{A}_t] = X_t$ for all $s > t$, Eq. (1.3) is proved.

If X has the representation (1.4) then it is uniformly integrable by Proposition 1.3. Conversely, if X is uniformly integrable then it has a separable modification and then (1.4) is true with $B = X_\infty$. This proves part (b).

By Eqs. (1.3) and (1.4),

$$E[B|\mathcal{A}_t] = E[X_\infty|\mathcal{A}_t] \quad \text{a.s. for all } t$$

Thus $E[(B - X_\infty)I_A] = 0$ for all A in the algebra $\bigcup_t \mathcal{A}_t$, and thus for all A in $\mathcal{A}_\infty$ by the extension theorem. Since X_∞ is $\mathcal{A}_\infty$ measurable, Eq. (1.5) follows. ∎

A nonnegative random variable $S, S \leq +\infty$, is called an $\mathcal{A}_.$ **stopping time** if

$$\{\omega: S(\omega) \leq t\} \in \mathcal{A}_t \quad \text{for all } t \geq 0$$

1. MARTINGALES

Equivalently, S is a stopping time if the zero-one valued random process U defined by $U_t = I_{\{S \leq t\}}$ is an adapted random process.

Given a stopping time S define the σ-algebra $\mathcal{Q}_S$ as follows:

$$E \in \mathcal{Q}_S \text{ if and only if } E \cap \{\omega: S(\omega) \leq t\} \in \mathcal{Q}_t, \text{ all } t$$

If $S = s_0$ is a deterministic time this definition causes no ambiguity. (Since we assume that $\mathcal{Q}_{t+} = \mathcal{Q}_t$, stopping times here are the same as Markov times defined in Section 5.3.)

The martingale property persists at stopping times:

Proposition 1.5 (Doob's Optional Sampling Theorem). Let $X = (X_t: t \geq 0)$ (or $X = (X_n: n \in \mathbb{Z})$) be a corlol submartingale, and let S and R be two stopping times such that $S \leq R$. Then

$$E[X_R | \mathcal{Q}_S] \geq X_S \quad \text{a.s.} \tag{1.6}$$

with equality in case X is a martingale, if R and S are bounded above by a constant. If X is a uniformly integrable martingale then equality holds in (1.6), even if R and S are not bounded.

Proof: The proposition is not difficult to verify if R and S each have a finite number of possible values (see Exercise 4), and the general case is proved by an approximation argument as in the proof of Proposition 3.2 in Chapter 2. ∎

Lemma. Let $(Y_k, \mathcal{Q}_k, k = 1, \ldots, n)$ be a positive submartingale. Define $Y_n^*(\omega) = \max\{Y_1(\omega), \ldots, Y_n(\omega)\}$. Then

(a) $\mathcal{P}[Y_n^* \geq \lambda]\lambda \leq E[I_{\{Y_n^* \geq \lambda\}} Y_n]$ for $\lambda > 0$, and
(b) $E[(Y_n^*)^2] \leq 4E(Y_n^2)$.

Proof: Let $\lambda > 0$ and define a stopping time τ to be the first k such that $Y_k \geq \lambda$, and $\tau = n$ if there is no such k. By the optional sampling theorem, $Y_\tau \leq E[Y_n | \mathcal{Q}_\tau]$ so that

$$Y_\tau I_{\{Y_\tau \geq \lambda\}} \leq E\left[Y_n I_{\{Y_\tau \geq \lambda\}} | \mathcal{Q}_\tau\right] \quad \text{a.s.}$$

and so

$$\mathcal{P}[Y_\tau \geq \lambda]\lambda \leq E\left[Y_\tau I_{\{Y_\tau \geq \lambda\}}\right] \leq E\left[Y_n I_{\{Y_\tau \geq \lambda\}}\right]$$

Since the event that $Y_\tau \geq \lambda$ is the same as the event that $Y_n^* \geq \lambda$, this proves (a).

Suppose that $E(Y_n^2)$ is finite, since (b) is trivial otherwise. Define $F(\lambda) = \mathcal{P}(Y_n^* < \lambda)$. Then, using integration by parts and part (a),

$$E\left[(Y_n^*)^2\right] = \int_0^\infty \lambda^2 F(d\lambda)$$

$$= \lim_{c \to \infty} \left[\int_0^c 2\lambda(1 - F(\lambda))\, d\lambda - \lambda^2(1 - F(\lambda))|_{\lambda=0}^c\right]$$

$$\leq \int_0^\infty 2\lambda(1 - F(\lambda))\, d\lambda$$

$$\leq 2\int_0^\infty E\left[I_{\{Y^* \geq \lambda\}} Y_n\right] d\lambda = 2E\left[Y_n^* Y_n\right]$$

$$\leq 2E\left[(Y_n^*)^2\right]^{1/2} E\left[Y_n^2\right]^{1/2}$$

Now $Y_n^* \leq Y_1 + \cdots + Y_n$ and $E[Y_i^2] \leq E[Y_n^2] < +\infty$ so that $E[(Y_n^*)^2]$ is finite, so (b) is implied. ∎

Proposition 1.6. (Doob's L^2 Inequality). Let M be a separable martingale and $T < +\infty$. Then

$$E\left[\sup_{0 \leq t \leq T} |M_t|^2\right] \leq 4E\left[|M_T|^2\right] \quad (1.7)$$

and if M is uniformly integrable then the inequality persists if $T = +\infty$.

Proof: Suppose $s_1, s_2, \ldots$ is a separating set for the interval $[0, T]$ with $s_1 = T$. For $n \geq 1$, let $t_1^n, \ldots, t_n^n$ be $s_1, \ldots, s_n$ arranged in increasing order. Then $(|M_{t_k^n}|, \mathcal{C}_{t_k^n} : 1 \leq k \leq n)$ is a positive submartingale so, by the lemma,

$$E \max_{k \leq n} |M_{s_k}|^2 \leq 4E\left[|M_T|^2\right] \quad (1.8)$$

As n tends to infinity the left-hand side of (1.8) converges to the left-hand side of (1.7) by the monotone convergence theorem, and the proposition is proved. ∎

Martingales are ideal for studying differentiation of probability measures, as we show next. Let Ω be a nonempty set and let $\mathcal{C}$ be a σ-algebra of subsets of Ω. The pair $(\Omega, \mathcal{C})$ is called a measurable space. Now consider two nonnegative σ-additive set functions M and M_0 defined on sets in $\mathcal{C}$ such that $M(\Omega)$ and $M_0(\Omega)$ are finite. We call M and M_0 finite measures on $(\Omega, \mathcal{C})$. For a pair of finite measures on a measurable space we define the

1. MARTINGALES

concepts of absolute continuity and singularity as follows:

M is said to be **absolutely continuous** with respect to M_0 if for every A in $\mathcal{C}$ such that $M_0(A) = 0$, we have $M(A) = 0$. In symbols we write $M \ll M_0$.

M and M_0 are said to be **mutually absolutely continuous** if $M \ll M_0$ and $M_0 \ll M$. In symbols we write $M \equiv M_0$.

M and M_0 are said to be **singular** if there exists a set A in $\mathcal{C}$ such that $M(A) = 0$ and $M_0(\Omega - A) = 0$. We denote this by $M \perp M_0$.

Two basic theorems related to these properties of measures are the Lebesgue decomposition theorem and the Radon-Nikodym theorem. These can be combined and stated for the case of finite measures as follows.

Proposition 1.7. Let $(\Omega, \mathcal{C})$ be a measurable space, and let M_0 and M be two finite measures defined on $(\Omega, \mathcal{C})$. Then there exists a nonnegative, $\mathcal{C}$-measurable function Λ and a finite measure μ such that for every A in $\mathcal{C}$,

$$M(A) = \int_A \Lambda(\omega) M_0(d\omega) + \mu(A)$$

where μ is singular with respect to M_0. Λ is unique up to a set of M_0 measure zero, and μ is unique.

Remark: It is clear that $M \ll M_0$ if and only if μ is the zero measure. In that case we call Λ the **Radon-Nikodym derivative** of M with respect to M_0 and denote Λ by

$$\Lambda = \frac{dM}{dM_0}$$

If $M \ll M_0$ and $\mathcal{B}$ is a sub-σ-algebra of $\mathcal{C}$ then the restriction of M to $\mathcal{B}$ is always absolutely continuous with respect to the corresponding restriction of M_0. It is easy to verify this fact and the formula (see Exercise 6)

$$\frac{dM^{\mathcal{B}}}{dM_0^{\mathcal{B}}} = E_0\left[\frac{dM}{dM_0}\bigg|\mathcal{B}\right] \quad \text{a.s. } M_0 \qquad (1.9)$$

where $M^{\mathcal{B}}$ and $M_0^{\mathcal{B}}$ denote the restrictions of M and M_0 to $\mathcal{B}$, respectively, and E_0 indicates that the conditional expectation is with respect to M_0 measure. Although our definition for conditional expectation given in Section 1.7 was for probability measures, extension to finite measures causes no difficulty.

We shall now restrict ourselves to probability measures which are generated by stochastic processes. Let T be a possibly infinite subinterval of

$\mathbb{R}$ and let $(X_t: t \in T)$ be a family of m-vector valued functions on Ω such that for each ω fixed, $X_t(\omega)$ is right continuous in t. Let $\mathcal{C}$ denote the σ-algebra generated by $\{X_t: t \in T\}$. We call the pair $(\Omega, \mathcal{C})$ the *sample space* of $(X_t: t \in T)$. Let (T_n) be an increasing sequence of finite subsets of T such that T_n becomes dense in T as n tends to infinity, and T_n contains the largest t in T, if such t exists. Then let $\mathcal{C}_n$ denote the σ-algebra generated by $\{X_t: t \in T_n\}$ and let $\mathcal{C}_\infty$ denote the smallest σ-algebra containing $\bigcup_n \mathcal{C}_n$. It is clear that $\mathcal{C} \supset \mathcal{C}_\infty$. On the other hand, by the right continuity of $X_t(\omega)$ in t, each random variable X_t is $\mathcal{C}_\infty$ measurable and so $\mathcal{C} = \mathcal{C}_\infty$.

Now suppose that $\mathcal{P}$ and $\mathcal{P}_0$ are two probability measures on $(\Omega, \mathcal{C})$ such that $\mathcal{P} \ll \mathcal{P}_0$. For each n, let

$$L_n = E_0\left[\frac{d\mathcal{P}}{d\mathcal{P}_0}\bigg|\mathcal{C}_n\right]$$

Then L_n is a nonnegative uniformly integrable martingale, and by Proposition 1.4 we have

$$\lim_{n \to \infty} L_n = \frac{d\mathcal{P}}{d\mathcal{P}_0} \qquad \text{a.s. and in 1-mean,} \qquad \mathcal{P}_0$$

This equation provides a useful method for computing $d\mathcal{P}/d\mathcal{P}_0$ since L_n can often be computed directly, as we now demonstrate.

First, by (1.9) we have

$$L_n = \frac{d\mathcal{P}^{\mathcal{C}_n}}{d\mathcal{P}_0^{\mathcal{C}_n}}$$

Now suppose that T_n contains n points. Then, since X_t is m-vector valued, the ordered collection $(X_t: t \in T_n)$ is mn-vector valued. Let $\mathbf{P}_{T_n}$ and $\mathbf{P}_{T_n^0}$ denote the probability measures, defined on the Borel subsets of $\mathbb{R}^{mn}$, which are induced by $(X_t, t \in T_n)$ under $\mathcal{P}$ and $\mathcal{P}_0$, respectively. Then, with $\mathcal{P}_0$ probability one,

$$L_n(\omega) = \frac{d\mathbf{P}_{T_n}}{d\mathbf{P}_{T_n^0}}(X_t(\omega), t \in T_n) \tag{1.10}$$

where the Radon-Nikodym derivative $d\mathbf{P}_{T_n}/d\mathbf{P}_{T_n^0}$ is a Borel measurable function on $\mathbb{R}^{mn}$. If both $\mathbf{P}_{T_n}$ and $\mathbf{P}_{T_n^0}$ are absolutely continuous with respect to some measure (e.g., Lebesgue measure) then (1.10) can be written in terms of the density functions p_{T_n} and $p_{T_n^0}$ as

$$L_n(\omega) = \frac{p_{T_n}}{p_{T_n^0}}(X_t(\omega), t \in T_n)$$

2. SAMPLE PATH INTEGRALS

For a specific example, consider the following. Suppose that under $\mathcal{P}_0, \{X_t, 0 \leq t \leq 1\}$ is a Brownian motion and under $\mathcal{P}, \{X_t - t, 0 \leq t \leq 1\}$ is a Brownian motion. Let $T_n = \{k/2^n, k = 0, 1, \ldots, 2^n\}$. Then,

$$\frac{d\mathbf{P}_{T_n}}{d\mathbf{P}_{T_n^0}}(x_0, x_1, \ldots, x_{2^n}) = \frac{p_{T_n}(x_0, x_1, \ldots, x_{2^n})}{p_{T_n^0}(x_0, x_1, \ldots, x_{2^n})}$$

$$= \frac{\prod_{\nu=0}^{2^n-1} \left(1/\sqrt{2\pi 2^{-n}}\right) e^{-\frac{1}{2}(2^n)(x_{\nu+1} - x_\nu - 2^{-n})^2}}{\prod_{\nu=0}^{2^n-1} \left(1/\sqrt{2\pi 2^{-n}}\right) e^{-\frac{1}{2}(2^n)(x_{\nu+1} - x_\nu)^2}}$$

$$= \exp\left[+\sum_{\nu=0}^{2^n-1}(x_{\nu+1} - x_\nu)\right] e^{-\frac{1}{2}}$$

$$= e^{-\frac{1}{2}} e^{(x_1 - x_0)}$$

Therefore, in this somewhat trivial example, we have with $\mathcal{P}_0$-measure 1,

$$\frac{d\mathcal{P}}{d\mathcal{P}_0}(\omega) = e^{-\frac{1}{2}} e^{X_1(\omega) - X_0(\omega)} = e^{-\frac{1}{2}} e^{X_1(\omega)}$$

As a check, we can verify that $E_0 \, d\mathcal{P}/d\mathcal{P}_0 = 1$ by computing

$$E_0 e^{-\frac{1}{2}} e^{X_1} = e^{-\frac{1}{2}} \frac{1}{\sqrt{2\pi}} \int_{-\infty}^{\infty} e^x e^{-\frac{1}{2}x^2} \, dx = \frac{1}{\sqrt{2\pi}} \int_{-\infty}^{\infty} e^{-\frac{1}{2}(x-1)^2} \, dx = 1$$

2. SAMPLE PATH INTEGRALS

Consider the integral

$$\int_0^t \phi(s, \omega) \, dV(s, \omega)$$

which we often write as

$$\int_0^t \phi(s) \, dV_s, \quad \text{or simply as} \quad \phi \cdot V_t$$

for random processes ϕ and V. Such an integral can in some cases be defined as an integral involving deterministic functions (namely the sample paths of ϕ and V) for each ω. Before defining the integral as a Lebesgue-Stieltjes integral, we will briefly review the theory of Lebesgue-Stieltjes integrals for deterministic functions.

Let G be a bounded increasing function on $\mathbb{R}_+$ which is right continuous. Then there exists a unique measure μ defined on the Borel subsets of $\mathbb{R}_+$

such that

$$\mu((a,b]) = G(b) - G(a)$$

Then the Lebesgue-Stieltjes integral of a (deterministic) nonnegative Borel function ϕ with respect to G,

$$\int_0^\infty \phi(s)\, dG(s)$$

is defined to be equal to the Lebesgue integral of ϕ with respect to μ:

$$\int_0^\infty \phi(s)\mu(ds)$$

(The integral we defined in Section 1.5 is the Lebesgue integral.)

Given a and b with $a < b$, let π_{ab} denote the set of vectors $\underline{t}$ such that, for some n, $\underline{t} = (t_0, t_1, \ldots, t_n)$, where $a \leq t_0 < t_1 \cdots < t_n \leq b$. Then given a real function v on $\mathbb{R}_+$, the variation of v over a finite interval $[a,b]$ is

$$\int_a^b |dv_s| \triangleq \sup_{\underline{t} \in \pi_{ab}} \sum_{k=0}^{n-1} |v(t_{k+1}) - v(t_k)| \leq +\infty$$

and the variation over $\mathbb{R}_+$ is equal to

$$\int_0^\infty |dv_s| = \lim_{t \to \infty} \int_0^t |dv_s| \leq +\infty$$

We say that v has finite variation if its variation over $\mathbb{R}_+$ is finite.

A real function v has finite variation if and only if v can be written as

$$v(t) = v(0) + v_1(t) - v_2(t) \qquad t \in \mathbb{R}_+ \tag{2.1}$$

where v_1 and v_2 are bounded increasing functions. If v has such a representation then there is a unique choice of v_1 and v_2 (we call this (v_1, v_2) canonical) such that $v_1(0) = v_2(0) = 0$ and

$$\int_0^t |dv_s| = v_1(t) + v_2(t) \tag{2.2}$$

Henceforth we suppose that v is right continuous. Then its variation over intervals $[0, t]$ and the canonical (v_1, v_2) can be found by taking limits (use

2. SAMPLE PATH INTEGRALS

$\delta_k^n = 2^{-n}k$, $u_+ = u \vee 0$ and $u_- = (-u)_+$):

$$\int_0^t |dv_s| = \lim_{n \to \infty} \sum_{k=0}^{2^n-1} |v(t\delta_{k+1}^n) - v(t\delta_k^n)| \tag{2.3}$$

$$v_1(t) = \lim_{n \to \infty} \sum_{k=0}^{2^n-1} (v(t\delta_{k+1}^n) - v(t\delta_k^n))_+ \tag{2.4}$$

and

$$v_2(t) = \lim_{n \to \infty} \sum_{k=0}^{2^n-1} (v(t\delta_{k+1}^n) - v(t\delta_k^n))_- \tag{2.5}$$

We suppose, furthermore, that v has finite variation. Then v_1 and v_2 are bounded right-continuous increasing functions on $\mathbb{R}_+$.

Then we define the Lebesgue-Stieltjes integral of ϕ with respect to v by

$$\int_0^\infty \phi\, dv_s = \int_0^\infty \phi_+\, dv_1(s) - \int_0^\infty \phi_-\, dv_1(s)$$
$$- \int_0^\infty \phi_+\, dv_2(s) + \int_0^\infty \phi_-\, dv_2(s)$$

The integral is well defined and finite if the integral of $|\phi|$ with respect to the variation of v, defined by

$$\int_0^\infty |\phi||dv_t| = \int_0^\infty |\phi|\, dv_1(t) + \int_0^\infty |\phi|\, dv_2(t) \tag{2.6}$$

is finite. The integral of ϕ with respect to v over a compact interval $[0, t]$ is defined by

$$\int_0^t \phi_s\, dv_s = \int_0^\infty \phi_s I_{[0,t]}\, ds \tag{2.7}$$

If the left-hand side of (2.6) is finite then the integral in (2.7) is finite, and as a function of t the integral in (2.7) is right continuous and its total variation is equal to the left-hand side of (2.6). It is convenient to use the notation $\phi \cdot v$ to denote the integral as a function of t—thus

$$\phi \cdot v_t = \int_0^t \phi_s\, dv_s$$

A function v is said to have locally finite variation if the variation over each compact interval $[0, t]$ is finite. Then for any $T > 0$ the function v^T defined by $v^T(t) = v(t \wedge T)$ has finite variation. We can still define $\phi \cdot v_t$ for each t by letting it equal

$$\int_0^t \phi_s\, dv_s^T \qquad (\text{for some } T, T \geq t) \tag{2.8}$$

because the value of the integral in (2.8) is the same for all T exceeding t. Then $\phi \cdot v_t$ is well defined and finite for all t if

$$\int_0^t |\phi_s||dv_s| < +\infty \qquad \text{for all } t \tag{2.9}$$

and the integral in (2.9) is the variation of $\phi \cdot v$ over the interval $[0, t]$.

Suppose now that $(\Omega, \mathcal{C}, \mathcal{P})$ is a probability space equipped with an increasing family of sub-σ-algebras of $\mathcal{C}, \mathcal{C}. = (\mathcal{C}_t, t \geq 0)$. Suppose that $V = (V_t: t \geq 0)$ is a real corlol random process which is $\mathcal{C}.$ adapted. We then define $\int_0^t |dV_s|$, $V_1(t)$ and $V_2(t)$ by the same equations, Eqs. (2.3)–(2.5), which were used for deterministic functions v. Then, for example, for each ω fixed,

$$\int_0^t |dV_s|(\omega) = \lim_{n \to \infty} \sum_{k=0}^{2^n - 1} |V(t\delta_{k+1}^n, \omega) - V(t\delta_k^n, \omega)|$$

V is said to have locally finite variation if

$$\mathcal{P}\left(\int_0^t |dV_s| < +\infty\right) = 1 \qquad \text{for each } t$$

For the remainder of this section we will assume that V has locally finite variation.

The corresponding facts for deterministic processes immediately imply that $\int_0^t |dV_s|$, $V_1(t)$ and $V_2(t)$ are increasing and right continuous in t and that Eqs. (2.1) and (2.2) hold, for a.e. ω. On the other hand, for each t fixed these quantities are defined as (pointwise) limits of sequences of $\mathcal{C}_t$ measurable random variables. Therefore $\int_0^t |dV_s|$, V_1 and V_2 are adapted random processes.

Next, if $\phi(\omega, t)$ is a real Borel measurable function of t for each ω then we define $\phi \cdot V_t(\omega)$ for each ω to be the Lebesgue-Stieltjes integral of the sample function $s \rightsquigarrow \phi(s, \omega)$ with respect to the sample function $s \rightsquigarrow V(s, \omega)$.

In order that the resulting integral be measurable and adapted, a condition on ϕ as a function of ω is needed. The appropriate condition is that of progressive measurability.

Definition. $\phi = (\phi(t, \omega))$ is an $\mathcal{C}.$ **progressively measurable** random process if, for each $t \geq 0$, $\phi(s, \omega)$ restricted to $[0, t] \times \Omega$ is a $\mathcal{B}([0, t]) \times \mathcal{C}_t$ measurable function of (s, ω).

2. SAMPLE PATH INTEGRALS 221

Proposition 2.1. Let V be a corlol, $\mathcal{A}_{\cdot}$-adapted random process and let ϕ be an $\mathcal{A}_{\cdot}$-progressively measurable process. Then the sample path Lebesgue-Stieltjes integral

$$\int_0^t |\phi(s,\omega)||dV(s,\omega)| \tag{2.10}$$

is an adapted random process which may be $+\infty$ for some (t,ω). If it is a.s. finite for each t then

$$\int_0^t \phi(s,\omega)\, dV(s,\omega) \tag{2.11}$$

is an adapted corlol random process, the variation of which is given by (2.10).

Proof: The only part of the proposition that does not follow immediately from the facts given above for deterministic functions is that the integrals in (2.10) and (2.11) are $\mathcal{A}_t$ measurable for each t. It is sufficient to consider the case when ϕ is bounded, because in the general case ϕ is the pointwise limit of a sequence of bounded ϕ_n with $|\phi_n| \leq |\phi|$, and the integrals of the ϕ_n converge pointwise to those of ϕ by the dominated convergence theorem. Fix t with $t \geq 0$. It suffices to prove (the stronger result) that the random variable U defined by

$$U(\omega) = \int_0^t G(s,\omega)\, dV(s,\omega) \qquad (t \text{ fixed})$$

is $\mathcal{A}_t$ measurable whenever G is a (not necessarily adapted) bounded, $\mathcal{B}([0,t]) \times \mathcal{A}_t$ measurable function of (s,ω). Let $\mathcal{H}$ denote the collection of such G for which U is $\mathcal{A}_t$ measurable. Clearly $\mathcal{H}$ contains functions of the form $I_{[a,b] \times D}$ where $0 \leq a < b \leq t$ and D is in $\mathcal{A}_t$. The class of such simple functions is closed under multiplication. Using the dominated convergence theorem it is also clear that $\mathcal{H}$ is closed under uniform and bounded monotone limits. Thus, by the monotone class theorem, $\mathcal{H}$ contains all bounded $\mathcal{B}([0,t]) \times \mathcal{A}_t$ measurable functions, as was to be proved. ∎

A random **counting process** $(C_t : t \geq 0)$ is defined to be an increasing right-continuous integer-valued random process such that the jumps $\Delta C_t = C_t - C_{t^-}$ are either zero or one for all ω and t. A Poisson random process is a counting process (N_t) such that $N_0 = 0$,

$$N_{t_2} - N_{t_1}, N_{t_3} - N_{t_2}, \ldots, N_{t_n} - N_{t_{n-1}}$$
are independent if $0 \leq t_1 < \cdots < t_n$,

$N_t - N_s$ is a Poisson random variable with mean $t - s$ whenever $t > s$

A Poisson random process is called an $\mathcal{Q}_*$ Poisson process if it is $\mathcal{Q}_*$ adapted and if $\{(N_s - N_t): s \geq t\}$ is independent of $\mathcal{Q}_t$ for each $t > 0$. Given an $\mathcal{Q}_*$ Poisson process N, we define $n_t = N_t - t$. Then for $s \geq t$, $n_s - n_t$ has mean zero and is independent of $\mathcal{Q}_t$ so that

$$E[n_s - n_t | \mathcal{Q}_t] = E[n_s - n_t] = 0$$

Therefore n is a martingale. The variation of n is given by

$$\int_0^t |dn_s| = N_t + t$$

If ϕ is a progressively measurable process then we compute that

$$\int_0^t \phi_s \, dN_s = \sum_{\substack{s \leq t \\ \Delta N_s = 1}} \phi_s$$

and

$$\int_0^t \phi_s \, dn_s = \sum_{\substack{s \leq t \\ \Delta N_s = 1}} \phi_s - \int_0^t \phi_s \, ds$$

For example,

$$\int_0^t N_s \, dN_s = \sum_{\substack{s \leq t \\ \Delta N_s = 1}} N_s = 1 + 2 + \cdots + N_t = \frac{N_t(N_t + 1)}{2}$$

and, since $\Delta N_s = 0$ for all but countably many values of s,

$$\int_0^t \Delta N_s \, dn_s = \sum_{\substack{s \leq t \\ \Delta N_s = 1}} \Delta N_s - \int_0^t \Delta N_s \, ds = N_t \tag{2.12}$$

Equation (2.12) shows that the integral of a bounded progressively measurable (in particular, adapted) random process with respect to a martingale is not necessarily a martingale. The reason is that the integrand and integrator can simultaneously jump in a positively correlated way. Integrals with respect to martingales are martingales if the integrands are "predictable" as we see in Section 4.

We defined the Lebesgue-Stieltjes integral $\phi \cdot V$ in this section only when ϕ and V are real valued. The definitions extend naturally to complex-valued ϕ and V by requiring the integral to be linear in ϕ and linear in V.

3. PREDICTABLE PROCESSES

A submartingale $(B_t, \mathcal{Q}_t : t \geq 0)$ has nonnegative drift by its very definition:

$$E[B_t - B_s | \mathcal{Q}_s] \geq 0 \quad \text{a.s. for } t \geq s$$

3. PREDICTABLE PROCESSES

Clearly an adapted increasing process A with $A_0 = 0$ has the same drift as B, i.e.,

$$E[B_t - B_s | \mathcal{A}_s] = E[A_t - A_s | \mathcal{A}_s] \qquad \text{a.s. when } t \geq s$$

if and only if M defined by $M_t = B_t - A_t$ is a martingale. Such an A gives rise to a representation of B as the sum of an increasing process A and a martingale M:

$$B_t = A_t + M_t \tag{3.1}$$

The representation (3.1) of B as the sum of an increasing process and a martingale is not unique. For example, a Poisson process N is a submartingale relative to its own family of σ-algebras $(\mathcal{A}_t^N)$, and $t + (N_t - t)$ and $N_t + 0$ are each representations of N with the form (3.1). In order to make the representation (3.1) unique, we need to further restrict A.

To understand how a canonical version of A can be identified, we will briefly consider discrete-time processes. Suppose that $(\mathcal{A}_n : n \in \mathbb{Z}_+)$ is an increasing family of sub-σ-algebras of $\mathcal{A}$. A discrete-time process $(A_n : n \in \mathbb{Z}_+)$ is called *predictable* if A_{n+1} is $\mathcal{A}_n$ measurable for each n. A predictable process is adapted, but the converse is not generally true. Suppose now that $(B_n, \mathcal{A}_n : n \in \mathbb{Z}_+)$ is a submartingale and that

$$B_n = A_n + M_n \qquad n \in \mathbb{Z}_+ \tag{3.2}$$

where

$(M_n, \mathcal{A}_n : n \in \mathbb{Z}_+)$ is a martingale, and $\qquad (3.3)$

A is predictable and $A_0 = 0$ $\qquad (3.4)$

Then

$$E[B_{n+1} - B_n | \mathcal{A}_n] = E[A_{n+1} - A_n | \mathcal{A}_n] + E[M_{n+1} - M_n | \mathcal{A}_n]$$
$$= A_{n+1} - A_n \geq 0 \qquad \text{a.s.}$$

Thus, it must be that A is an increasing process and

$$A_n = \sum_{k=0}^{n-1} E[B_{k+1} - B_k | \mathcal{A}_k] \qquad \text{a.s.} \tag{3.5}$$

Therefore, the submartingale B and conditions (3.2)–(3.4) uniquely determine A and M. In addition, given a submartingale B, if A is defined by Eq. (3.5) and M is then defined by Eq. (3.2), then A is increasing and conditions (3.3) and (3.4) are true.

The predictability (rather than merely the adaptedness) requirement on A was essential for making A unique. A similar concept plays an analogous role for continuous-time processes, which we turn to next.

Let $(\Omega, \mathcal{A}, \mathcal{P})$ be a probability space equipped with an increasing family $(\mathcal{A}_t : t \geq 0)$ of sub-σ-algebras of $\mathcal{A}$ satisfying the usual conditions.

Definition. Σ_p is the σ-algebra of subsets of $\mathbb{R}_+ \times \Omega$ generated by random processes ϕ (viewed as functions on $\mathbb{R}_+ \times \Omega$) of the form

$$\phi(t, \omega) = U(\omega) I_{(a, b]}(t) \tag{3.6}$$

where a and b are deterministic and U is an $\mathcal{A}_a$-measurable random variable. Sets in Σ_p are called **predictable** (or $\mathcal{A}_-$ predictable). A random process $H = (H(t, \omega))$ is called **predictable** if it is a Σ_p-measurable function of (t, ω).

The processes ϕ in this definition are $\mathcal{A}_-$ adapted, and it is crucial that they are left continuous—left continuity implies that $\phi(t, \omega)$ is determined by $\{\phi(s, \omega) : s < t\}$.

Proposition 3.1.

(a) Suppose that H is an $\mathcal{A}_-$-adapted process which is left continuous in t for each ω. Then H is predictable.
(b) Suppose that $\varepsilon > 0$ and that K is a corlol random process such that K_t is $\mathcal{A}_{t-\varepsilon}$ measurable for all t (set $\mathcal{A}_{t-\varepsilon} = \mathcal{A}_0$ if $t < \varepsilon$). Then K is predictable.

Proof: For each n let

$$H^n(t, \omega) = \sum_{k=0}^{n^2} H\left(\frac{k}{n}, \omega\right) I_{(k/n, (k+1)/n]}(t).$$

Then H^n is clearly predictable for each n and

$$\lim_{n \to \infty} H^n(t, \omega) = H(t, \omega) \qquad (t, \omega) \in \mathbb{R}_+ \times \Omega.$$

This proves part (a).

If $0 < \delta < \varepsilon$ then the process $K_{(t+\delta)-}$ is predictable by part (a). This process converges pointwise to the process K as δ decreases to zero, and this establishes part (b). ∎

Definition. Given any process X adapted to $\mathcal{A}_-$, if there exists a process A such that $A_0 = 0$, A is predictable, A has corlol sample paths of locally finite variation, and $X_t - A_t$ is a martingale relative to $(\mathcal{A}_-, \mathcal{P})$, then A is called the **predictable compensator** of X relative to $(\mathcal{A}_-, \mathcal{P})$.

3. PREDICTABLE PROCESSES

A corlol submartingale $(B_t, \mathcal{Q}_t : t \geq 0)$ is class D if the collection of random variables

$$\{ B_\tau : \tau \text{ is an } \mathcal{Q}_\cdot\text{-stopping time}\}$$

is uniformly integrable. The following is a celebrated theorem of P. A. Meyer (the second part of which was more or less conjectured earlier by Doob on the basis of the discrete-time analogue and many examples) about representations of the form (3.1).

Proposition 3.2.

(a) If a random process X has a predictable compensator, then it is unique in the sense that any two predictable compensators are equal to each other for all t, with probability one.

(b) If B is a corlol class D submartingale, then B has a predictable compensator A, and the sample paths of A are increasing with probability one.

Proof: Rather than give a proof (which can be found in [Dellacherie and Meyer, 1982] and in [Doob, 1984]) we only attempt to make plausible the existence of predictable compensators for a class D submartingale B. Given a positive integer M, define a process A^m by (note the similarity to Eq. (3.5)):

$$A_t^m = 0 \quad \text{for } 0 \leq t < 2^{-m}$$

$$A_t^m = \sum_{k=0}^{n-1} E\left[B_{(k+1)2^{-m}} - B_{k2^{-m}} | \mathcal{Q}_{k2^{-m}} \right]$$

$$\text{for } n2^{-m} \leq t < (n+1)2^{-m} \text{ and } n \geq 1$$

Then A^m has right-continuous increasing sample paths with probability one, and for each t, either $A_t^m = 0$ or A_t^m is $\mathcal{Q}_{t-2^{-m}}$ measurable. Thus, A^m is predictable. Furthermore, by the reasoning given for discrete-time processes,

$$(B_t - A_t^m, \mathcal{Q}_t : t = n2^{-m}, \quad n = 0,1,2, \ldots)$$

is a martingale. The assumption that B is class D implies that A^m converges in an appropriate sense to a process A which satisfies the required conditions. A proof of (a variation of) Proposition 3.2 which is based on this idea was given by K. M. Rao (1969). ∎

Remark: If A and A' are each predictable compensators of X, then their difference $(A_t - A'_t)$ is predictable, is a martingale, has initial value zero, and has corlol locally finite variation sample paths. Thus, assertion (a) of the proposition is equivalent to the fact that such martingales are equal to zero for all t, with probability one.

Meyer's original proof, as well as those of Doléans (1968) and Rao (1969) were based on a different characterization of compensator—they were called "natural increasing processes." Doléans (1967) first established equivalence of the two characterizations (see [Dellacherie and Meyer 1982, p. 126] and [Doob, 1984, p. 483]).

Example 1. Let C be a counting process adapted to $\mathcal{A}_\cdot$ such that EC_t is finite for each t. We will show that C has a predictable compensator. First, for fixed n, the process $(C_{t \wedge n})$ is a class D submartingale since the family of random variables

$$\{|C_{\tau \wedge n}| : \tau \text{ a stopping time}\}$$

is dominated by the integrable random variable $|C_0| + |C_n|$. Hence, $(C_{t \wedge n})$ has a predictable compensator A^n for each n—in particular $C_{t \wedge n} - A_t^n$ is a martingale. If $m < n$ then $C_{t \wedge m} - A_{t \wedge m}^n$ is clearly also a martingale, so by the uniqueness of the compensator of $(C_{t \wedge m})$ we have $A_t^m = A_{t \wedge m}^n$ for all t, with probability one or, in other words, $A_t^m = A_t^n$ for all t in $[0, m \wedge n]$, with probability one. Thus, a process A is well defined up to a set of zero probability by

$$A_t = \lim_{n \to \infty} A_t^n \quad \text{for all } t, \text{ a.s.}$$

and such a process A is the predictable compensator of C. If A has the representation

$$A_t = \int_0^t \lambda_s \, ds$$

for some nonnegative $\mathcal{A}_\cdot$ progressive process λ, then λ is called the **intensity** of C with respect to $(\mathcal{A}_\cdot, \mathcal{P})$.

Example 2. Recall that if N is an $\mathcal{A}_\cdot$ Poisson process then $N_t - t$ is a martingale, so that $A_t = t$ is the $\mathcal{A}_\cdot$ predictable compensator of N. Equivalently, the counting process N has intensity $\lambda = 1$, relative to $(\mathcal{A}_\cdot, \mathcal{P})$.

On the other hand, if $\mathcal{F}_t = \mathcal{A}$ for all t, then N is an $\mathcal{F}_\cdot$ predictable increasing process and $N_t - N_t$ is a martingale (trivially) relative to $(\mathcal{F}_\cdot, \mathcal{P})$. Thus, the unique $\mathcal{F}_\cdot$ predictable compensator of N is N itself, and N does not have an intensity relative to $\mathcal{F}_\cdot$ unless delta functions are allowed. Thus, the predictable compensator of a process depends crucially on the family of σ-algebras used in the definition.

4. ISOMETRIC INTEGRALS

Recall that a corlol martingale $(M_t: t \geq 0)$ is uniformly integrable if and only if there is an $\mathcal{A}_\infty$-measurable random variable M_∞ such that $M_t = E[M_\infty | \mathcal{A}_t]$. Let $\mathcal{M}^2$ denote the space of complex-valued corlol uniformly integrable martingales M such that $E|M_\infty|^2$ is finite. For M and N in $\mathcal{M}^2$ we define their inner product to be $EM_\infty \overline{N}_\infty$, which is simply the inner product of M_∞ and N_∞ in the space $L^2(\mathcal{A}_\infty)$ of square-integrable $\mathcal{A}_\infty$-measurable random variables. The correspondence between a process M in $\mathcal{M}^2$ and the random variable M_∞ in $L^2(\mathcal{A}_\infty)$ gives a one-to-one isometric mapping of $\mathcal{M}^2$ onto $L^2(\mathcal{A}_\infty)$. Since $L^2(\mathcal{A}_\infty)$ is a Hilbert space, so is $\mathcal{M}^2$.

If $M \in \mathcal{M}^2$ then it is not difficult to show that for $t > s$,

$$E[|M_t - M_s|^2 | \mathcal{A}_s] = E[|M_t|^2 | \mathcal{A}_s] - |M_s|^2 \quad \text{a.s.} \tag{4.1}$$

Since the left-hand side of this equation is nonnegative we conclude that $(|M_t|^2: t \geq 0)$ is an $\mathcal{A}_*$ submartingale. In addition, by Doob's L^2 inequality (Proposition 1.6), $\sup_t |M_t|^2$ has finite mean, and this random variable dominates

$$\{|M_T|^2: T \text{ is stopping time}\}$$

Therefore, $|M|^2$ is a class D submartingale so that it has a predictable compensator which we denote by $\langle M, M \rangle$.

If M and N are in $\mathcal{M}^2$ we define $\langle M, N \rangle$ to be the predictable compensator of $(M_t \overline{N}_t)$. Since

$$M\overline{N} = \tfrac{1}{4}\{|M + N|^2 - |M - N|^2 + i|-iM + N|^2 - i|iM + N|^2\} \tag{4.2}$$

the compensator $\langle M, N \rangle$ exists and

$$\langle M, N \rangle = \tfrac{1}{4}\{\langle M + N, M + N \rangle - \langle M - N, M - N \rangle$$
$$+ i\langle -iM + N, -iM + N \rangle - i\langle iM + N, iM + N \rangle\}$$

Note from the definition that $\langle N, M \rangle = \overline{\langle M, N \rangle}$.

For all complex λ,

$$0 \leq \langle M + \lambda N, M + \lambda N \rangle$$
$$= \langle M, M \rangle + 2\operatorname{Re}(\lambda \langle M, N \rangle) + |\lambda|^2 \langle N, N \rangle$$

The fact that the right-hand side is non-negative for all complex λ implies the useful inequality:

$$|\langle M, N \rangle_t|^2 \leq \langle M, M \rangle_t \langle N, N \rangle_t \quad \text{all } t, \text{ a.s.} \tag{4.3}$$

We now define a stochastic integral with integrator M in $\mathfrak{M}^2$ and integrand ϕ in $\mathcal{L}^2(M)$ where

$$\mathcal{L}^2(M) = \left\{ \phi : \phi \text{ is predictable and } \|\phi\|^2 = E\int_0^\infty |\phi_s|^2 \, d\langle M, M \rangle_s < +\infty \right\}$$

The result is a random process in $\mathfrak{M}^2$ which we denote by $\phi \cdot M$—that is,

$$\phi \cdot M_t = \int_0^t \phi_s \, dM_s$$

The construction closely parallels the construction of integrals with respect to the Wiener process, but now $\langle M, M \rangle$ plays the important role that $\langle w, w \rangle$ did for Wiener integrals.

Proposition 4.1. There is a unique mapping $\phi \rightsquigarrow \phi \cdot M$ from $\mathcal{L}^2(M)$ to $\mathfrak{M}^2$ which satisfies properties (a)–(c):

(a) If ϕ is a function in $\mathcal{L}^2(M)$ such that

$$\phi_t = \begin{cases} \phi_{t_k+} & \text{if } t_k < t \le t_{k+1}, \text{ some } k \\ 0 & \text{otherwise} \end{cases} \quad (4.4)$$

for some $t_0, \ldots, t_n$ (such a ϕ is called a **step function**) then

$$\phi \cdot M_\infty = \sum_{k=0}^{n-1} \phi_{t_k+} \left(M_{t_{k+1}} - M_{t_k} \right) \quad (4.5)$$

(b) (Linearity)

$$(\phi + \psi) \cdot M_\infty = \phi \cdot M_\infty + \psi \cdot M_\infty \quad (4.6)$$

(c) (Isometry)

$$E|\phi \cdot M_\infty|^2 = \|\phi\|^2$$

The additional properties hold

(c') (Isometry)

$$E\phi \cdot M_\infty \overline{\psi \cdot M_\infty} = E\int_0^\infty \phi_s \overline{\psi}_s \, d\langle M, M \rangle_s$$

(d) $\phi \cdot M_t = (I_{(0,t]}\phi) \cdot M_\infty$ a.s. for each t

(e) $\Delta(\phi \cdot M)_t = \phi_t \Delta M_t$ for all t, with probability one, where $\Delta Z_t = Z_t - Z_{t-}$ is the jump of a process Z at time t.

4. ISOMETRIC INTEGRALS

(*f*) Assume that M has locally finite variation and that $M \in \mathcal{M}^2$. Suppose that ϕ is a predictable process such that $\phi \cdot M$ is well defined as an isometric integral and as a random Lebesgue-Stieltjes integral. Then the two integrals agree for all t, with probability one.

(*g*) For M in $\mathcal{M}^2$ and ϕ in $\mathcal{L}^2(M)$, $\phi \cdot M$ is the unique process in $\mathcal{M}^2$ with initial value zero such that

$$\langle \phi \cdot M, N \rangle = \phi \cdot \langle M, N \rangle \tag{4.7}$$

for all N in $\mathcal{M}^2$. (The integral on the right-hand side of (4.7) is a Lebesgue-Stieltjes sample-path integral, and we consider two processes which agree for all t with probability one to be the same.)

Proof: If ϕ is a step function then the set $t_0, \ldots, t_n$ in (4.4) is not unique. For example, if (4.4) is true for $t_0, \ldots, t_n$ then it is also true for any sequence $t'_0, t'_1, \ldots, t'_{n'}$ for which $t_0, \ldots, t_n$ is a subsequence. In spite of this nonuniqueness, the sum in (4.5) does not depend on the choice of $t_0, \ldots, t_n$, so we can use (*a*) to define $\phi \cdot M_\infty$ for a step function ϕ. If ψ is another step function in $\mathcal{L}^2(M)$, then a single sequence $t_0, \ldots, t_n$ can be found so that (4.3) and the corresponding statement for ψ are both true. The linearity property is easily established in this case. Next,

$$|\phi \cdot M_\infty|^2 = \sum_{k=0}^{n-1} \sum_{j=0}^{n-1} \phi_{t_k+} \bar{\phi}_{t_j+} (M_{t_{k+1}} - M_{t_k})\overline{(M_{t_{j+1}} - M_{t_j})}$$

Let D_{kj} denote the kjth term on the right-hand side. The fact that ϕ_{t_k+} is $\mathcal{Q}_{t_k}$ measurable for each k implies that for $k > j$,

$$ED_{kj} = E\left[\phi_{t_k+} \bar{\phi}_{t_j+} \overline{(M_{t_{j+1}} - M_{t_j})} E\left[M_{t_{k+1}} - M_{t_k} | \mathcal{Q}_{t_k} \right] \right] = 0$$

and similarly $ED_{kj} = 0$ for $k < j$. Also, using (4.1),

$$ED_{kk} = E\left[|\phi_{t_k+}|^2 E\left[|M_{t_{k+1}} - M_{t_k}|^2 | \mathcal{Q}_{t_k} \right] \right]$$
$$= E\left[|\phi_{t_k+}|^2 E\left[\langle M, M \rangle_{t_{k+1}} - \langle M, M \rangle_{t_k} | \mathcal{Q}_{t_k} \right] \right]$$
$$= E\left[|\phi_{t_k+}|^2 (\langle M, M \rangle_{t_{k+1}} - \langle M, M \rangle_{t_k}) \right].$$

Therefore

$$E|\phi \cdot M_\infty|^2 = E \sum_{k=0}^{n-1} |\phi_{t_k+}|^2 (\langle M, M \rangle_{t_{k+1}} - \langle M, M \rangle_{t_k})$$
$$= E \int_0^\infty |\phi_t|^2 \, d\langle M, M \rangle_t = \|\phi\|^2.$$

Thus, properties (*b*) and (*c*) are both true when ϕ and ψ are step functions.

So far, we have only defined a random variable $\phi \cdot M_\infty$ for each step function ϕ. However, because of the one-to-one correspondence between $\mathfrak{M}^2$ and the space of square integrable $\mathcal{A}_\infty$-measurable random variables discussed at the beginning of the section, this serves to define the process $\phi \cdot M$ in $\mathfrak{M}^2$ for any step function ϕ. The case of general ϕ will be considered next.

Let L denote the set of functions ϕ in $\mathcal{L}^2(M)$ such that there exists a sequence of step functions ϕ^n in $\mathcal{L}^2(M)$ such that

$$\|\phi - \phi^n\|^2 \xrightarrow[n \to \infty]{} 0.$$

It is not difficult to verify that L is a closed subspace of $\mathcal{L}^2(M)$. In particular, L is closed under both uniform and bounded monotone convergence, so by the monotone class theorem, L contains all of the bounded real-valued functions in M. From this it follows that $L = \mathcal{L}^2(M)$.

We have thus proven that for any ϕ in $\mathcal{L}^2(M)$, there exists a sequence of step functions ϕ^n in $\mathcal{L}^2(M)$ such that $\|\phi^n - \phi\| \to 0$. In order that the stochastic integral satisfy properties (a)–(c) we are forced to have

$$E\big[|\phi \cdot M_\infty - \phi^n \cdot M_\infty|^2\big] = E\big[|(\phi - \phi^n) \cdot M_\infty|^2\big] = \|\phi - \phi^n\|^2 \xrightarrow[n \to \infty]{} 0$$

and thus, if $\phi \cdot M$ exists, we must have

$$\phi \cdot M_\infty = \lim \text{ in q.m. } \phi^n \cdot M_\infty \tag{4.8}$$

This proves the assertion that the map satisfying properties (a)–(c) is unique, if it exists.

Continuing now with the existence proof, we attempt to define $\phi \cdot M$ by Eq. (4.8). To see that this works we need only check two things: First, (use $|a + b|^2 \leq 2|a|^2 + 2|b|^2$),

$$E\big[|\phi^n \cdot M_\infty - \phi^m \cdot M_\infty|^2\big] = \|\phi^n - \phi^m\|^2$$
$$\leq 2\big(\|\phi^n - \phi\|^2 + \|\phi^m - \phi\|^2\big) \xrightarrow[m,n \to \infty]{} 0$$

so that the sequence of random variables $\phi^n \cdot M_\infty$ is Cauchy in 2-mean and hence the limit in (4.8) exists. Second, if (ψ^n) is another sequence of step functions in $\mathcal{L}^2(M)$ such that $\|\phi - \psi^n\|$ tends to zero as n tends to infinity, then

$$E\big[|\psi^n \cdot M_\infty - \phi \cdot M_\infty|^2\big] \leq 2E\big[|\psi^n \cdot M_\infty - \phi^n \cdot M_\infty|^2$$
$$+ |\phi^n \cdot M_\infty - \phi \cdot M_\infty|^2\big] \tag{4.9}$$

and since

$$E\big[|\psi^n \cdot M_\infty - \phi^n \cdot M_\infty|^2\big] = \|\psi^n - \phi^n\|^2 \leq 2\big(\|\psi^n - \phi\|^2 + \|\phi^n - \phi\|^2\big)$$

4. ISOMETRIC INTEGRALS

the left-hand side of (4.9) tends to zero as n tends to infinity. Thus $\psi^n \cdot M$ converges in 2-mean to $\phi \cdot M$, so our definition of $\phi \cdot M$ does not depend on the sequence ϕ^n chosen. Thus $\phi \cdot M$ is well defined for any ϕ in $\mathcal{L}^2(M)$.

If ϕ^n and ψ^n are step functions then

$$(\phi^n + \psi^n) \cdot M_\infty = \phi^n \cdot M_\infty + \psi^n \cdot M_\infty \tag{4.10}$$

and if the functions are chosen so that $\|\phi - \phi^n\|$, $\|\psi - \psi^n\|$, and hence also $\|\phi + \psi - \phi^n - \psi^n\|$ tend to zero as n tends to infinity, we obtain (4.6) from (4.10) by taking the limit in 2-mean of each term in (4.10). Property (c) is easily verified for any ϕ in $\mathcal{L}^2(M)$ by a similar argument.

The isometry property (c') is implied by property (c) and the identities obtained by substituting $\phi \cdot M$ and $\psi \cdot M$ or ϕ and ψ in for M and N in Eq. (4.2).

For any s and t,

$$E[M_s | \mathcal{C}_t] = M_{s \wedge t} \quad \text{a.s.}$$

Property (d) is thus easy to establish for a step function ϕ:

$$\phi \cdot M_t \triangleq E[\phi \cdot M_\infty | \mathcal{C}_t] = \sum_{k=0}^{n-1} \phi_{t_k +} (M_{t_{k+1} \wedge t} - M_{t_k \wedge t}) \quad \text{a.s.}$$
$$= (I_{(0,t]} \phi) \cdot M_\infty \quad \text{a.s.}$$

To prove (d) in general we chose a sequence of step functions ϕ^n with $\|\phi - \phi^n\|$ tending to zero. Then

$$E[|\phi \cdot M_t - \phi^n \cdot M_t|^2] \leq E[|\phi \cdot M_\infty - \phi^n \cdot M_\infty|^2] = \|\phi - \phi^n\|^2$$

so that

$$\phi \cdot M_t = \lim_{n \to \infty} \text{in q.m.} \, \phi^n \cdot M_t = \lim_{n \to \infty} \text{in q.m.} \, I_{(0,t]} \phi^n \cdot M_\infty$$
$$= I_{(0,t]} \phi \cdot M_\infty \quad \text{a.s.}$$

which establishes (d).

Property (e) is certainly true if ϕ is a step function. Suppose that (e) is true for each ϕ^n in a sequence such that ϕ^n converges either montonically or uniformly to a function ϕ in $\mathcal{L}^2(M)$. Then

$$\sup_t |\Delta(\phi \cdot M)_t - \Delta(\phi^n \cdot M)_t| \leq 2 \sup_t |\phi^n \cdot M_t - \phi \cdot M_t| \to 0 \quad \text{a.s.}$$

and

$$\phi_t^n \Delta M_t \to \phi_t \Delta M_t \quad \text{for all } t, \omega$$

Thus, by the monotone class theorem, (e) is true for all bounded real functions ϕ in $\mathcal{L}^2(M)$. Appealing to the approximation argument again, one easily sees that (e) is true for all ϕ in $\mathcal{L}^2(M)$.

Property (f) is proved by the same method used to prove property (e). To prove Eq. (4.7) we only need to establish that U defined by

$$U_t = (\phi \cdot M_t)\overline{N}_t - \phi \cdot \langle M, N \rangle_t$$

is a martingale. Since $M\overline{N} - \langle M, N \rangle$ is a martingale it is easy to verify that U^n defined by

$$U_t^n = (\phi^n \cdot M_t)\overline{N}_t - \phi^n \cdot \langle M, N \rangle_t$$

is a martingale for each n, where ϕ^n is a sequence of bounded step functions converging in $\mathcal{L}^2(M)$ to ϕ. Now by the Schwarz inequality and (4.3),

$$E|U_t - U_t^n| \le E|((\phi - \phi^n) \cdot M_t)\overline{N}_t| + E|(\phi - \phi^n) \cdot \langle M, N \rangle_t|$$
$$\le \left(E|(\phi - \phi^n) \cdot M_t|^2 E|N_t|^2\right)^{1/2}$$
$$+ E\left[\left(|\phi - \phi^n|^2 \cdot \langle M, M \rangle_t \langle N, N \rangle_t\right)^{1/2}\right]$$
$$\le 2\|\phi - \phi^n\|\left(E|N_\infty^2|\right)^{1/2}$$

Thus, U_t^n converges to U_t in 1-mean for each t. To see that this implies that U is a martingale, note that for $s > t$

$$E|E[U_s|\mathcal{C}_t] - U_t^n| = E|E[U_s - U_s^n|\mathcal{C}_t]|$$
$$\le E|U_s - U_s^n| \xrightarrow[n \to \infty]{} 0$$

so that U_t^n converges to $E[U_s|\mathcal{C}_t]$, as well as to U_t, in 1-mean. Thus U_t is a.s. equal to $E[U_s|\mathcal{C}_t]$, so U is a martingale and (4.7) is established.

Finally, if Z is any process in $\mathcal{M}^2$ with $Z_0 = 0$ such that $\langle Z, N \rangle = \phi \cdot \langle M, N \rangle$ for all N in $\mathcal{M}^2$, then in view of (4.7),

$$\langle \phi \cdot M - Z, N \rangle = \phi \cdot \langle M, N \rangle - \phi \cdot \langle M, N \rangle = 0$$

for all N in $\mathcal{M}^2$. Choosing $N = \phi \cdot M - Z$ we have $\langle \phi \cdot M - Z, \phi \cdot M - Z \rangle = 0$ so $Z = \phi \cdot M$. This completes the proof of (g), and hence the proof of the proposition. ∎

We can apply similar reasoning to Stieltjes-Lebesgue integrals:

Proposition 4.2. Let V be a (not necessarily square-integrable) martingale and a process with integrable variation and let H be a bounded predictable process. Then

$$\int_0^t H_s \, dV_s$$

defined as a Lebesgue-Stieltjes integral for each (t, ω), is a martingale.

Proof: The result is obvious if H is a step function. In general there is a sequence H^n of step functions such that

$$E \int_0^\infty |H_s^n - H_s| |dV_s| \xrightarrow[n \to \infty]{} 0.$$

Then

$$E \left| \int_0^t H_s \, dV_s - \int_0^t H_s^n \, dV_s \right| \leq E \int_0^t |H_s - H_s^n| |dV_s| \xrightarrow[n \to \infty]{} 0.$$

Thus, the martingales $\int_0^t H_s^n \, dV_s$ converge to $\int_0^t H_s \, dV_s$ in 1-mean for each t, which yields the desired result. ∎

5. SEMIMARTINGALE INTEGRALS

Let X have a representation

$$X_t = X_0 + M_t + A_t$$

where $M \in \mathcal{M}^2$ and A is an adapted corlol process with finite variation. This representation is not unique, but given any other such representation $X_t = X_0 + M_t' + A_t'$, the process $M - M'$ is equal to $A' - A$ and therefore it is in $\mathcal{M}^2$ and is also a process with finite variation. Given a bounded predictable process ϕ we define the stochastic integral $\phi \cdot X$ by

$$\phi \cdot X_t = \phi \cdot M_t + \phi \cdot A_t.$$

By part (f) of Proposition 4.1 this definition is not ambiguous even though the representation for X is not unique.

The class of processes X with the representation above is nearly closed under sufficiently smooth nonlinear transformations. The main reason it is not is that $f(X_t)$ need not be integrable, even though X_t is. In this section we introduce the technique of localizing processes to a random time interval, and this will allow us to present an important nonlinear transformation formula in the next section.

Given a random process $X = (X_t: t \geq 0)$ and a stopping time T, we define X^T, the process X stopped at time T, by

$$X_t^T = \begin{cases} 0 & \text{if } T = 0 \\ X_t & \text{if } 0 \leq t \leq T \quad \text{and } T > 0 \\ X_T & \text{if } t \geq T \quad \text{and } T > 0 \end{cases}$$

or equivalently, $X_t^T = I_{\{T > 0\}} X_{t \wedge T}$. A random process M is called a **local martingale** (resp. **local square integrable martingale**) if it is corlol and if there is a sequence (T_n) of stopping times which is increasing (i.e., $T_n(\omega) \leq T_{n+1}(\omega)$ a.s.) and which satisfies $T_n \xrightarrow{n \to \infty} +\infty$ a.s., such that M^{T_n} is a martingale (resp. square integrable martingale) for each n. Similarly, a random process H is called locally bounded if there is a sequence $T_n \uparrow +\infty$ of stopping times such that H^{T_n} is a bounded random process for each n.

Definition. A **semimartingale** relative to the family $(\mathcal{Q}_t)$ is any random process X which can be written as

$$X_t = X_0 + M_t + A_t \qquad \text{for } t \geq 0, \text{ a.s.} \tag{5.1}$$

where X_0 is an $\mathcal{Q}_0$-measurable random variable, M is a local square integrable martingale with $M_0 = 0$, and A is a right-continuous adapted process whose paths are of locally finite variation with $A_0 = 0$. We call (5.1) a semimartingale representation of X.

Remark: It is more common in the literature to only require that M be a local martingale (rather than a local square integrable martingale) in the definition above. However, local martingales are semimartingales as we have defined them [Delacherie and Meyer, 1982] so our definition of a semimartingale is equivalent to the common definition. Note that semimartingales have corlol sample paths since we require local square integrable martingales to have corlol sample paths.

Let X be a semimartingale and let H be a locally bounded predictable process. We call a stopping time R good relative to (X, H) if H is bounded on the interval $(0, R]$ and if X stopped at R has a representation

$$X_t^R = X_0 + M_t + A_t \qquad \text{for } t \geq 0, \text{ a.s.}$$

where $M \in \mathcal{M}^2$ and A has finite variation. For any good stopping time R, the integral $H \cdot X^R$ of H with respect to X^R can thus be defined by the procedure given at the beginning of the section. If S is another good stopping time, then so is $R \wedge S$ and

$$H \cdot X_t^S = H \cdot X_t^{R \wedge S} = H \cdot X_t^R \qquad \text{for } t \leq R \wedge S, \text{ a.s.}$$

5. SEMIMARTINGALE INTEGRALS

There exists a sequence $R_n \uparrow \infty$ of good stopping times, and then

$$(H \cdot X^{R_{n+1}})^{R_n} = H \cdot X^{R_n} \qquad \text{for } t \leq R_n, \text{a.s.}$$

Thus, the processes $H \cdot X^{R_n}$ paste together into a unique process $H \cdot X$ such that

$$H \cdot X_t = H \cdot X_t^{R_n} \qquad 0 \leq t \leq R_n, \text{a.s.}$$

If $S_n \uparrow \infty$ is another sequence of good stopping times then

$$H \cdot X_t = H \cdot X_t^{R_n} = H \cdot X_t^{S_n} \qquad \text{for } t \leq R_n \wedge S_n$$

so that the definition of $H \cdot X$ does not depend on which particular sequence of good stopping times is used.

So defined, the semimartingale integral $H \cdot X$ inherits many of the properties which are common to the isometric and Lebesgue-Stieltjes integrals. For example,

(a) $\Delta(H \cdot X)_t = H_t \Delta X_t$, all t, a.s.
(b) If X is a local martingale or a process of finite variation, then $H \cdot X$ has the same property.
(c) We can obtain the integral $H \cdot X$ as a limit of sums (see [Dellacherie and Meyer, 1982] for more information):

Proposition 5.1. Let X be a semimartingale and let H^n be predictable processes such that $H^n(t, \omega)$ converges to zero for each (t, ω) as n tends to infinity and such that there exists a sequence of stopping times $T_k \uparrow +\infty$ and constants c_k such that $|H_t^n| \leq c_k$ for $0 \leq t < T_k$ and all n. Then

$$\sup_{0 \leq t \leq T} |H^n \cdot X_t| \xrightarrow[n \to \infty]{\text{in p.}} 0 \qquad (5.2)$$

for any finite T. If U is an adapted corlol process then for each $t > 0$ (write δ_k^n for $k2^{-n}$):

$$\int_0^t U_{s-} \, dX_s = \lim_{n \to \infty} \text{in p.} \sum_{0 \leq k < 2^n} U_{t\delta_k^n}(X_{t\delta_{k+1}^n} - X_{t\delta_k^n}) \qquad (5.3)$$

Proof: First suppose that the H^n are uniformly bounded by a constant c and that $X = X_0 + M + A$ where $M \in \mathfrak{M}^2$ and A is an adapted corlol process with finite variation. Then by Doob's L^2 inequality,

$$E \sup_t |H^n \cdot M_t|^2 \leq 4E \int_0^\infty |H_t^n|^2 \, d\langle M, M \rangle_t \xrightarrow[n \to \infty]{} 0$$

so that $\sup_t |H^n \cdot M_t|$ converges to zero in 2-mean, and

$$\sup_t |H^n \cdot A_t| \le \int_0^\infty |H^n_t||dA_t| \xrightarrow[n\to\infty]{} 0 \quad \text{a.s.}$$

We thus have that $\sup_t |H^n \cdot X_t|$ converges to zero in probability as n tends to infinity.

In the general case there are stopping times $R^k \uparrow \infty$ such that

$$X_{t \wedge R_k} = A_t^k + M_t^k \quad \text{for } t \ge 0, \text{ a.s.}$$

where A^k has finite variation and $M^k \in \mathfrak{M}^2$. Then $H^n \cdot X_t = H^n \cdot X_{t \wedge R_k}$ for $t < R_k$ so that

$$\sup_{0 \le t < T_k \wedge R_k} H^n \cdot X_t \xrightarrow[n\to\infty]{\text{p.}} 0$$

by our conclusion in the special case. This implies Eq. (5.2). Applying what was just proved to

$$H_s^n = U_{s-} I_{\{s \le t\}} - \sum_{0 \le k < 2^n} U_{t\delta_k^n} I_{(t\delta_k^n, t\delta_{k+1}^n]}(s),$$

$c_k = 2k$ and $T_k = \inf\{t : |U_t| \ge k\}$ yields Eq. (5.3). ∎

6. QUADRATIC VARIATION AND THE CHANGE OF VARIABLE FORMULA

Let X and Y be semimartingales. We define their co-quadratic variation process $[X, Y]$ by

$$[X, Y]_t = \lim_{n\to\infty} \text{in p. } X_0 \overline{Y}_0 + \sum_{0 \le k < 2^n} \left(X_{t\delta_{k+1}^n} - X_{t\delta_k^n} \right) \left(\overline{Y}_{t\delta_{k+1}^n} - \overline{Y}_{t\delta_k^n} \right) \tag{6.1}$$

To see that the limit exists, note that

$$\left(X_{t\delta_{k+1}^n} - X_{t\delta_k^n} \right) \left(\overline{Y}_{t\delta_{k+1}^n} - \overline{Y}_{t\delta_k^n} \right)$$
$$= \left(X_{t\delta_{k+1}^n} \overline{Y}_{t\delta_{k+1}^n} - X_{t\delta_k^n} \overline{Y}_{t\delta_k^n} \right) - X_{t\delta_k^n} \left(\overline{Y}_{t\delta_{k+1}^n} - \overline{Y}_{t\delta_k^n} \right)$$
$$- \overline{Y}_{t\delta_k^n} \left(X_{t\delta_{k+1}^n} - X_{t\delta_k^n} \right)$$

Thus, the right-hand side of (6.1) is the limit of

$$X_t \overline{Y}_t - \sum_{0 \le k < 2^n} X_{t\delta_k^n} \left(\overline{Y}_{t\delta_{k+1}^n} - \overline{Y}_{t\delta_k^n} \right) - \sum_{0 \le k < 2^n} \overline{Y}_{t\delta_k^n} \left(X_{t\delta_{k+1}^n} - X_{t\delta_k^n} \right)$$

as n tends to infinity. Proposition 5.1 implies that the limit exists in the

6. QUADRATIC VARIATION AND THE CHANGE OF VARIABLE FORMULA

sense of convergence in probability, and it is given by

$$[X,Y]_t = X_t \overline{Y}_t - \int_0^t X_{s-} \, d\overline{Y}_s - \int_0^t \overline{Y}_{s-} \, dX_s \qquad (6.2)$$

This equation shows that we can choose a corlol version of $[X,Y]$, which we always will. Since Eq. (6.1) defines $[X,Y]_t$ up to a set of zero probability for each t, and since a corlol process is determined by its values at a countable set of times, any two corlol versions of $[X,Y]$ are equal for all t, with probability one.

The process $[X,Y]$ is linear in X and $[X,Y] = \overline{[Y,X]}$, so (ignoring a set of zero probability)

$$[X,Y] = \tfrac{1}{4}\{[X+Y, X+Y] - [X-Y, X-Y]$$
$$+ i[-iX+Y, -iX+Y] - i[iX+Y, iX+Y]\} \qquad (6.3)$$

Processes of the form $[Z,Z]_t$ are increasing in t, and thus $[X,Y]$ is a locally finite variation process. The continuous part, $[X,Y]^c$, of the process $[X,Y]$ is defined by

$$[X,Y]_t^c = [X,Y]_t - [X,Y]_0 - \sum_{0 < s \le t} \Delta[X,Y]_s$$

Proposition 6.1. Let X and Y be semimartingales.

(a) $\Delta[X,Y]_t = \Delta X_t \Delta \overline{Y}_t$ for all t, with probability one.
(b) $[X,Y]_t^c = [X,Y]_t - X_0 \overline{Y}_0 - \sum_{0 < s \le t} \Delta X_s \Delta \overline{Y}_s$ for all t, with probability one.
(c) If either X or Y have locally finite variation sample paths then

$$[X,Y]_t = X_0 \overline{Y}_0 + \sum_{0 < s \le t} \Delta X_s \Delta \overline{Y}_s$$

and $[X,Y]^c = 0$.
(d) If H is a locally bounded predictable process then $[H \cdot X, Y] = H \cdot [X,Y]$ and $[H \cdot X, Y]^c = H \cdot [X,Y]^c$.
(e) If X and Y are in $\mathcal{M}^2$ (resp. are local martingales) then $X\overline{Y} - [X,Y]$ is a uniformly integrable martingale (resp. a local martingale).

Proof: By Eq. (6.2),

$$\Delta[X,Y]_t = \Delta(X_t \overline{Y}_t) - X_{t-} \Delta \overline{Y}_t - \overline{Y}_{t-} \Delta X_t = \Delta X_t \Delta \overline{Y}_t$$

which proves (a), and (b) is an immediate consequence. The first equation in (c) can be readily deduced from the definition (6.1) of $[X,Y]$, and then the fact that $[X,Y]^c = 0$ is an immediate consequence. It is easy to verify that

$[H \cdot X, Y] = H \cdot [X, Y]$ when H is a bounded predictable step function and the general case can then be proved by an approximation argument similar to those of the previous section. Then

$$\Delta[H \cdot X, Y]_t = \Delta(H \cdot [X, Y]_t) = H_t \Delta[X, Y]_t$$

and so the second assertion in part (d) is true.

If X and Y are each local martingales, then the stochastic integrals in Eq. (6.2) are also local martingales, and so $X\overline{Y} - [X, Y]$ is also a local martingale. Now, suppose that X is in $\mathfrak{M}^2$. Then by what was just proved, $|X|^2 - [X, X]$ is a local martingale. Thus, there exist stopping times $R_n \uparrow \infty$ so that for each n,

$$|X_{t \wedge R_n}|^2 - [X, X]_{t \wedge R_n} \tag{6.4}$$

is a uniformly integrable martingale. It has initial value zero so that

$$E|X_{t \wedge R_n}|^2 = E[X, X]_{t \wedge R_n} \tag{6.5}$$

Since $|X|^2$ is a uniformly integrable submartingale the variables $|X_{t \wedge R_n}|^2$ converge in 1-mean to $|X_t|^2$ as n tends to infinity for fixed $t, t \leq +\infty$. By Eq. (6.5), $[X, X]_{t \wedge R_n}$ also converges in 1-mean, to $[X, X]_t$. Thus, for each $t, t \leq +\infty$, the variable in (6.4) converges in 1-mean to $|X_t|^2 - [X, X]_t$. Hence, this limit process is also a uniformly integrable martingale. Summarizing, if X is in $\mathfrak{M}^2$ then $|X_t|^2 - [X, X]_t$ is a uniformly integrable martingale. The analogous fact for $X\overline{Y} - [X, Y]$ can now be deduced using (6.3), and the proposition is proved. ∎

If M is in $\mathfrak{M}^2$ then both $|M^2| - [M, M]$ and $|M^2| - \langle M, M \rangle$ are martingales. Thus $[M, M] - \langle M, M \rangle$ is a martingale. Therefore, $\langle M, M \rangle$ is the (unique) predictable compensator of the process $[M, M]$. So far we have defined $\langle X, X \rangle$ only for X in $\mathfrak{M}^2$.

To define $\langle X, X \rangle$ more generally we can extend the definition of the predictable compensator A of a process X by relaxing the condition that $X - A$ be a martingale to the condition that $X - A$ be a local martingale. With this more general definition, the predictable compensator of a process X continues to be unique (if it exists). A sufficient condition for the existence of a predictable compensator for X is that X be a corlol process with locally integrable variation.

If X and Y are semimartingales, we define $\langle X, Y \rangle$ to be the predictable compensator, if it exists, of $[X, Y]$. The process $\langle X, Y \rangle$ exists, for example, if X and Y are locally square integrable, for then $[X, Y]$ is locally integrable. Note that $[X, Y] = \langle X, Y \rangle$ whenever $[X, Y]$ is predictable and $X_0 Y_0 = 0$—this is true, for example, whenever $[X, Y]$ is continuous and $X_0 Y_0 = 0$.

6. QUADRATIC VARIATION AND THE CHANGE OF VARIABLE FORMULA

Examples. Let w be a Wiener martingale. Then, as shown in Proposition 2.3.4, $[w, w]_t = t$. Since this is a predictable increasing function with initial value zero, it is equal to its predictable compensator. Thus, $\langle w, w \rangle_t = t$.

For another example, let N be an $\mathscr{Q}_*$ Poisson process and let n be the martingale $n_t = N_t - t$. Then both N and n have locally finite variation sample paths, $\Delta N_s = \Delta n_s$ for all s, and ΔN_s takes values zero and one. Therefore

$$[N, N]_t = [n, n]_t = \sum_{0 < s \leq t} |\Delta N_s|^2 = \sum_{0 < s \leq t} \Delta N_s = N_t$$

Since the predictable compensator of N is t, it follows that $\langle N, N \rangle_t = \langle n, n \rangle_t = t$.

Equation (6.2) (which is virtually the definition of $[X, Y]$) yields that for semimartingales X and Y,

$$X_t Y_t = \int_0^t X_{s-} \, dY_s + \int_0^t Y_{s-} \, dX_s + [X, \overline{Y}]_t \tag{6.6}$$

We call Eq. (6.6) *Itô's product formula*. It is an important special case of a generalization of the next proposition to functions f on $\mathbb{C}^n$.

Proposition 6.2 (Generalized Itô's Formula). Let f be a twice continuously differentiable function on $\mathbb{R}$ and let X be a real semimartingale, or let f be an analytic function and let X be a possibly complex-valued semimartingale. Then $f(X_t)$ is also a semimartingale and

$$\begin{aligned} f(X_t) = f(X_0) &+ \int_0^t f'(X_{s-}) \, dX_s \\ &+ \sum_{s \leq t} \left(f(X_s) - f(X_{s-}) - f'(X_{s-}) \Delta X_s \right) \\ &+ \frac{1}{2} \int_0^t f''(X_s) \, d[X, \overline{X}]_s^c \end{aligned} \tag{6.7}$$

Proof: Since $f(X_{s-}) + f'(X_{s-}) \Delta X_s$ is the first-order Taylor's approximation to $f(X_s)$, Taylor's approximation formula yields that for $t \geq 0$,

$$\sum_{s \leq t} |f(X_s) - f(X_{s-}) - f'(X_{s-}) \Delta X_s| \leq K(X, f, t) \sum_{s \leq t} |\Delta X_s|^2 \quad \text{a.s.} \tag{6.8}$$

where K is random and is defined by

$$K(X, f, t) = \tfrac{1}{2} \max \left\{ |f''(u)| : |u| \leq \sup_{s \leq t} |X_s| \right\}$$

Thus, the sum of jumps term on the right-hand side of Eq. (6.7) converges absolutely for each finite t, with probability one, and its variation is locally finite. The right-hand side of Eq. (6.7) is therefore a semimartingale.

Equation (6.7) is certainly true if $f(x) = 1$ or if $f(x) = x$. The equation when $f(x) = x^n$ becomes

$$X_t^n = X_0^n + n\int_0^t X_{s-}^{n-1} dX_s + \sum_{s \leq t} \left(X_s^n - X_{s-}^n - nX_{s-}^{n-1}\Delta X_s \right)$$
$$+ \frac{n(n-1)}{2}\int_0^t X_{s-}^{n-2} d[X, \overline{X}]_s^c \tag{6.9}$$

For $n = 2$ this equation reduces to the product formula (6.6) and it is hence true in that case. Now, argue by induction and suppose that for some $n, n \geq 2$, that Eq. (6.9) for X^n is correct. Then, since $X^{n+1} = X(X^n)$, the product formula yields

$$X_t^{n+1} = \int_0^t X_{s-} dX_s^n + \int_0^t X_{s-}^n dX_s + [X, \overline{X^n}]_t \tag{6.10}$$

Now, using Eq. (6.9),

$$[X, \overline{X^n}]_t = X_0^{n+1} + n\int_0^t X_{s-}^{n-1} d[X, \overline{X}]_s$$
$$+ \sum_{s \leq t} \Delta X_s \left(X_s^n - X_{s-}^n - nX_{s-}^{n-1}\Delta X_s \right)$$
$$= X_0^{n+1} + n\int_0^t X_{s-}^{n-1} d[X, \overline{X}]_s^c + \sum_{s \leq t} \Delta X_s \left(X_s^n - X_{s-}^n \right)$$

and

$$\int_0^t X_{s-} dX_s^n = n\int_0^t X_{s-}^n dX_s + \sum_{s \leq t} X_{s-} \left(X_s^n - X_{s-}^n - nX_{s-}^{n-1}\Delta X_s \right)$$
$$+ \frac{n(n-1)}{2}\int_0^t X_{s-}^{n-1} d[X, \overline{X}]_s^c$$

Substituting these expressions into the right-hand side of Eq. (6.10), we obtain that Eq. (6.9) is true for n replaced by $n + 1$. Therefore, Eq. (6.9) is true for nonnegative integers n.

Since each side of Eq. (6.7) is linear in f, it is thus true whenever f is a polynomial. In the general case, there exists a sequence of polynomials f_n so that f_n and the first and second derivatives of f_n converge to f and its first and second derivatives, uniformly on each bounded set. Then Eq. (6.7) is valid for each f_n and, by Proposition 5.1 and the estimate (6.8) with f replaced by $f - f_n$, we can take a term by term limit in probability to deduce that Eq. (6.7) is valid for f. ∎

7. SEMIMARTINGALE EXPONENTIALS AND APPLICATIONS

Given a semimartingale X, the exponential process of X, $\mathcal{E}(X)$, is defined by

$$\mathcal{E}(X)_t = \exp\left(X_t - \tfrac{1}{2}[X,\overline{X}]_t^c\right) \prod_{s \leq t} (1 + \Delta X_s) e^{-\Delta X_s} \tag{7.1}$$

Given any $D > 0$, there is a constant K so that

$$|(1+z)e^{-z} - 1| \leq K|z|^2 \quad \text{if } |z| \leq D$$

and since the sum of $|\Delta X_s|^2$ over s in any finite interval is a.s. finite, the product term in (7.1) converges absolutely for all t, with probability one. The product term thus represents a process with locally finite variation, so the process $\mathcal{E}(X)$ is itself a semimartingale. Note that $\mathcal{E}(X)_t = \exp(t)$ if $X_t \equiv t$.

Using the generalized Itô's formula, it is not hard (although it is tedious) to show that

$$\mathcal{E}(X)_t = \exp(X_0) + \int_0^t \mathcal{E}(X)_{s-} \, dX_s \tag{7.2}$$

Equation (7.2) can be viewed as a stochastic integral equation for the semimartingale $\mathcal{E}(X)$, for X given. It can be shown [C. Doleans-Dade, 1970] by methods similar to those in Section 4.4 that $\mathcal{E}(X)$ is the *unique*, up to a set of zero probability, solution of the equation. (If we had provided a proof of this fact or used it, we would have defined $\mathcal{E}(X)$ using Eq. (7.2).) An important consequence of Eq. (7.2) is that if X is a local martingale, then so is $\mathcal{E}(X)$.

It is easy to verify directly from the definitions that if X and Y are each semimartingales then

$$\mathcal{E}(X)\mathcal{E}(Y) = \mathcal{E}(X + Y + [X,\overline{Y}]) \tag{7.3}$$

and that if X is a real semimartingale with $\Delta X_s \geq -1$ for all s then $\mathcal{E}(X)_t \geq 0$ for all t.

The first application of martingale exponentials is to prove a proposition which combines a characterization of the Wiener process due to P. Lévy with a characterization of the Poisson process due to S. Watanabe.

Proposition 7.1 (Characterization Theorem). Let w and n be real local martingales where w is sample continuous with $[w,w]_t = t$ and N defined by $N_t = n_t + t$ is a counting process with $N_0 = 0$. Then w is an $\mathcal{C}_{\cdot}$ Wiener martingale, N is an $\mathcal{C}_{\cdot}$ Poisson process, and for any $s \geq 0$ the following three σ-algebras are independent:

$$\sigma\{w_t - w_s : t \geq s\}, \quad \sigma\{N_t - N_s : t \geq s\}, \quad \mathcal{C}_s$$

Proof: Since w and N are adapted, it suffices to prove that for any s, t with $0 \leq s < t$,

$w_t - w_s$ is Gaussian, mean 0, variance $t - s$;
$N_t - N_s$ is Poisson, mean $t - s$; and
$w_t - w_s, N_t - N_s$ and $\mathcal{Q}_s$ are independent $\qquad(7.4)$

Choose real constants u and v and define a local martingale X by

$$X_t = iuw_t + (e^{iv} - 1)n_t$$

$\mathcal{E}(X)$ is also a local martingale, and

$$[X, \overline{X}]_t^c = -u^2 t, \quad \Delta X_t = (e^{iv} - 1)\Delta N_t \quad \text{and} \quad \Delta N_t \in \{0, 1\}$$

so we obtain

$$\mathcal{E}(X)_t = \exp\bigl(iuw_t + ivN_t + t(u^2/2 + 1 - e^{iv})\bigr)$$

Since the local martingale $\mathcal{E}(X)$ is bounded on finite intervals independently of ω, it is a martingale. Thus, for $s \leq t$,

$$E[\mathcal{E}(X)_t | \mathcal{Q}_s] = \mathcal{E}(X)_s \qquad \text{a.s.}$$

or, equivalently,

$$E[\mathcal{E}(X)_t / \mathcal{E}(X)_s | \mathcal{Q}_s] = 1 \qquad \text{a.s.}$$

This translates to

$$E\bigl[\exp(iu(w_t - w_s) + iv(N_t - N_s)) | \mathcal{Q}_s\bigr]$$
$$= \exp\bigl(-(t - s)(u^2/2 + 1 - e^{iv})\bigr)$$

As a function of (u, v) the right-hand side of this equation is the joint characteristic function of a pair of independent random variables, the first of which is Gaussian with mean 0 and variance $t - s$, and the second of which is Poisson with mean $t - s$. Thus, by the uniqueness of characteristic functions, the conditions in (7.4) must be true. ∎

The next application of semimartingale exponentials is to the description of local martingales when the probability measure on the underlying probability space is changed. We start out with a probability measure $\mathcal{P}_0$ on $(\Omega, \mathcal{Q})$. Suppose that M is a local martingale relative to $(\mathcal{Q}_t)$ on the probability space $(\Omega, \mathcal{Q}, \mathcal{P}_0)$. For short, we say that M is a $\mathcal{P}_0$ local martingale. Assume also that M is real valued, that $M_0 = 0$, and that $\Delta M_t > 1$ for all t, and let $L = \mathcal{E}(M)$. Then $L_0 = 1$, and L is a positive $\mathcal{P}_0$ local martingale. Hence, L is also a $\mathcal{P}_0$ supermartingale (see Exercise 8) and the a.s. limit L_∞

7. SEMIMARTINGALE EXPONENTIALS AND APPLICATIONS

exists. We assume that

$$E_0 L_\infty = 1$$

where the subscript "0" on the expectation indicates that it is defined using measure $\mathcal{P}_0$. Then L is a uniformly integrable $\mathcal{P}_0$ martingale. We can define a new probability measure $\mathcal{P}$ on $(\Omega, \mathcal{A})$ which is equivalent to $\mathcal{P}_0$ (i.e., $\mathcal{P}$ and $\mathcal{P}_0$ assign zero probability to the same sets) by specifying the Radon-Nikodym derivative:

$$\frac{d\mathcal{P}}{d\mathcal{P}_0} = L_\infty \qquad (7.5)$$

We will use "E" without a subscript to denote expectations defined using measure $\mathcal{P}$.

Lemma 7.1. For any random variable D with $E|D|$ finite and any sub-σ-algebra $\mathcal{F}$ of $\mathcal{A}$,

$$E[D|\mathcal{F}] = \frac{E_0[L_\infty D|\mathcal{F}]}{E_0[L_\infty|\mathcal{F}]} \qquad \mathcal{P} \text{ a.s.}$$

Proof: We will first show that

$$E[D|\mathcal{F}] E_0[L_\infty|\mathcal{F}] = E_0[L_\infty D|\mathcal{F}] \qquad \mathcal{P}_0 \text{ a.s.} \qquad (7.6)$$

Let V denote the left-hand side of this equation. To show that it is equal to the right-hand side we will show that V has the two properties which characterize the right-hand side. First, V is $\mathcal{F}$ measurable. Second, we must prove that

$$E_0[ZL_\infty D] = E_0[ZV] \qquad \text{all bounded } \mathcal{F} \text{ measurable } Z \qquad (7.7)$$

Now

$$E_0[ZL_\infty D] = E[ZD] = E[ZE[D|\mathcal{F}]]$$
$$= E_0[ZL_\infty E[D|\mathcal{F}]] = E_0[E_0[ZL_\infty E[D|\mathcal{F}]|\mathcal{F}]]$$
$$= E_0[ZE[D|\mathcal{F}] E_0[L_\infty|\mathcal{F}]] = E_0[ZV]$$

which proves (7.7) and hence also (7.6).

Since $\mathcal{P} \ll \mathcal{P}_0$, the equality in (7.6) is also true $\mathcal{P}$ a.s., and since

$$\mathcal{P}(E_0[L_\infty|F] = 0) = E_0\left[L_\infty I_{\{E_0[L_\infty|\mathcal{F}]=0\}}\right]$$
$$= E_0\left[E_0[L_\infty|\mathcal{F}] I_{\{E_0[L_\infty|\mathcal{F}]=0\}}\right] = 0$$

the lemma follows. ∎

Lemma 7.2. Suppose that U is an adapted random process such that $L_t U_t$ is a $\mathcal{P}_0$ martingale (resp. $\mathcal{P}_0$ local martingale). Then U is a $\mathcal{P}$ martingale (resp. $\mathcal{P}$ local martingale).

Proof: Suppose that $L_t U_t$ is a $\mathcal{P}_0$ martingale. Then, by Lemma 7.1, for $s > t$,

$$E[U_s|\mathcal{Q}_t] = \frac{E_0[L_\infty U_s|\mathcal{Q}_t]}{L_t} = \frac{L_t U_t}{L_t} = U_t \quad \mathcal{P} \text{ a.s.}$$

and the first assertion is proved. If, instead, $L_t U_t$ is only a $\mathcal{P}_0$ local martingale, then there exist stopping times $R_n \uparrow \infty$ such that $(LU)_{t \wedge R_n}$ is a $\mathcal{P}_0$ martingale for each n. Then, by the argument above, $U_{t \wedge R_n}$ is a $\mathcal{P}$ martingale for each n, which implies that U is a $\mathcal{P}$ local martingale. ∎

Proposition 7.2 (Abstract Girsanov's Theorem, [Girsanov, 1960], [Van Schuppen and Wong, 1974]). Let Y be a $\mathcal{P}_0$ local martingale. If $\langle Y, M \rangle$ exists (computed under measure $\mathcal{P}_0$), then $Y - \langle Y, M \rangle$ is a $\mathcal{P}$ local martingale.

Proof: By Lemma 7.2, it suffices to prove that $(Y - \langle Y, M \rangle)L$ is a $\mathcal{P}_0$ local martingale. By Itô's product formula,

$$(Y_t - \langle Y, M \rangle_t) L_t = -\int_0^t L_{s-} \, d\langle Y, M \rangle_s + [Y - \langle Y, M \rangle, L]_t$$
$$+ \text{ a } \mathcal{P}_0 \text{ local martingale} \quad (7.8)$$

Since $\langle Y, M \rangle$ has locally finite variation and is predictable and since $dL_t = L_{t-} \, dM_t$,

$$[Y - \langle Y, M \rangle, L]_t = [Y, L]_t - \int_0^t \Delta \langle Y, M \rangle_s \, dL_s$$
$$= \int_0^t L_{s-} \, d[Y, M]_s + \text{ a } \mathcal{P}_0 \text{ local martingale}$$

Finally, since $\langle Y, M \rangle - [Y, M]$ is a $\mathcal{P}_0$ local martingale, the right-hand side of (7.8) is a $\mathcal{P}_0$ local martingale. ∎

Example. Suppose that w is an $(\mathcal{Q}_\cdot, \mathcal{P}_0)$ Wiener martingale and that N is a counting process with predictable intensity λ_t, where $\lambda_t > 0$ a.s. Let h and ρ be predictable processes such that the equation

$$M_t = \int_0^t h_s \, dw_s + \int_0^t \frac{\lambda_s - \rho_s}{\lambda_s} (dN_s - \lambda_s \, ds)$$

defines a local martingale M. Suppose that $E_0 L_\infty = 1$ where $L = \mathcal{E}(M)$

7. SEMIMARTINGALE EXPONENTIALS AND APPLICATIONS

and that $\mathcal{P}$ is defined by Eq. (7.5). Then, computing under measure $\mathcal{P}_0$,

$$\langle M, w \rangle_t = \int_0^t h_s \, ds$$

and

$$\left\langle M, N - \int_0^\cdot \lambda_s \, ds \right\rangle_t = \langle M, N \rangle_t = \int_0^t \frac{\lambda_s - \rho_s}{\lambda_s} d\langle N, N \rangle_s$$

$$= \int_0^t (\lambda_s - \rho_s) \, ds$$

Therefore, under measure $\mathcal{P}$,

$$w_t - \int_0^t h_s \, ds \quad \text{and} \quad N_t - \int_0^t \rho_s \, ds$$

are each local martingales. Since the quadratic variation process of the first of these is t, the first of these processes is in fact a $\mathcal{P}$ Wiener process. The intensity of N under measure $\mathcal{P}$ is ρ.

The final application of martingale exponentials we give is a useful martingale representation theorem. Let w denote a Wiener process and let N denote a Poisson process, both defined on a complete probability space $(\Omega, \mathcal{C}, \mathcal{P}_0)$. Define an increasing family of σ-algebras $(\mathcal{O}_t)$ by

$$\mathcal{O}_t = \sigma(w_s, N_s : s \leq t) \vee \mathcal{P}_0\text{-null sets}$$

The following lemma ensures that the family $(\mathcal{O}_t)$ satisfies the usual conditions, so that, for example, $(\mathcal{O}_t, \mathcal{P}_0)$ martingales have corlol modifications.

Lemma 7.3 (Conditional Zero-One Law). $\mathcal{O}_t = \mathcal{O}_{t^+}$.

Proof. (Based on an idea in [Doob, 1953]): For $t \geq 0$ define

$$\mathcal{U}_t = \sigma(w_s - w_t, N_s - N_t : s \geq t)$$

Fix t and let X be a bounded $\mathcal{O}_{t^+}$-measurable random variable. Since $\mathcal{O}_{t^+}$ is independent of $\mathcal{U}_{t+\frac{1}{n}}$ for each $n \geq 1$ we have

$$E_0\left[X | \mathcal{U}_{t+\frac{1}{n}}\right] = E_0 X \quad \text{a.s.}$$

The martingale convergence theorem ensures that the left-hand side converges a.s. to $E_0[X|\mathcal{U}_t]$ so that $E_0[X|\mathcal{U}_t] = E_0 X$ a.s. By considering all possible X we conclude that the σ-algebras $\mathcal{U}_t$ and $\mathcal{O}_{t^+}$ are independent. Now if Z is a sum of finitely many terms $A_i B_i$ where the A_i are bounded $\mathcal{O}_t$-measurable random variables and the B_i are bounded $\mathcal{U}_t$-measurable random variables, then

$$E_0\big[(X - E_0[X|\mathcal{O}_t])Z\big] = \sum_{i=1}^{n} E_0\big[(X - E_0[X|\mathcal{O}_t])A_i\big] E_0 B_i = 0$$

By the monotone class theorem, the left-most expectation in this equation is zero for all bounded $\mathcal{O}_\infty$-measurable random variables Z. Thus, $X = E_0[X|\mathcal{O}_t]$ a.s. which, since $\mathcal{O}_t$ contains all $\mathcal{P}_0$ null sets, implies that X is $\mathcal{O}_t$ measurable. Since X was an arbitrary bounded $\mathcal{O}_{t+}$-measurable random variable, the lemma follows. ∎

Proposition 7.3 (Martingale Representation Theorem). Let Y be any (corlol) local martingale relative to $(\mathcal{O}_t, \mathcal{P}_0)$. Then Y has the representation

$$Y_t = Y_0 + \int_0^t H_s\, dw_s + \int_0^t K_s\, dn_s \tag{7.9}$$

where $n_t = N_t - t$, and H and K are predictable with

$$\int_0^t |H_s|^2 + |K_s|\, ds < +\infty \qquad \mathcal{P}_0 \text{ a.s. for } t < +\infty \tag{7.10}$$

In the special case that Y is a martingale relative to $(\mathcal{O}_t, \mathcal{P}_0)$ with $E_0|Y_t|^2 < +\infty$,

$$E_0|Y_t|^2 = E_0|Y_0|^2 + E_0 \int_0^t |H_s|^2 + |K_s|^2\, ds \tag{7.11}$$

Proof: Assume that Y is a real bounded corlol $(\mathcal{O}_\cdot, \mathcal{P}_0)$ martingale and choose T with $T > 0$. Then by part (g) of Proposition 4.1, inequality (4.3), and the fact that $\langle w, w \rangle_t = t$,

$$\left| E_0 \int_0^T C(t, \omega)\, d\langle Y, w \rangle_t \right| = |E_0 \langle Y, C \cdot w \rangle_T|$$

$$\leq E_0 \left[\left(\langle Y, Y \rangle_T \int_0^T C_s^2\, ds \right)^{1/2} \right]$$

for real bounded predictable processes C. This implies the following absolute continuity condition for the indicated measures on the predictable subsets of $[0, T] \times \Omega$:

$$d\langle Y, w \rangle_t \mathcal{P}_0(d\omega) \ll dt\, \mathcal{P}_0(d\omega)$$

Let H denote the corresponding Radon-Nikodym derivative—so, by definition,

$$E_0 \int_0^T C_t\, d\langle Y, w \rangle_t = E_0 \int_0^T C_t H_t\, dt$$

7. SEMIMARTINGALE EXPONENTIALS AND APPLICATIONS

for all bounded predictable processes C. By choosing C of the form $C_t(\omega) = Z(\omega) I_{\{a < t \le b\}}$ where Z is $\mathcal{C}_a$ measurable, we conclude that

$$\langle Y, w \rangle_t - \int_0^t H_s \, ds$$

is a martingale. It is also predictable and has corlol locally finite variation sample paths. It is thus zero for all t, with probability one, so

$$\langle Y, w \rangle_t = \int_0^t H_s \, ds \qquad 0 \le t \le T \text{ a.s.}$$

Similarly, there is a predictable process K so that

$$\langle Y, n \rangle_t = \int_0^t K_s \, ds \qquad 0 \le t \le T \text{ a.s.}$$

We will prove that Eqs. (7.9) and (7.11) hold.

Choose $c > 0$ and define the predictable process U by

$$U_s = I_{\{|H_s| \le c \text{ and } |K_s| \le c\}}$$

and define martingales $\tilde{Y}$ and $\tilde{\tilde{Y}}$ by

$$\tilde{Y}_t = Y_0 + \int_0^t U_s \, dY_s$$

and

$$\tilde{\tilde{Y}}_t = Y_0 + \int_0^t H_s U_s \, dw_s + \int_0^t K_s U_s \, dn_s$$

We will show that $\tilde{Y}_t = \tilde{\tilde{Y}}_t$. First define $z_t = \tilde{Y}_t - \tilde{\tilde{Y}}_t$ and note that $\langle z, w \rangle = \langle z, n \rangle = 0$. The jumps of z are uniformly bounded so that we can choose $\lambda > 0$ so that $\lambda \Delta z_t > -1$ for all t. Then $\mathcal{E}(\lambda z)$ is a positive local martingale so there exist $\mathcal{O}_.$ stopping times $R_k \uparrow T$ such that $L^k = \mathcal{E}(\lambda z_{t \wedge R_k})$ is a martingale for each k. Let $\mathcal{P}^k$ denote the probability measure on $(\Omega, \mathcal{C})$ which is absolutely continuous with respect to $\mathcal{P}_0$ such that $d\mathcal{P}^k = L_T^k \, d\mathcal{P}_0$. Since $\langle z, w \rangle = \langle z, n \rangle = 0$, Proposition 7.1 yields that w is a $\mathcal{P}^k$ Wiener martingale and N is a $\mathcal{P}^k$ Poisson process. Thus $\mathcal{P}^k$ and $\mathcal{P}_0$ are identical on $\mathcal{O}_T$, which implies that $L_T^k = 1$, $\mathcal{P}_0$ a.s. Thus, $z_t = 0$ and $\tilde{Y}_t = \tilde{\tilde{Y}}_t$ for $0 \le t \le R_k$. Since k was arbitrary, $\tilde{Y}_t = \tilde{\tilde{Y}}_t$ for $0 \le t \le T$. Now, $\tilde{Y}_t$ converges in 2-mean to Y_t as $c \to +\infty$ and so Eqs. (7.9) and (7.11) are obtained in the limit. By an easy patching-together argument, Y has a representation of the form (7.9) for all $t \ge 0$.

Next, suppose that Y is a square integrable $(\mathcal{O}_t, \mathcal{P}_0)$ martingale, and let $T > 0$. Then there exists a sequence of bounded $\mathcal{O}_T$-measurable random variables Y_T^k converging in 2-mean to Y_T. Let Y^k denote the corlol martingale on $[0, T]$ defined by $Y_t^k = E_0[Y_t^k | \mathcal{O}_t]$. Then Y^k is bounded so that

$$Y_t^k = Y_0^k + \int_0^t H_s^k \, dw_s + \int_0^t K_s^k \, dn_s \qquad t \leq T$$

for some predictable H^k and K^k. Now

$$E_0|Y_0^k - Y_0^m|^2 + E_0 \int_0^T |H_s^k - H_s^m|^2 + |K_s^k - K_s^m|^2 \, ds = E_0|Y_T^k - Y_T^m|^2$$

and the right-hand side of this equation converges to zero as m and k tend to infinity, so there must exist predictable functions H and K so that

$$E_0 \int_0^T |H_s^k - H_s|^2 + |K_s^k - K_s|^2 \, ds \xrightarrow[k \to \infty]{} 0$$

Equations (7.9) and (7.11) thus hold for $0 \leq t \leq T$, and it is then an easy matter to show that there exist H and K so that Eqs. (7.9) and (7.11) hold for $0 \leq t \leq T$.

The general case of the proposition can be proved by a more elaborate approximation argument [Liptser and Shiryaev, 1976], which we omit. See [Jacod, 1979] for a variety of extensions. ∎

EXERCISES

1. Verify Eq. (4.1).

2. Suppose that S and T are stopping times. Show that their minimum $S \wedge T$ and maximum $S \vee T$ are stopping times. If (T_n) is a sequence of stopping times show that $T^* = \sup_n T_n$ is a stopping time.

3. Use the optional sampling theorem to show that if M is a corlol martingale and T is a stopping time then M^T defined by $M_t^T = M_{t \wedge T}$ is also a martingale.

4. Prove the optional sampling theorem (Proposition 1.5) in the special case that there are two nonrandom times $t_1 < t_2$ such that $S, R \in \{t_1, t_2\}$ a.s.

5. Let $U_1, U_2, \ldots$ be independent and suppose that each U_i is uniformly distributed over the interval $[0, 2]$. Let $Z_n = U_1 U_2 \ldots U_n$. Show that Z is a martingale relative to $\mathcal{O}Z$ and discuss the limiting behavior of Z_n as n tends to infinity (Hint: Consider $\ln Z_n$).

6. Verify Eq. (1.9).

7. Show that a positive local martingale M is a supermartingale.

EXERCISES

8. Let h be a predictable process such that $\int_0^\infty h_s^2\, ds \leq K$ a.s., let w be a Wiener martingale, and let $p > 1$. Use the fact that $E\mathcal{E}(ph \cdot w)_\tau \leq 1$ for all stopping times τ to obtain an upper bound on $E[\mathcal{E}(h \cdot w)_\tau^p]$ which is uniform in τ. Then deduce that $\mathcal{E}(h \cdot w)$ is class D, and finally, that $E\mathcal{E}(h \cdot w)_\infty = 1$.

9. Let $(\mathcal{C}_n)$ be an increasing family of sub-σ-algebras of $\mathcal{C}$, let $\mathcal{C}_\infty = \vee \mathcal{C}_n$, let $\mathcal{P}_0$ and $\mathcal{P}$ be probability measures on $\mathcal{C}_\infty$, and let $\mathcal{P}_0^n$ and $\mathcal{P}^n$ denote the restrictions of $\mathcal{P}_0$ and $\mathcal{P}$ to $\mathcal{C}_n$. Suppose that $\mathcal{P}_0^n \ll \mathcal{P}^n$ for each n and let $L_n = d\mathcal{P}_0^n/d\mathcal{P}^n$. Then L is a positive $\mathcal{P}_0$ martingale, so that $L_\infty = \lim_{n\to\infty}$ a.s. L_n exists. Show that $\mathcal{P} \ll \mathcal{P}_0$ if and only if $E_0 L_\infty = 1$. (Hint: $E_0 L_\infty = 1$ is equivalent to L being uniformly integrable.)

10. [Kakutani, 1948] Let $\Omega = \{(x_1, x_2, \ldots) : x_i \in \mathbb{R}\}$ and let X_i denote the function on Ω defined by $X_i(x) = x_i$. Let $\mathcal{C}_n = \sigma(X_1, \ldots, X_n)$ and $\mathcal{C}_\infty = \vee_n \mathcal{C}_n$. Let $\mathcal{P}_0$ and $\mathcal{P}$ be probability measures on $(\Omega, \mathcal{C}_\infty)$ so that $X_1, X_2, \ldots$ are independent unit variance Gaussian random variables under $\mathcal{P}_0$ and under $\mathcal{P}$. Assume that $E_0 X_i = 0$ and $E X_i = a_i$ for all i. Let $S = \sum_{i=1}^\infty a_i^2$. Using the previous problem, show that $\mathcal{P}_0 \equiv \mathcal{P}$ if $S < +\infty$ and that $\mathcal{P}_0 \perp \mathcal{P}$ otherwise. Show also that if $\mathcal{P}_0 \equiv \mathcal{P}$ then the Radon-Nikodym derivative of $\mathcal{P}$ with respect to $\mathcal{P}_0$ is given by

$$\frac{d\mathcal{P}}{d\mathcal{P}_0} = \exp\left(\sum_{i=1}^\infty A_i X_i - \frac{1}{2}S\right)$$

where the sum converges $\mathcal{P}_0$ a.s.

11. Find the predictable compensators for w_t^4 and for N_t^4, where w is a Wiener martingale and N is an $\mathcal{C}$. Poisson process.

12. Let $(V_n, \mathcal{C}_n : n \in \mathbb{Z}_+)$ be a martingale with $V_0 = 0$, let $B_n = V_{n+1} - V_n$ for $n \geq 1$, and suppose that $|B_n| = 1$ for all n and ω.

(*a*) Show that B_{n+1} is independent of $\mathcal{C}_n$ for each n (so that $B_1, B_2, \ldots$ are mutually independent) and that $\mathcal{P}(B_i = 1) = \mathcal{P}(B_i = -1) = 0.5$.

(*b*) Show that any martingale (M_n) relative to the σ-algebras $(\mathcal{C}_n^V)$ generated by V has a representation

$$M_n = M_0 + \sum_{k=1}^n H_k B_k \qquad n \geq 1$$

where H is a discrete-time predictable process.

7
Detection and Filtering

1. INTRODUCTION

In both the detection problem and the filtering problem a sample path from some random process O is observed. In the detection problem, it is postulated that one of two known probability measures $\mathcal{P}$ or $\mathcal{P}_0$ is given on the underlying space $(\Omega, \mathcal{A})$. Presumably the distribution of the observed process is different under $\mathcal{P}$ and $\mathcal{P}_0$. The problem is to intelligently guess which of the two measures is in effect, on the basis of the observation. In the filtering problem a random process ξ, which may be a coordinate of some other random process, is given and the problem is to produce at each time t an estimate of ξ_t.

The detection and filtering problems are closely related. On the one hand, the solution of the detection problem can involve estimating processes under the assumption that one of the given probabilities is in effect. On the other hand, one treatment of filtering is to introduce a probability measure in addition to the one given, thus leading to the situation encountered in the detection problem.

1. INTRODUCTION

For both the detection and the filtering problem the role of the observations is to provide information, and this information is summarized by the increasing family of σ-algebras $\mathcal{O}_. = (\mathcal{O}_t)$ generated by the observations:

$$\mathcal{O}_t = \sigma(O_s : 0 \le s \le t)$$

We shall be largely interested in computing $E[\xi_t|\mathcal{O}_t]$, which is known to be the minimum mean square error estimate of ξ_t given $\mathcal{O}_t$. This expectation is defined only up to sets of zero probability which can depend on t, but a nice version of the process always exists:

Proposition 1.1. Let $\mathcal{O}_.$ be an increasing family of sub-σ-algebras of A (not necessarily satisfying the usual conditions) and let $\xi = (\xi(t, \omega))$ be a $\mathcal{B}(\mathbb{R}_+) \times \mathcal{A}$-measurable random process. Then there exists an $\mathcal{O}_.$ progressively measurable process $\hat{\xi}$ such that $\hat{\xi}_t = E[\xi_t|\mathcal{O}_t]$ a.s. for all t such that $E\xi_t^+$ or $E\xi_t^-$ is finite.

Proof: Consider first the case that $\xi(t, \omega) = f(t)U(\omega)$ where f is a bounded Borel-measurable function and U is a bounded random variable. Let (M_t) denote a separable version of the martingale $E[U|\mathcal{O}_t]$. There exists a countable subset D of $\mathbb{R}_+$ such that for each t not in D, M is a.s. left continuous at t [Doob 1953, Th. 11.2, p. 358]. Then the process $\hat{U}$ defined by

$$\hat{U}_t = \begin{cases} \limsup_{n \to \infty} M_{\lfloor tn \rfloor /n} & \text{if } t \notin D \\ M_t & \text{if } t \in D \end{cases}$$

is a progressively measurable process (it is a limit of such processes) and it is a modification of M, and therefore of $E[U_t|\mathcal{O}_t]$. Thus $\hat{\xi}$ defined by $\hat{\xi}_t = f(t)\hat{U}_t$ is a progressively measurable version of $E[\xi_t|\mathcal{O}_t]$.

Let $\mathcal{H}$ denote the vector space of all bounded measurable processes ξ such that there exists a progressively measurable version of $E[\xi_t|\mathcal{O}_t]$. Then $\mathcal{H}$ includes the algebra of all finite linear combinations of processes of the form $f(t)U(\omega)$ as above.

Next, suppose that ξ is a bounded measurable process and that ξ_n is a sequence of processes in $\mathcal{H}$ such that either ξ_n converges to ξ uniformly, or $\xi_n(t, \omega)$ increases to $\xi(t, \omega)$ for all t, ω, as n tends to infinity. Then $\hat{\xi}$ defined by

$$\hat{\xi}(t, \omega) = \limsup_{n \to \infty} \hat{\xi}_n(t, \omega) \tag{1.1}$$

is a progressively measurable version of $E[\xi_t|\mathcal{O}_t]$. Thus $\xi \in \mathcal{H}$. $\mathcal{H}$ is therefore closed under uniform limits and under bounded monotone limits. By the monotone class theorem, $\mathcal{H}$ contains all bounded measurable processes.

Finally, given an arbitrary measurable process, ξ, ξ is the pointwise limit of a sequence of bounded measurable processes ξ_n with $|\xi_n| \le |\xi|$. Then $\hat{\xi}$ defined by (1.1) satisfies the requirements of the proposition. ∎

If $\hat{\xi}_t$ is redefined on a (t, ω) set of $dt\, d\mathcal{P}(\omega)$ measure zero, then a predictable process can be obtained:

Proposition 1.2. Let $\xi = (\xi(t, \omega))$ be a $\mathcal{B}(\mathbb{R}_+) \times \mathcal{C}$-measurable random process. Then there exists an $\mathcal{O}_\cdot$ predictable process $\hat{\hat{\xi}}$ and a subset D of $\mathbb{R}$ with Lebesgue measure zero (in fact D can be chosen to be countable) such that $\hat{\hat{\xi}}_t = E[\xi_t | \mathcal{O}_t]$ a.s. for all t such that t is not contained in D and either $E\xi_t^+$ or $E\xi_t^-$ is finite.

Proof: Let $\mathcal{H}$ denote the space of bounded ξ for which Proposition 1.2 is true. If $\xi(t, \omega) = f(t)U(\omega)$ and M and D are as in the proof of Proposition 1.1, then $\hat{\hat{\xi}}$ defined by

$$\hat{\hat{\xi}}_t = \begin{cases} f(t) \limsup_{n \to \infty} M_{\lfloor t(n-1) \rfloor / n} & \text{if } t \notin D \\ 0 & \text{if } t \in D \end{cases}$$

(with the convention that $M_t = 0$ for $t < 0$) and D satisfy the conclusion of Proposition 1.2. Thus ξ is in $\mathcal{H}$, and $\mathcal{H}$ also includes finite linear combinations of such processes. The rest of the proof is nearly identical to that of Proposition 1.1. ∎

Remark: $\hat{\xi}$ and $\hat{\hat{\xi}}$ are a poor person's version of the so-called optional projection and predictable projection, respectively, of ξ onto the family $(\mathcal{O}_t)$ (see [Dellacherie and Meyer, 1982, Appendix I, Th. 6]).

In order to describe the dynamics of processes which are not "observed," we assume that there is a second increasing family $\mathcal{C}_\cdot$ of sub-σ-algebras of $\mathcal{C}$ such that $\mathcal{O}_t \subset \mathcal{C}_t$ for each t.

Proposition 1.3 (Representation of Projections onto $\mathcal{O}_\cdot$). Let Z be an $(\mathcal{C}_\cdot, \mathcal{P})$ semimartingale with representation

$$Z_t = Z_0 + \int_0^t f_s\, ds + m_t$$

where $E|Z_0|$ is finite, f is an $\mathcal{C}_\cdot$ progressive process with

$$E\int_0^t |f_s|\, ds < +\infty$$

and m is an $(\mathcal{C}_\cdot, \mathcal{P})$ martingale. Then there exists an $(\mathcal{O}_\cdot, \mathcal{P})$ martingale

1. INTRODUCTION

D so that for each t,

$$\hat{Z}_t = \hat{Z}_0 + \int_0^t \hat{f}_s\, ds + D_t \qquad \mathcal{P} \text{ a.s.} \tag{1.2}$$

and

$$\hat{Z}_t = \hat{Z}_0 + \int_0^t \hat{\hat{f}}_s\, ds + D_t \qquad \mathcal{P} \text{ a.s.} \tag{1.3}$$

Proof: If we use Eq. (1.2) to define D and use the representation for Z and the definition of $\hat{Z}_t$ we obtain

$$D_t = E[m_t|\mathcal{O}_t] + E[Z_0|\mathcal{O}_t] - E[Z_0|\mathcal{O}_0] + \int_0^t E[f_s|\mathcal{O}_t] - E[f_s|\mathcal{O}_s]\, ds$$

where $\mathcal{O}_{\cdot}$ progressive versions are chosen for conditional expectations where appropriate. Now for $0 \le a < t$,

$$E[E[m_t|\mathcal{O}_t]|\mathcal{O}_a] = E[m_t|\mathcal{O}_a]$$
$$= E[E[m_t|\mathcal{Q}_a]|\mathcal{O}_a] = E[m_a|\mathcal{O}_a]$$

and

$$E\left[\int_0^t E[f_s|\mathcal{O}_t] - E[f_s|\mathcal{O}_s]\, ds \Big| \mathcal{O}_a\right] = \int_0^t E[f_s|\mathcal{O}_a] - E[f_s|\mathcal{O}_{s \wedge a}]\, ds$$
$$= \int_0^a E[f_s|\mathcal{O}_a] - E[f_s|\mathcal{O}_s]\, ds$$

which shows that D is indeed a martingale. Equation (1.3) is a consequence of Eq. (1.2) since $\hat{f} = \hat{\hat{f}}$ except on a set of $dt\, d\mathcal{P}(\omega)$ measure zero. ∎

Suppose now that the observed process is given by $O_t = (y_t, N_t)$ and that

$$y_t = \int_0^t h_s\, ds + w_t, \tag{1.4}$$

where w is an $(\mathcal{Q}_{\cdot}, \mathcal{P})$ Wiener martingale, $\tag{1.5}$

h is $\mathcal{Q}_{\cdot}$ progressive, $E\int_0^t |h_s|\, ds < +\infty$ $\tag{1.6}$

and

$$N_t = \int_0^t \lambda_s\, ds + M_t, \tag{1.7}$$

where N is a counting process, $\tag{1.8}$

$$M \text{ is an } (\mathcal{Q}_\cdot, \mathcal{P}) \text{ martingale,} \tag{1.9}$$

$$\lambda \text{ is } \mathcal{Q}_\cdot \text{ predictable, } E\int_0^t |\lambda_s|\, ds < +\infty \tag{1.10}$$

The main point of conditions (1.7)–(1.10) is that N is a counting process with $\mathcal{Q}_\cdot$ intensity λ.

Since y and N are $\mathcal{O}_\cdot$ progressive, we have $y = \hat{y}$ and $N = \hat{N}$. Thus, Proposition 1.3 yields that $\tilde{y}$ and $\tilde{N}$ defined by

$$\tilde{y}_t = y_t - \int_0^t \hat{h}_s\, ds \quad \text{and} \quad \tilde{N}_t = N_t - \int_0^t \hat{\lambda}_s\, ds \tag{1.11}$$

are $\mathcal{O}_\cdot$ martingales. The processes $\tilde{y}$ and $\tilde{N}$ are called the **innovations processes** corresponding to the observation process y and N.

2. LIKELIHOOD RATIO REPRESENTATION

Consider a simple hypothesis testing problem for which a process ($O_s : 0 \leq s \leq T$) is observed and the problem is to decide which of two given probability measures $\mathcal{P}$ or $\mathcal{P}_0$ is in effect on the underlying space $(\Omega, \mathcal{Q})$. Associated with any decision procedure we can identify two conditional error probabilities:

$$p_e^I = \mathcal{P}[\text{decide } \mathcal{P}_0 \text{ is in effect}]$$
$$p_e^{II} = \mathcal{P}_0[\text{decide } \mathcal{P} \text{ is in effect}]$$

The σ-algebra of the observations is $\mathcal{O}_T$ where we now let

$$\mathcal{O}_t = \sigma(O_s : 0 \leq s \leq t) \vee \{\mathcal{P}_0 \text{ null sets}\}$$

We also assume that $\mathcal{O}_t = \mathcal{O}_{t+}$ and that $\mathcal{O}_0$ consists only of $\mathcal{P}_0$ null sets and their complements. Let $\mathcal{P}_0^t$ and $\mathcal{P}^t$ denote the restrictions of $\mathcal{P}_0$ and $\mathcal{P}$ to $\mathcal{O}_t$. We assume for simplicity that $\mathcal{P}_0^T$ and $\mathcal{P}^T$ are mutually absolutely continuous and define Λ_t for $0 \leq t \leq T$ by

$$\Lambda_t = \frac{d\mathcal{P}^t}{d\mathcal{P}_0^t}$$

We observe that $\Lambda_t = E_0[\Lambda_T | \mathcal{O}_t]$ so that Λ is a $\mathcal{P}_0$ martingale. By our assumptions on $\mathcal{O}_\cdot$, $\mathcal{P}_0(\Lambda_0 = 1) = 1$ and we can choose a modification of Λ which is corlol.

The likelihood ratio test with threshold γ and randomization probability α for deciding which probability measure is in effect is the following:

2. LIKELIHOOD RATIO REPRESENTATION

compute Λ_T and

if $\Lambda_T > \gamma$ decide $\mathcal{P}$

if $\Lambda_T = \gamma$ decide $\mathcal{P}$ with probability α, otherwise decide $\mathcal{P}_0$

if $\Lambda_T < \gamma$ decide $\mathcal{P}_0$

Then, for example,

$$p_e^{II} = \mathcal{P}_0(\Lambda_T > \gamma) + \alpha \mathcal{P}_0(\Lambda_T = \gamma)$$

and for any ε between zero and one there is a choice of the parameters γ and α such that $p_e^{II} = \varepsilon$. By the well-known Neyman-Pearson lemma of statistics, the resulting likelihood ratio test achieves the minimum p_e^I over all decision rules with $p_e^{II} = \varepsilon$.

The key to implementing and evaluating a likelihood ratio test is to compute the likelihood ratio Λ_T, and the key to understanding the likelihood ratio process (Λ_t) is to connect it to the behavior of (O_t) under $\mathcal{P}$ and $\mathcal{P}_0$. We will first represent Λ as a martingale exponential.

Define, for $n \geq 1$,

$$R_n = \inf\{s : 0 \leq s \leq T \text{ and } \Lambda_s \leq 1/n, \text{ or } s = +\infty\}$$

Then R_n is increasing in n for each ω so there exists a limit R of R_n as n tends to infinity. Since Λ has right-continuous sample paths, $\Lambda_{R_n} \leq 1/n$ if $R_n \leq T$. By this fact and the optional sampling theorem, we have (using I_n for the indicator of the event that $R_n \leq T$):

$$E_0 \Lambda_T I_{\{R \leq T\}} \leq E_0 \Lambda_T I_n = E_0\left(E_0\left[\Lambda_T I_n | \mathcal{O}_{R_n}\right]\right)$$

$$= E_0\left(E_0\left[\Lambda_T | \mathcal{O}_{R_n}\right] I_n\right) = E_0 \Lambda_{R_n} I_n \leq 1/n$$

Thus $E_0 \Lambda_T I_{\{R \leq T\}} = 0$, and since $\mathcal{P}_0(\Lambda_T > 0) = 1$ it follows that $\mathcal{P}_0(R \leq T) = 0$. Therefore Λ_{s-}^{-1} is locally bounded, and Λ_{s-} is locally bounded away from zero. We can thus define

$$X_t = \int_0^t \Lambda_{s-}^{-1} d\Lambda_s$$

and apply Itô's formula to obtain

$$\ln \Lambda_t = \int_0^t \frac{1}{\Lambda_{s-}} d\Lambda_s - \frac{1}{2} \int_0^t \frac{1}{\Lambda_{s-}^2} d[\Lambda, \Lambda]_s^c$$

$$+ \sum_{s \leq t} \ln\left(1 + \frac{\Delta \Lambda_s}{\Lambda_{s-}}\right) - \frac{\Delta \Lambda_s}{\Lambda_{s-}}$$

or

$$\Lambda_t = \exp\left(X_t - \frac{1}{2}[X,X]_t^c + \sum_{s \le t} \ln(1 + \Delta X_s) - \Delta X_s \right)$$

Equivalently, Λ has the representation $\Lambda = \mathcal{E}(X)$, and X is a $\mathcal{P}_0$ local martingale since Λ is one as well.

Our goal now is to relate X (rather than Λ directly) to the behavior of O under $\mathcal{P}$ and under $\mathcal{P}_0$. We now restrict our attention to the case that $O_t = (y_t, N_t)$. We suppose that (y_t, N_t) satisfies conditions (1.4)–(1.10) for some λ and h—these are conditions on (y, N) under the measure $\mathcal{P}$. We suppose, on the other hand, that

y is an $(\mathcal{O}_{\cdot}, \mathcal{P}_0)$ Wiener martingale, and

N is an $(\mathcal{O}_{\cdot}, \mathcal{P}_0)$ Poisson process

By the Levy-Watanabe characterization theorem, Proposition 6.7.1, this completely specifies the distribution of (y, N) under measure $\mathcal{P}_0$.

We can now invoke the martingale representation theorem under measure $\mathcal{P}_0$ to deduce that (let $n_t = N_t - t$)

$$X_t = \int_0^t \phi_s \, dy_s + \int_0^t \psi_s \, dn_s \qquad 0 \le t \le T$$

for some $\mathcal{O}_{\cdot}$ predictable processes ϕ and ψ such that

$$\int_0^T \phi_s^2 + |\psi_s| \, ds < +\infty \qquad \mathcal{P}_0 \text{ a.s.}$$

To see how ϕ and ψ are related to $\mathcal{P}$, first apply the abstract Girsanov's theorem, Proposition 6.7.2, to deduce that

$$y_t - \int_0^t \phi_s \, ds \text{ and } N_t - \int_0^t \psi_s \, ds \text{ are } (\mathcal{O}_{\cdot}, \mathcal{P}) \text{ martingales}$$

Comparing this with the fact that the innovations processes $\tilde{y}$ and $\tilde{N}$ in (1.11) are also $(\mathcal{O}_{\cdot}, \mathcal{P})$ martingales and using the fact that the $(\mathcal{O}_{\cdot}, \mathcal{P})$ predictable compensators of y and N are unique, we conclude that

$$\phi_t(\omega) = \hat{h}_t(\omega) \quad \text{and} \quad 1 + \psi_t(\omega) = \hat{\lambda}_t(\omega) \qquad dt\, d\mathcal{P}(\omega) \text{ a.e.}$$

Hence the representation for X becomes

$$X = \hat{h} \cdot y + (\hat{\lambda} - 1) \cdot n$$

3. FILTER REPRESENTATION — CHANGE OF MEASURE DERIVATION

so that

$$\Lambda = \mathcal{E}(X) = \mathcal{E}(\hat{h} \cdot y)\mathcal{E}((\hat{\lambda} - 1) \cdot n) \tag{2.1}$$

where

$$\mathcal{E}(\hat{h} \cdot y)_t = \exp\left(\int_0^t \hat{h}_s \, dy_s - \tfrac{1}{2}\int_0^t \hat{h}_s^2 \, ds\right)$$

and

$$\mathcal{E}((\hat{\lambda} - 1) \cdot n)_t = \exp\left(\int_0^t \ln(\hat{\lambda}_s) \, dN_s - \int_0^t (\hat{\lambda}_s - 1) \, ds\right)$$

$$= \exp\left(-\int_0^t \hat{\lambda}_s - 1 \, ds\right) \prod_{\substack{s \le t \\ \Delta N_s = 1}} \hat{\lambda}_s$$

Examination of Eq. (2.1) reveals that the optimal solution to the detection problem cleanly separates into a pure estimation part and a pure detection part. More precisely, for the "pure detection case" in which λ and h are known nonrandom functions, Eq. (2.1) with $\hat{h} = h$ and $\hat{\lambda} = \lambda$, together with the likelihood ratio test, prescribes the optimal strategy. In the general case, the first part of the solution is to compute the estimates $\hat{h}$ and $\hat{\lambda}$ and the second part is to proceed as in the pure detection case.

Remark: The likelihood ratio representation (2.1) can be traced back to work of Sosulin and Stratonovich (1965), Duncan (1968), and Kailath (1969). For an account of its history for point process observations, see [Brémaud (1981)].

3. FILTER REPRESENTATION — CHANGE OF MEASURE DERIVATION

Let $(\Omega, \mathcal{A}, \mathcal{P})$ be a complete probability space and let $\mathcal{S}_\cdot = (\mathcal{S}_t)$ and $\mathcal{O}_\cdot = (\mathcal{O}_t)$ be two increasing families of sub-σ-algebras of $\mathcal{A}$ which satisfy the usual conditions. The σ-algebra $\mathcal{S}_t$ represents "state information" up to time t and $\mathcal{O}_t$ represents "observed information" up to time t. It is useful to define a third increasing family $\mathcal{A}_\cdot$ by $\mathcal{A}_t = \mathcal{S}_t \vee \mathcal{O}_t$. We will assume that the observations σ-algebras are generated by a pair of processes y and N:

$$\mathcal{O}_t = \sigma(y_s, N_s : 0 \le s \le t) \vee \{\mathcal{P} \text{ null sets}\} \tag{3.1}$$

A random process ξ adapted to $\mathcal{S}_\cdot$ represents the state, or one coordinate of the state, to be estimated. We assume that ξ, y, and N have the following representations:

State Coordinate: $\quad \xi_t = \xi_0 + \int_0^t \beta_s \, ds + m_t \quad 0 \le t \le T \quad (3.2)$

Observation:
$$\begin{cases} y_t = \int_0^t h_s \, ds + w_t & (3.3) \\ N_t = \int_0^t \lambda_s \, ds + M_t & (3.4) \end{cases}$$

where

ξ_0 is $\mathcal{S}_0$ measurable, $E|\xi_0| < +\infty$,

β is $\mathcal{S}_{\cdot}$ predictable, $E\int_0^T |\beta_s| \, ds < +\infty$,

m is an $\mathcal{S}_{\cdot}$ martingale,

h is $\mathcal{Q}_{\cdot}$ progressive, $\int_0^T h_s^2 \, ds < +\infty$ a.s., $\quad (3.5)$

w is an $\mathcal{O}_t \vee \mathcal{S}_T$ Wiener martingale, $\quad (3.6)$

N is an $\mathcal{O}_t \vee \mathcal{S}_T$ adapted counting process, $\quad (3.7)$

λ is $\mathcal{Q}_{\cdot}$ predictable, $E\int_0^T \lambda_s \, ds < +\infty, \lambda_s > 0,\quad (3.8)$

M is an $\mathcal{O}_t \vee \mathcal{S}_T$ martingale $\quad (3.9)$

Remark: If we specialize the conditions to require h to be $\mathcal{S}_{\cdot}$ progressive and λ to be $\mathcal{S}_{\cdot}$ predictable then by a slight extension of the Levy-Watanabe characterization theorem, w is independent of $\mathcal{S}_T$ and, given $\mathcal{S}_T$, N is an independent increment process with Poisson distributed increments. The model is called doubly stochastic in this special case.

The purpose of this section is to derive a representation for $\hat{\xi}_t = E[\xi_t|\mathcal{O}_t]$ by carrying out calculations under a second probability measure $\mathcal{P}_0$. This approach finds its roots in the work of Duncan (1968), Mortensen (1966), and Zakai (1969). To begin, define a $\mathcal{P}$ local martingale U by

$$U = \mathcal{E}\left(-h \cdot w - \left(\frac{\lambda - 1}{\lambda}\right) \cdot M\right)$$

We invoke the following assumption:

Assumption: $\quad EU_T = 1 \quad (3.10)$

Define a new measure $\mathcal{P}_0$ on $(\Omega, \mathcal{Q})$ by

$$\frac{d\mathcal{P}_0}{d\mathcal{P}} = U_T$$

3. FILTER REPRESENTATION — CHANGE OF MEASURE DERIVATION

Proposition 3.1.

(a) $\mathcal{P} \ll \mathcal{P}_0$ and (with $n_t = N_t - t$) $d\mathcal{P}/d\mathcal{P}_0 = L_T$ where $L = \mathcal{E}(h \cdot y + (\lambda - 1) \cdot n)$. ($L$ is given by the right-hand side of Eq. (2.1) with $\hat{h}$ and $\hat{\lambda}$ replaced by h and λ.)

(b) The restrictions of $\mathcal{P}$ and $\mathcal{P}_0$ to $\mathcal{S}_T$ are the same.

(c) y is an $(\mathcal{O}_., \mathcal{P}_0)$ Wiener martingale.

(d) N is an $(\mathcal{O}_., \mathcal{P}_0)$ Poisson process.

(e) For each $s \geq 0$, the three σ-algebras

$$\mathcal{S}_T \vee \mathcal{O}_s, \quad \sigma(w_t - w_s : t \geq s) \quad \text{and} \quad \sigma(N_t - N_s : t \geq s)$$

are $\mathcal{P}_0$ independent. In particular (take $s = 0$), $\mathcal{S}_T$ and $\mathcal{O}_T$ are $\mathcal{P}_0$ independent.

Proof: Using the identity (6.7.3) we easily check that $L_t U_t = 1$, which implies (a). Next, U is an $(\mathcal{S}_T \vee \mathcal{O}_t, \mathcal{P})$ martingale with $U_0 = 1$ so that $E[U_T|\mathcal{S}_T] = 1$. Thus, for any bounded $\mathcal{S}_T$-measurable random variable X,

$$E_0 X = E[XU_T] = E[E[XU_T|\mathcal{S}_T]]$$
$$= E[XE[U_T|\mathcal{S}_T]] = EX$$

so that (b) is true. Now, computing under measure $\mathcal{P}$,

$$w_t - \langle -h \cdot w + \frac{\lambda - 1}{\lambda} \cdot M, w \rangle_t = y_t$$

and

$$M_t - \langle -h \cdot w + \frac{\lambda - 1}{\lambda} \cdot M, M \rangle_t = N_t - t = n_t$$

so by the abstract Girsanov's theorem, y and n are each $(\mathcal{O}_. \vee \mathcal{S}_T, \mathcal{P}_0)$ local martingales. Moreover, y is sample continuous with $[y, y]_t = t$ and N is a counting process. Properties (c)–(e) are then implied by the Levy-Watanabe characterization theorem, Proposition 6.7.1. ∎

Remark: We will no longer appeal directly to assumption (3.10), but will only use the fact that $\mathcal{P}_0$ satisfying the conclusions of Proposition 3.1 exists. This is important since Proposition 3.1 can be established under much less restrictive assumptions. In particular, it is not necessary to have $\mathcal{P}_0 \ll \mathcal{P}$.

By Lemma 6.7.1, if η is a measurable process with $E(\eta_t)$ finite for all t, then

$$\hat{\eta}_t = \check{\eta}_t / \check{1}_t \quad \mathcal{P} \text{ a.s. for each } t$$

where $\check{\eta}$ is any $\mathcal{O}_*$ progressive process such that

$$\check{\eta}_t = E_0[L_t \eta_t | \mathcal{O}_t] \tag{3.11}$$

Note that

$$\check{1}_t = E_0[L_t | \mathcal{O}_t] = \Lambda_t$$

where Λ is the likelihood ratio process for $\mathcal{P}$ versus $\mathcal{P}_0$ relative to $\mathcal{O}_*$.

It follows that if we define $\hat{\eta}$ to be any predictable process such that $\hat{\eta}_t = \eta_t$ $\mathcal{P}_0$ a.s. unless t is in some Borel subset D of $\mathbb{R}_+$ with Lebesgue measure zero, then we have

$$\hat{\eta}_t = \check{n}_t / \check{1}_t \qquad dt\, d\mathcal{P}(\omega) \qquad \text{a.e.}$$

We make the additional assumptions

$$E_0 \int_0^T (\xi_s L_s h_s)^2 \, ds < +\infty$$

$$E_0 \int_0^T |\xi_{s-} L_{s-}(\lambda_s - 1)| \, ds < +\infty$$

$$E_0 \int_0^T L_{s-}^2 \, d\langle m, m \rangle_s < +\infty \qquad \text{or} \qquad E_0 \int_0^T |L_{s-}| |dm_s| < +\infty$$

Proposition 3.2. Under the above assumptions,

$$\xi_t = \xi_0 + \int_0^t (\xi_s h_s)^{\check{}} \, dy_s + \int_0^t (\xi_{s-}(\lambda_s - 1))^{\check{}} \, dn_s + \int_0^t \check{\beta}_s \, ds \qquad \text{a.s. for } t \geq 0$$

Lemma 3.1. $[y, m]_t = [n, m]_t = 0$ for $t \geq 0$, $\mathcal{P}_0$ a.s.

Proof of Lemma 3.1: y, and therefore also $[y, m]$, is sample continuous so that

$$[y, m]_t = \langle y, m \rangle_t \qquad (\mathcal{Q}_*, \mathcal{P}_0)$$

If $s < t$ then $y_t - y_s$ and $m_t - m_s$ under $\mathcal{P}_0$ are conditionally independent and each has conditional mean zero given $\mathcal{Q}_t$. Thus $m_t y_t$ is an $(\mathcal{Q}_*, \mathcal{P}_0)$ martingale so that $\langle y, m \rangle = 0$, and thus $[y, m] = 0$.

Since n has locally finite variation,

$$[n, m]_t = \sum_{s \leq t} \Delta n_s \Delta m_s \tag{3.12}$$

Since the probability that n has a jump at any fixed time is zero and since n and m are independent under $\mathcal{P}_0$, it follows from (3.12) that $[n, m] = 0$. ∎

3. FILTER REPRESENTATION — CHANGE OF MEASURE DERIVATION

Lemma 3.2.

(a) $E_0[\int_0^t \beta_s L_s \, ds | \mathcal{O}_t] = \int_0^t \check{\beta}_s \, ds.$

Let ϕ be $\mathcal{Q}_\cdot$ predictable. Then

(b) $E_0[\int_0^t \phi_s L_{s-} \, dy_s | \mathcal{O}_t] = \int_0^t \check{\phi}_s \, dy_s$ if $E_0 \int_0^T |\phi_s L_s|^2 \, ds < +\infty,$

(c) $E_0[\int_0^t \phi_s L_{s-} \, dn_s | \mathcal{O}_t] = \int_0^t \check{\phi}_s \, dn_s$ if $E_0 \int_0^T |\phi_s L_s| \, ds < +\infty,$

(d) $E_0[\int_0^t \phi_s \, dm_s | \mathcal{O}_t] = 0$ if $E_0 \int_0^T \phi_s^2 \, d\langle m, m \rangle_s < +\infty$ or if $E_0 \int_0^T |\phi_s| |dm_s| < +\infty.$

Proof of Lemma 3.2: First,

$$E_0\left[\int_0^t \beta_s L_s \, ds | \mathcal{O}_t\right] = \int_0^t E_0[\beta_s L_s | \mathcal{O}_t] \, ds \tag{3.13}$$

if a jointly measurable version of the integrand on the right-hand side is chosen. Now for $s \leq t$,

$$\mathcal{O}_t = \mathcal{O}_s \vee \sigma\{w_u - w_s, N_u - N_s : s \leq u \leq t\}$$

and the second σ-algebra on the right-hand side is $\mathcal{P}_0$ independent of $\mathcal{Q}_s$. Since $\beta_s L_s$ is $\mathcal{Q}_s$ measurable and $\mathcal{Q}_s \supset \mathcal{O}_s$, we claim that

$$E_0[\beta_s L_s | \mathcal{O}_t] = E_0[\beta_s L_s | \mathcal{O}_s] \quad \text{a.s.} \tag{3.14}$$

To establish Eq. (3.14) it suffices to prove that

$$E_0[\beta_s L_s Z] = E_0[E_0[\beta_s L_s | \mathcal{O}_s] Z]$$

for all bounded $\mathcal{O}_t$-measurable random variables Z, which is easily done by considering Z of a special form and applying the monotone class theorem. Now, the right-hand side of (3.14) is, by definition, equal to $\check{\beta}_s$. Equations (3.13) and (3.14) thus imply part (a) of the lemma.

Part (b) is proved similarly:

$$E_0\left[\int_0^t \phi_s L_{s-} \, dy_s | \mathcal{O}_t\right] = \int_0^t E_0[\phi_s L_{s-} | \mathcal{O}_s] \, dy_s$$
$$= \int_0^t \check{\phi}_s \, dy_s$$

The first equality can be proved by reducing it to the case that $\gamma_s L_{s-}$ is replaced by a bounded $\mathcal{Q}_\cdot$ predictable step function, and the second equality is true by the definition of $\check{\phi}_s$. The proof of part (c) is similar.

The fact that m is an $(\mathcal{S}_\cdot, \mathcal{P}_0)$ martingale and that $\mathcal{O}_T$ is $\mathcal{P}_0$ independent of $\mathcal{S}_T$ implies that m, and therefore $(\phi \cdot m_t)$, is a $(\mathcal{P}_0, \mathcal{S}_t \vee \mathcal{O}_T)$

martingale. Thus,

$$E_0[\phi \cdot m_t|\mathcal{O}_T] = \phi \cdot m_0 = 0$$

and Lemma 3.2 is proved.

Proof of Proposition 3.2: By the stochastic integral equation for semimartingale exponentials,

$$L_t = 1 + \int_0^t L_{s-} h_s \, dy_s + \int_0^t L_{s-} (\lambda_s - 1) \, dn_s$$

so that, using Lemma 3.1,

$$[L, \xi]_t = \int_0^t L_{s-} h_s \, d[y, m]_s + \int_0^t L_{s-} (\lambda_s - 1) \, d[n, m]_s = 0$$

Using this and Itô's product formula we have

$$L_t \xi_t = \xi_0 + \int_0^t \xi_{s-} \, dL_s + \int_0^t L_{s-} \, d\xi_s$$

or

$$L_t \xi_t = \xi_0 + \int_0^t \xi_{s-} L_{s-} h_s \, dy_s + \int_0^t \xi_{s-} L_{s-} (\lambda_s - 1) \, dn_s$$
$$+ \int_0^t L_{s-} \beta_s \, ds + \int_0^t L_{s-} \, dm_s \qquad (3.15)$$

The proof is completed by taking the conditional $\mathcal{P}_0$ expectation given $\mathcal{O}_t$ of each term in (3.15), and using Lemma 3.2. ■

4. FILTER REPRESENTATION — INNOVATIONS DERIVATION

The so-called "innovations method" is applied to this section to derive another representation of $\hat{\xi}$. In this section we assume once again that $\mathcal{O}_{\cdot}$ is generated by processes y and N, that ξ is a process to be estimated, and that $\mathcal{Q}_{\cdot}$ is an increasing family of σ-algebras satisfying the usual conditions with $\mathcal{O}_t \subset \mathcal{Q}_t$. (The σ-algebras $\mathcal{S}_{\cdot}$ of the last section are not used in this section.) We assume that ξ, y, and N have a representation of the form (3.2)–(3.4), where now we assume that

ξ_0 is $\mathcal{Q}_0$ measurable, $E|\xi_0| < +\infty$

β is $\mathcal{Q}_{\cdot}$ predictable, $E \int_0^T (\beta_s)^2 \, ds < +\infty$

m is an $\mathcal{Q}_{\cdot}$ martingale, $Em_T^2 < +\infty$

4. FILTER REPRESENTATION — INNOVATIONS DERIVATION

$$\mathcal{O}_t = \sigma(y_s, N_s : s \le t) \vee \{\mathcal{P} \text{ null sets}\}$$

h is $\mathcal{O}_\cdot$ progressive, $E\int_0^T h_s^2\, ds < +\infty$

w is an $\mathcal{O}_\cdot$ Wiener martingale

N is an $\mathcal{O}_\cdot$ adapted counting process

λ is $\mathcal{O}_\cdot$ predictable, $E\int_0^T \lambda_s\, ds < +\infty,\ \lambda_s > 0$

M is an $\mathcal{O}_\cdot$ martingale

We will also make the following assumptions, which can be relaxed.

$\lambda_t \ge \varepsilon \quad$ for all (t,ω), where $\varepsilon > 0$

$E\int_0^T \xi_t^2 \lambda_t\, dt < +\infty \quad$ (e.g., λ bounded)

By the argument used in the first part of the proof of Proposition 6.7.3, there exist $\mathcal{O}_\cdot$ predictable processes ϕ and ψ so that

$$\langle m, w \rangle_t = \int_0^t \phi_s\, ds$$
$$\langle m, M \rangle_t = \int_0^t \psi_s\, ds \qquad 0 \le t \le T,\ \mathcal{P} \text{ a.s.}$$

Proposition 4.1. Under the above assumptions,

$$\hat{\xi}_t = \hat{\xi}_0 + \int_0^t \hat{\beta}_s\, ds + \int_0^t H_s\, d\tilde{y}_s + \int_0^t K_s\, d\tilde{N}_s \tag{4.1}$$

where H and K are the predictable processes given by

$$H_s = ((\xi_s - \hat{\xi}_s)(h_s - \hat{h}_s))\hat{}\ + \hat{\phi}_s \qquad ds\, d\mathcal{P}(\omega) \text{ a.e.} \tag{4.2}$$

$$K_s = \frac{((\xi_s - \hat{\xi}_s)(\lambda_s - \hat{\lambda}_s))\hat{}\ + \hat{\psi}_s}{\hat{\lambda}_s} \qquad ds\, d\mathcal{P}(\omega) \text{ a.e.} \tag{4.3}$$

and $\tilde{y}$ and $\tilde{N}$ are the innovations processes

$$\tilde{y}_t = y_t - \int_0^t \hat{h}_s\, ds \quad \text{and} \quad \tilde{N}_t = N_t - \int_0^t \hat{\lambda}_s\, ds$$

The truth of the proposition is directly implied by the next three lemmas. Our proof is essentially that of Fujisaki, Kallianpur, and Kunita (1972).

Lemma 4.1. There exists an $(\mathcal{O}_\cdot, \mathcal{P})$ martingale D with $D_0 = 0$ and

$ED_T^2 < +\infty$ such that

$$\hat{\xi}_t = \hat{\xi}_0 + \int_0^t \hat{\beta}_s \, ds + D_t \qquad 0 \leq t \leq T, \text{a.s.}$$

Proof: This is simply an application of the projection result, Proposition 1.3. ∎

Lemma 4.2. There exist predictable processes H and K such that

$$\int_0^T H_s^2 + K_s^2 \, ds < +\infty \qquad \mathcal{P} \text{ a.s.}$$

and

$$D_t = \int_0^t H_s \, d\tilde{y}_s + \int_0^t K_s \, d\tilde{N}_s \qquad 0 \leq s \leq T, \mathcal{P} \text{ a.s.}$$

Proof: Define $X = -\hat{h} \cdot \tilde{y} - (1 - 1/\hat{\lambda}) \cdot \tilde{N}$ and let $L = \mathcal{E}(X)$, or

$$L_t = \exp\left(-\hat{h} \cdot \tilde{y}_s - \int_0^s \left(\frac{1}{2}\hat{h}_s^2 + \hat{\lambda}_s - 1\right) ds\right) \prod_{\substack{s \leq t \\ \Delta N_s = 1}} \frac{1}{\hat{\lambda}_s}$$

Let $\tau_n = \inf\{t: L_t \geq n \text{ or } \langle X, X \rangle_t \geq n\}$ and define X^n and L^n by $X_t^n = X_{t \wedge \tau_n}$ and $L_t^n = L_{t \wedge \tau_n}$. Since $L_t < n$ for $t < \tau_n$ and $L_t \leq L_{t^-}/\varepsilon$, we have that $L_t^n \leq n/\varepsilon$ for all t. By computation we see that $\langle X, X \rangle$ is sample continuous, also $\langle X^n, X^n \rangle = \langle X, X \rangle_{t \wedge \tau_n}$. Thus, both L^n and $\langle X^n, X^n \rangle$ are bounded (uniformly in t, ω). Since L^n is a bounded local $\mathcal{O}_\cdot$ martingale it is actually a martingale, so EL_t^n is equal to one. Define a probability measure $\mathcal{P}^n$ by

$$d\mathcal{P}^n = L_T^n \, d\mathcal{P}$$

and then by the abstract Girsanov's theorem,

$$\tilde{y} - \langle \tilde{y}, X^n \rangle, \quad \tilde{N} - \langle \tilde{N}, X^n \rangle \quad \text{and} \quad D - \langle D, X^n \rangle$$

are $\mathcal{P}^n$ local martingales. Now

$$\tilde{y}_t - \langle \tilde{y}, X^n \rangle_t = w_t \qquad \text{for } t \leq \tau_n$$

and

$$\tilde{N}_t - \langle \tilde{N}, X^n \rangle_t = N_t - t \qquad \text{for } t \leq \tau_n$$

Thus, under measure $\mathcal{P}^n$, w up to time τ_n is an $\mathcal{O}_\cdot$ Wiener martingale and N up to time τ_n is an $\mathcal{O}_\cdot$ Poisson process.

4. FILTER REPRESENTATION — INNOVATIONS DERIVATION

By inequality (6.4.3),

$$\langle D, X^n \rangle_t^2 \leq \langle X^n, X^n \rangle_t \langle D, D \rangle_t$$
$$\leq (\text{const.}) \langle D, D \rangle_t$$

so that $\langle D, X^n \rangle_t^2$, as well as D_t^2, has finite expectation under $\mathcal{P}$. Since L^n is bounded, both $\langle D, X^n \rangle_t^2$ and D_t^2 have finite means under $\mathcal{P}^n$ as well. Thus, $D - \langle D, X^n \rangle$ is an $(\mathcal{O}_\cdot, \mathcal{P}^n)$ square integrable martingale with initial value zero, so by the representation theorem of Section 6.7,

$$D_t - \langle D, X^n \rangle_t = \int_0^t H_s \, dy_s + \int_0^t K_s \, d(N_s - s) \qquad \text{for } t \leq \tau_n$$

where H and K are predictable with

$$E^n \int_0^T H_s^2 + K_s^2 \, ds < +\infty$$

Therefore, for $t \leq \tau_n$,

$$D_t = \int_0^t H_s \, d\tilde{y}_s + \int_0^t K_s \, d\tilde{N}_s$$
$$+ \left\{ \langle D, X \rangle_t + \int_0^t H_s \hat{h}_s \, ds + \int_0^t K_s (\hat{\lambda}_s - 1) \, ds \right\}$$

Now the sum of terms in brackets is an $(\mathcal{O}_\cdot, \mathcal{P}^n)$ martingale up to time τ_n (since the other terms are) and it is predictable and has locally finite variation. It is thus zero, so we have

$$D_t = \int_0^t H_s \, d\tilde{y}_s + \int_0^t K_s \, d\tilde{N}_s \qquad t \leq \tau_n$$

Now, letting n tend to infinity and patching together processes H and K, we obtain

$$D_t = \int_0^t H_s \, d\tilde{y}_s + \int_0^t K_s \, d\tilde{N}_s \qquad 0 \leq t \leq T$$

where H and K are predictable processes with

$$\int_0^T H_s^2 \, ds < +\infty \quad \text{and} \quad \int_0^T K_s^2 \, ds < +\infty \qquad \text{a.s.} \quad \blacksquare$$

Lemma 3. *H and K in Lemma 4.2 are given by Eqs. (4.2) and (4.3).*

Proof: (See the remark below for a heuristic approach.) Define a sequence

(S_n) of $\mathcal{O}$. stopping times by

$$S_n = \inf\{t : |y_t| \geq n\}$$

Then, since y has continuous sample paths, $|y_t|$ is bounded above by n for $0 \leq t \leq S_n$. In what follows we will use μ to denote Lebesgue measure so that, for example, $y\beta \cdot \mu$ represents the process

$$y\beta \cdot \mu_t = \int_0^t y_s \beta_s \, ds$$

Applying Itô's product formula, using the representations (3.2) and (3.3) for ξ and y, and using the fact that $[m, w] = \langle m, w \rangle$ since w is sample continuous, we obtain

$$\xi y = y\beta \cdot \mu + y \cdot m + \xi h \cdot \mu + \xi \cdot w + \langle m, w \rangle \tag{4.4}$$

so that

$$(\xi y)_{t \wedge S_n} = (y\beta + \xi h + \phi) \cdot \mu_{t \wedge S_n} + \text{an } \mathcal{C}. \text{ martingale}$$

Thus, projecting this process onto the observations σ-algebras (Proposition 1.3) yields

$$\widehat{(\xi y)_{t \wedge S_n}} = (y\beta + \xi h + \phi)\hat{\ } \cdot \mu_{t \wedge S_n} + \text{an } \mathcal{O}. \text{ martingale} \tag{4.5}$$

On the other hand, another application of Itô's product formula and use of the representation (4.1) for $\hat{\xi}$ (which is valid by Lemmas 4.1 and 4.2) gives

$$\hat{\xi} y = (y\hat{\beta} + \hat{\xi}\hat{h} + H) \cdot \mu + \text{an } \mathcal{O}. \text{ local martingale} \tag{4.6}$$

Using the fact that

$$(y\beta)\hat{\ } = y\hat{\beta} \quad \text{and} \quad \widehat{(\xi y)_{t \wedge S_n}} = (\hat{\xi} y)_{t \wedge S_n}$$

and comparing Eqs. (4.5) and (4.6) we find that

$$\left(\widehat{\xi h} - \hat{\xi}\hat{h} + \hat{\phi} - H\right) \cdot \mu_{t \wedge S_n}$$

is an $\mathcal{O}$. local martingale. It is also predictable, has initial value zero, and has locally finite variation corlol sample paths. It is hence zero for all t with probability one, so that H_t must be given by Eq. (4.2) for $t \leq S_n$. Since S_n increases to T as n tends to infinity, Eq. (4.2) is thus true in general.

The identification of K is similar: First define a sequence (T_n) of $\mathcal{O}$. stopping times by

$$T_n = \inf\{t : N_t \geq n\}$$

4. FILTER REPRESENTATION — INNOVATIONS DERIVATION

Since the jumps of N have size at most one, N_t is bounded above by n for $0 \leq t \leq T_n$. The analogue of Eq. (4.4) is

$$\xi N = N\beta \cdot \mu + N_- \cdot m + \xi\lambda \cdot \mu + \xi_- \cdot M$$
$$+ ([m, M] - \langle m, M \rangle) + \langle m, M \rangle$$

where N_- at s is equal to N_{s-} (and similarly for ξ_-). Since $[m, M] - \langle m, M \rangle$ is an $\mathcal{Q}_\cdot$ martingale, we have

$$(\widetilde{\xi N})_{t \wedge T_n} = (N\beta + \xi\lambda + \psi)\hat{\ } \cdot \mu_{t \wedge T_n} + \text{an } \mathcal{Q}_\cdot \text{ martingale}$$

which is analogous to Eq. (4.5). The rest of the proof for identifying K is so similar to that for identifying H that we omit it. ∎

Remark: Since the proof of Lemma 4.3 is somewhat mysterious, we will give a heuristic derivation of the equations for H and K by appealing directly to the "orthogonality principle." Fix t, let dt be a small positive number, and use the notation $d\xi_t = \xi_{t+dt} - \xi_t$, etc. We know that

$$E\big((\xi_{t+dt} - \hat{\xi}_{t+dt})Y\big) = 0$$

for any square integrable $\mathcal{O}_{t+dt}$-measurable random variable Y. Since $d\tilde{y}_t$ is $\mathcal{O}_{t+dt}$ measurable, this implies that

$$E\big[(\xi_{t+dt} - \hat{\xi}_{t+dt}) d\tilde{y}_t | \mathcal{O}_t \big] = 0 \quad \text{a.s.}$$

Using the relationships

$$d\tilde{y}_t = (h_t - \hat{h}_t) dt + dw_t + o(dt) = dw_t + O(dt)$$
$$\xi_{t+dt} - \hat{\xi}_{t+dt} = \xi_t - \hat{\xi}_t + d\xi_t - d\hat{\xi}_t$$
$$= \xi_t - \hat{\xi}_t + dm_t - K_t d\tilde{y}_t - H_t d\tilde{N}_t + O(dt)$$
$$E\big[dw_t \, d\tilde{y}_t | \mathcal{O}_t\big] = E\big[dw_t(dw_t + O(dt))|\mathcal{O}_t\big] = dt + o(dt)$$
$$E\big[dw_t \, dm_t | \mathcal{O}_t\big] = E\big[d\langle w, m \rangle_t | \mathcal{O}_t\big] = \hat{\phi}_t \, dt + o(dt)$$
$$E\big[(\xi_t - \hat{\xi}_t) dw_t + (h_t - \hat{h}_t)(dm_t - K_t d\tilde{y}_t - H_t d\tilde{N}_t)|\mathcal{O}_t\big] = o(dt)$$

then implies that

$$E\big[((\xi_t - \hat{\xi}_t)(h_t - \hat{h}_t) + \hat{\phi}_t - H_t) dt | \mathcal{O}_t\big] = o(dt)$$

which implies (heuristically) Eq. (4.2). This derivation can be made rigorous, but only with much perseverance. Equation (4.3) for K can be obtained similarly.

Example [Kalman and Bucy, 1961]. Let ξ be the unique solution to the linear stochastic differential equation

$$d\xi_t = a\xi_t dt + b\,dv_t; \quad \xi_0 \sim N(\mu_0, \sigma_0^2) \tag{4.7}$$

and let the observation process y be given by

$$y_t = \int_0^t \xi_s\,ds + w_t$$

where v and w are Wiener processes, ξ_0 is Gaussian, and ξ_0, v, and w are mutually independent. Suppose that there are no counting process observations. Then ξ is a Markov process and ξ and y are jointly Gaussian random processes. By Proposition 4.1

$$d\hat{\xi}_t = a\hat{\xi}_t dt + H_t(dy_t - \hat{\xi}_t dt); \quad \hat{\xi}_0 = \mu_0 \tag{4.8}$$

where

$$H_t = \left((\xi_t - \hat{\xi}_t)^2\right)^\wedge = \widehat{\xi_t^2} - (\hat{\xi}_t)^2 \tag{4.9}$$

By Itô's formula and Eq. (4.7),

$$d\xi_t^2 = 2\xi_t d\xi_t + b^2\,dt = \left(2a\xi_t^2 + b^2\right)dt + 2b\xi_t\,dv_t$$

so that by Proposition 4.1

$$d\widehat{\xi_t^2} = \left(2a\widehat{\xi_t^2} + b^2\right)dt + \left((\xi_t - \hat{\xi}_t)(\xi_t^2 - \widehat{\xi_t^2})\right)^\wedge d\tilde{y}_t \tag{4.10}$$

Itô's formula and Eq. (4.8) yield

$$d(\hat{\xi}_t)^2 = \left(2a(\hat{\xi}_t)^2 + H_t^2\right)dt + 2\hat{\xi}_t H_t\,d\tilde{y}_t \tag{4.11}$$

Thus, by Eqs. (4.9)–(4.11),

$$dH_t = \left(2aH_t + b^2 - H_t^2\right)dt + \left[(\xi_t - 3\hat{\xi}_t)(\xi_t^2 - \widehat{\xi_t^2})\right]^\wedge d\tilde{y}_t$$

The coefficient of $d\tilde{y}_t$ can be reexpressed as

$$\widehat{(\xi_t - \hat{\xi}_t)^3}$$

which is the conditional third central moment of ξ_t given $\mathcal{O}_t$. Since ξ_t is conditionally Gaussian, this moment is zero, so that H satisfies the

5. RECURSIVE ESTIMATION

deterministic differential equation

$$\frac{dH_t}{dt} = 2aH_t + b^2 - H_t^2; \quad H_0 = \sigma_0^2 \tag{4.12}$$

Equations (4.8) and (4.12) are a special case of the Kalman-Bucy filter and associated Riccati equation given by Eqs. (3.9.37) and (3.9.39). These equations provide a recursive method for computing $\hat{\xi}_t$ (see Section 5).

5. RECURSIVE ESTIMATION

The estimation equations of the previous two sections are specialized in this section to the case that a Markov process X represents an unobserved state, and conditional moments

$$\widehat{g(X)}_t = E[g(X_t)|\mathcal{O}_t]$$

are to be computed. The goal of recursive filtering is to find a process $(U_t: 0 \le t \le T)$ taking values in a possibly infinite dimensional space such that the following conditions are satisfied:

- U_{t+dt} depends on U_t and the new observations (dy_t, dN_t).
- For each t the desired conditional moments $\widehat{g(X)}_t$ are completely determined by U_t.

Let the families of σ-algebras $(\mathcal{S}_t)$ and $(\mathcal{O}_t)$ represent state and observation information, respectively, as in Section 7.3. Suppose that X is a Markov process with respect to $(\mathcal{S}_t)$ with a stationary transition function

$$P = (P_x(A, t) : x \in S, A \in \mathcal{B}(S), t \in [0, T])$$

that is, X is $\mathcal{S}_.$ adapted,

$$\mathcal{P}(X_{T+s} \in A|\mathcal{S}_t) = P_{X_t}(A, s) \quad \text{a.s. for } A \in \mathcal{B}(S)$$

and P satisfies the Chapman-Kolmogorov equation, Eq. (5.1.2). We assume that the state space S of X is a separable metric space (for example, S can be an interval of the line as in Chapter 5, a subset of Euclidean n space, or a finite or countably infinite set), we let $\mathcal{B}(S)$ denote the collection of Borel subsets of S, and we assume that the sample paths of X are corlol. The treatment of Markov processes in Sections 5.1–5.3 immediately carries over to this setting. In particular, the generator of X is a linear operator A, the domain of which, $\mathcal{D}_A$, is a subset of the Banach space of bounded $\mathcal{B}(S)$-measurable functions on S.

We assume that the observation σ-algebras ($\mathcal{O}_t$) are generated by y and N as in (3.1), where

$$y_t = \int_0^t h(X_s)\,ds + w_t \tag{5.1}$$

$$N_t = \int_0^t \lambda(X_s)\,ds + M_t \tag{5.2}$$

and we suppose that conditions (3.5)–(3.9) are true with $h_s = h(X_s)$ and $\lambda_s = \lambda(X_{s-})$.

To apply the estimation equations of Propositions 3.2 and 4.1 to processes ξ of the form $\xi_t = g(X_t)$, we must find a semimartingale representation of ξ as in Eq. (3.2). If g is in $\mathcal{D}_A$, then an easy modification of the proof of Dynkin's identity given in Chapter 5 yields that for $0 \le s \le t$,

$$E\left[\int_s^t Ag(X_u)\,du \mid \mathcal{S}_s\right] = E[g(X_t) \mid \mathcal{S}_s] - g(X_s) \quad \text{a.s.}$$

(since here s and t are not random, it is not necessary that X be strong Markov). Equivalently, C^g defined by

$$C_t^g = g(X_t) - \int_0^t Ag(X_s)\,ds \tag{5.3}$$

is an $\mathcal{S}_\cdot$ martingale for each g in $\mathcal{D}_A$.

Remark: The fact that C_t^g in Eq. (5.3) is a martingale for g in $\mathcal{D}_A$ characterizes Ag. Moreover, if we use such martingale property as a definition for Ag, we can extend the operator A to a much larger domain (possibly including some unbounded functions).

Thus, if $g \in \mathcal{D}_A$ then $\xi_t = g(X_t)$ has the desired representation (3.2) with $\beta_s = Ag(X_{s-})$ and $M_t = C_t^g$. Proposition 3.2 then yields the representation

$$\widehat{dg(X)}_t = \widehat{Ag(X)}_t\,dt + \widehat{g(X)h(X)}_t\,dy_t + \widecheck{(g(X)(\lambda(X)-1))}_t\,dn_t$$

where, as always, $n_t = N_t - t$. This equation is written in differential form, which is simply a shorthand way to write an integral equation. This equation implies that there is a corlol version of $\widehat{g(X)}_t$. Suppose that the product $g\lambda$ is in $\mathcal{D}$. Then there exists a corlol version of $\widecheck{(g(X)\lambda(X))}_t$ and so

$$\widecheck{(g(X)(\lambda(X)-1))}_t = \widecheck{(g(X)(\lambda(X)-1))}_{t-}$$

5. RECURSIVE ESTIMATION

In that case we have

$$\overline{dg(X)}_t = \overline{Ag(X)}_t dt + \overline{g(X)h(X)}_t dy_t + \overline{g(X)(\lambda(X)-1)}_{t-} dn_t \tag{5.4}$$

The right-hand side of this equation involves the observations y and N, and unnormalized moments corresponding to functions Ag, gh and $g(\lambda - 1)$. If $\mathcal{U}$ is a subset of $\mathcal{D}_A$ such that (using 1 to denote the function identically equal to one)

$$1, \quad gh, \quad g(\lambda - 1), \quad Ag \in \text{span}(\mathcal{U}) \quad \text{whenever } g \in \mathcal{U}$$

then a set of equations of the form (5.4), one for each g in $\mathcal{U}$, yields a stochastic differential equation for

$$U_t = \left(\overline{g(X)}_t : g \in \mathcal{U} \right)$$

The equation is driven by the observations y and N. Since

$$\widehat{g(X)}_t = \overline{g(X)}_t / \check{1}_t = \text{a function of } U_t$$

U satisfies the requirements of recursive filtering discussed at the beginning of the section. If $\mathcal{U}$ is sufficiently large, then as we see next, U_t determines the conditional distribution of X_t given $\mathcal{O}_t$, so that recursive equations can be obtained for that distribution.

Example [Wonham, 1964]. Suppose the statespace S is $\{1, 2, \ldots, n\}$. Functions and measures on S are represented by column and row vectors, respectively. The generator A is then represented by an $n \times n$ matrix which we also call A, so that for a function f on S, it is consistent to interpret Af as the product of a matrix A and a column vector f.

Let δ_i denote the function on S such that $\delta_i(z)$ is one for $z = i$ and is zero otherwise, and let Π denote the vector process defined by

$$\Pi_t = \left(\overline{\delta_1(X)}_t, \overline{\delta_2(X)}_t, \ldots, \overline{\delta_n(X)}_t \right)$$

For a function f on S we then have

$$\overline{f(X)}_t = \Pi_t f \quad \text{and} \quad \widehat{f(X)}_t = \frac{\Pi_t f}{\Pi_t 1} \tag{5.5}$$

where 1 denotes a column vector of all ones.

Equations (5.4) yield a representation for each of the coordinates of Π, and these representations can be written in vector form as

$$d\Pi_t = \Pi_t A\, dt + \Pi_t \,\mathrm{diag}(h)\, dy_t + \Pi_{t-}\,\mathrm{diag}(\lambda - 1)\, dn_t \tag{5.6}$$

where $\mathrm{diag}(v)$ denotes the $n \times n$ diagonal matrix with the elements of v down the diagonal.

Equation (5.6) shows that Π_t can be computed "recursively" from the observations processes y and N, and by Eq. (5.5) the conditional moments $\widehat{f(X)}_t$ are easily expressed in terms of Π.

We will now obtain an analogue of Eq. (5.6) for the general case. For disjoint sets A_n in $\mathcal{B}(S)$,

$$\sum_{n=1}^{\infty} \mathcal{P}[Z_t \in A_n | \mathcal{O}_t] = \mathcal{P}\left[Z_t \in \bigcup_n A_n | \mathcal{O}_t\right] \quad \text{a.s.} \tag{5.7}$$

and since S is separable there exist versions of the conditional expectations so that the exceptional set in (5.7) does not depend on the sets A_n [Doob 1953, p. 29]. With more care, a process Q can be constructed so that

$$Q(\cdot, t, \omega) \text{ is a positive measure for } t, \omega \text{ fixed}$$
$$Q(A, \cdot, \cdot) \text{ is } \mathcal{O}_{\cdot} \text{ progressive for each } A \in \mathcal{B}(S) \tag{5.8}$$
$$\widehat{f(X)}_t = \int_S f(x) Q_t(dx)$$

If there is a process $(q_t(x, \omega))$ and a (deterministic) measure l on S such that

$$\widehat{f(X)}_t = \int_S f(x) q_t(x, \omega) l(dx) \quad f \text{ bounded, } \mathcal{B}(S) \text{ measurable}$$

then q_t is the density of Q_t with respect to l.

Similarly, there exists a process Π which satisfies the same conditions (5.8) as Q but with $\widehat{f(X)}_t$ replaced by $\overline{f(X)}_t$. If there is a process $\pi_t(x, \omega)$ such that

$$\overline{f(X)}_t = \int_S f(x) \pi_t(x, \omega) l(dx) \quad f \text{ bounded, } \mathcal{B}(S) \text{ measurable}$$

then π_t is the density of Π_t with respect to l.

5. RECURSIVE ESTIMATION

The bilinear product of a measure μ on $\mathcal{B}(S)$ and a bounded $\mathcal{B}(S)$-measurable function f is defined by

$$(\mu, f) = \int_S f \, d\mu$$

The adjoint A^* of A relative to this product is characterized by

$$(\mu, Af) = (A^*\mu, f) \qquad f \in \mathcal{D}_A$$

(We will not enter into a discussion of the domain of A^*.) If l is a reference measure on $\mathcal{B}(S)$ we use $(\cdot, \cdot)_l$ to denote the usual inner product

$$(f, g)_l = \int_S f(x) g(x) l(dx)$$

The adjoint $\mathcal{L}^*$ of A relative to this product is characterized by

$$(g, Af)_l = (\mathcal{L}^* g, f)_l$$

For example, if X is a diffusion with drift term $m(x)$ and diffusion term $\sigma^2(x)$ as described in Section 4.7, then

$$Af(x) = m(x)\frac{\partial f}{\partial x}(x) + \frac{1}{2}\sigma^2(x)\frac{\partial^2 f}{\partial x^2}(x)$$

and, by an integration by parts, if g is a Schwartz function of rapid descent, then the adjoint $\mathcal{L}^*$ of A relative to Lebesgue measure satisfies

$$\mathcal{L}^* g(x) = -\frac{\partial (mg)}{\partial x}(x) + \frac{1}{2}\frac{\partial^2(\sigma^2 g)}{\partial x^2}(x)$$

Now, Eq. (5.4) can be written

$$(g, \Pi_t) = (g, \Pi_0) + \int_0^t (Ag, \Pi_s) \, ds + \int_0^t (gh, \Pi_s) \, dy_s$$
$$+ \int_0^t (g(\lambda - 1), \Pi_{s-}) \, dn_s$$

or, using M_f to denote the operator on measures defined by $M_f \mu(A) = \int_A f \, d\mu$, we have

$$(g, \Pi_t) = \left(g, \Pi_0 + \int_0^t A^*\Pi_s \, ds + \int_0^t M_h \Pi_s \, dy_s + \int_0^t M_{\lambda-1}\Pi_{s-} \, dn_s\right)$$

This equation is true for g in $\mathcal{D}_A$ and $\mathcal{D}_A$ is dense in the space of all bounded $\mathcal{B}(S)$-measurable functions so we conclude that

$$d\Pi_t = A^*\Pi_t \, dt + M_h \Pi_t \, dy_t + M_{\lambda-1}\Pi_{t-} \, dn_t \tag{5.9}$$

This is a stochastic differential equation (actually, an integral equation) for Π driven by the observations. Since Π_t determines all the moments $\widehat{f(X)}_t$, Eq. (5.9) represents a possibly infinite dimensional filter. If Π has a density π with respect to some fixed reference measure l, then this equation becomes

$$d\pi_t(x) = \mathcal{L}^*\pi_t(x) \, dt + h(x)\pi_t(x) \, dy_t + (\lambda(x) - 1)\pi_{t-}(x) \, dn_t \tag{5.10}$$

Similarly, starting with Proposition 4.1, we can obtain a recursive equation for the normalized conditional density q:

$$dq_t(x) = \mathcal{L}^*q_t(x) \, dt + q_t(x)\left[h(x) - \widehat{h(X)}_t\right]\left(dy_t - \widehat{h(X)}_t \, dt\right)$$

$$+ q_{t-}(x)\left[\frac{\lambda(x) - \widehat{\lambda(X)}_{t-}}{\widehat{\lambda(X)}_{t-}}\right](dN_t - \hat{\lambda}_t \, dt) \tag{5.11}$$

Ancestors of Eq. (5.11) were first given by Stratonovich (1960) and Kushner (1967). In the case that h and λ are identically constant functions (equivalent to no observations) this equation reduces to $dq_t/dt = \mathcal{L}^*q_t$, which is the Kolmogorov forward equation for the family of densities of X_t as t varies.

Remark [Davis, 1980]. Consider Eq. (5.10) for an unnormalized conditional density when $\lambda \equiv 1$ (equivalent to no counting process observations):

$$d\pi_t(x) = \mathcal{L}^*\pi_t(x) \, dt + h(x)\pi_t(x) \, dy_t \tag{5.12}$$

Seek a solution of the form

$$\pi_t(x) = r(t, x) \exp(h(x) y_t)$$

Then $r(t, x) = \pi_t(x) \exp(-h(x) y_t)$, so we can apply Itô's formula and

5. RECURSIVE ESTIMATION

Eq. (2.12) to deduce that

$$dr(t,x) = \left(B_t^* - \tfrac{1}{2}h(x)^2 \right) r(t,x)\, dt \tag{5.13}$$

where B_t^* is the (random) operator defined by

$$B^*f(x) = \exp(-h(x)y_t)\mathcal{L}^*(\exp(h(\cdot)y_t)f(\cdot))(t,x)$$

for sufficiently regular functions f on S. The important point is that there is no dy_t term in Eq. (5.13). This affords the possibility that Eq. (5.12) for r can be solved *pathwise*, and this then gives a method for constructing π_t, and therefore all conditional moments $\overline{f(X)}_t$, as pathwise functions of the process y.

A different tactic for finding recursive filtering equations is to work directly from the definition (3.11) of $\overline{f(X)}_t$. Let $\mathcal{P}_0$ and L be defined as in Section 7.3 and use $P_\mu(A, t)$ to denote $\mathcal{P}(X_t \in A)$. By Proposition 3.1,

$$P_\mu(A, t) = \mathcal{P}_0(X_t \in A) = \mathcal{P}_0(X_t \in A | \mathcal{O}_t)$$

and so for bounded $\mathcal{B}(S)$-measurable functions f,

$$\begin{aligned}
\overline{f(X)}_t &= E_0\left[E_0\left[f(X_t) L_t | \mathcal{O}_t, X_t \right] | \mathcal{O}_t \right] \\
&= \int_S f(x) E_0\left[L_t | \mathcal{O}_t, X_t = x \right] \mathcal{P}_0\left[X_t \in dx | \mathcal{O}_t \right] \\
&= \int_S f(x) \pi_t(x) P_\mu(dx, t)
\end{aligned}$$

where π_t is defined by

$$\pi_t(x) = E_0\left[L_t | \mathcal{O}_t, X_t = x \right] \tag{5.14}$$

Thus, π_t is the density of Π_t with respect to the unconditional distribution of X, $P_\mu(\cdot, t)$.

Example. Let $X_t = X$ for all t for some random variable X. Suppose that h and λ have the form

$$h_t = a(X)u(t) \quad \text{and} \quad \lambda_t = \exp(b(X)v(t))$$

for bounded measurable functions a, b, u, and v. Define random processes

$$\Phi_t = \int_0^t u(s)\, dy_s \quad \text{and} \quad \Psi_t = \int_0^t v(s)\, dN_s$$

and define a function $F: \mathbb{R}^4 \to \mathbb{R}$ by

$$F(x, \phi, \psi, t) = \exp\left(a(x)\phi + \frac{1}{2}\int_0^t a(x)^2 u(s)^2\, ds + b(x)\psi \right.$$
$$\left. + \int_0^t b(x)v(s) - 1\, ds\right)$$

Then Eq. (5.14) becomes

$$\pi_t(x) = F(x, \Phi_t, \Psi_t, t)$$

Note that π_t determines all the moments $\widehat{f(X)}_t$ and that Φ and Ψ can be recursively updated using the equations

$$d\Phi_t = u(t)\, dy_t, \qquad d\Psi_t = v(t)\, dN_t$$

Example [Beneš, 1981]. Suppose the state process X is the solution to the stochastic differential equation

$$dX_t = f(X_t)\, dt + dv_t; \qquad X_0 = x_0$$

for some constant x_0, and suppose the observation process y is given by

$$dy_t = x_t\, dt + dw_t; \qquad y_0 = 0$$

where v and w are independent Wiener processes. Using a variation of the change of measure method just explained, Beneš shows that if

$$f'(x) + f(x)^2 + x^2 = k^2 x^2 + bx + c \tag{5.15}$$

for some constants k, b, and c then an unnormalized conditional density ρ_t (relative to Lebesgue measure) of X_t given $\mathcal{O}_t$ is given by

$$\rho_t(x) = \exp\left(F(x) - F(x_0) - \frac{(x - \mu_t)^2}{s_t}\right)$$

where

$$F(x) = \int_0^x f(u)\, du \qquad -\infty < x < \infty$$

and μ and s can be recursively computed using the two equations

$$\frac{ds_t}{dt} = 1 - k^2 s_t^2; \qquad s_0 = 0,$$

$$d\mu_t = -k^2 s_t \mu_t \, dt - \tfrac{1}{2} s_t b \, dt + s_t \, dy_t; \qquad \mu_0 = x_0$$

The process ρ does not satisfy the nonlinear filtering equation (5.10). However, using direct computation and Eq. (5.15), one readily finds that

$$d\rho_t(x) = \mathcal{L}^* \rho_t(x) + x\rho_t(x) \, dy_t - \rho_t(x)\mu_t(dy_t + x\,dt) + U_t \, dt$$

where U is a random process (not depending on x). By Itô's product rule we then verify the $\tilde{\rho}$ defined by

$$\tilde{\rho}_t(x) = \rho_t(x) \exp\left(\int_0^t \mu_s \, dy_s + \int_0^t U_s \, ds \right)$$

satisfies the stochastic equation (5.10) for π. Therefore, if we can establish uniqueness of solutions, we can conclude that $\tilde{\rho} = \pi$. Since for each t, ρ_t differs from $\tilde{\rho}_t$ by a random factor (not depending on x), this would then imply that ρ is an unnormalized conditional density as claimed.

Remark: Other examples which lead to finite dimensional nonlinear filters, geometric arguments for why finite dimensional filters do not exist for certain (in fact, "most") examples, approximate filters, bounds on minimal error for nonlinear filtering, connections to physics, applications to control, and a variety of other topics can be found in [Hazewinkel and Willems (1981)] and [Kohlmann and Vogel (1979)].

EXERCISES

1. Let $z = (z_k: k \in Z_+)$ and $0 = (O_k: k \in Z_+)$ be random processes and let $\mathcal{O}_k = \sigma(O_0, \ldots, O_k)$. Suppose that each z_k takes values in $S = \{1, 2, \ldots, n\}$ and that O_k takes values in some finite set (not depending on k or ω). Suppose that for each possible value θ of O_k that there is an $n \times n$ matrix $R(\theta)$ such that

$$\mathcal{P}[z_{k+1} = j, O_{k+1} = \theta | z_k = i, \mathcal{O}_k] = R_{ij}(\theta)$$

(i.e., (z_{k+1}, O_{k+1}) is conditionally independent of $\mathcal{O}_k$ given z_k, and the transition mechanisms are time homogeneous.) Define Π_k to be the row vector with ith entry $\mathcal{P}(z_k = i | O_1, \ldots, O_k)$. Derive the recursive filtering equation,

$$\Pi_{k+1} = \frac{\Pi_k R(O_{k+1})}{\Pi_k R(O_{k+1}) e}$$

where e is the column vector of all ones. (Note that the numerator on the right-hand side is an unnormalized version of Π_{k+1}).

2. Find $E[(\xi_t - \hat{\xi}_t)^2]$ in terms of a and b for the example of Section 7.4 in the case that $\sigma_0^2 = 0$. Find its limiting value as t tends to infinity.

3. Let U be a positive random variable with density function f and distribution function F. Suppose that N is a counting process with intensity $\lambda = aI_{[0,U)} + bI_{[U,+\infty)}$ relative to the family of σ-algebras

$$\sigma(U) \vee \sigma(N_s : s \leq t) \qquad t \geq 0$$

Find a recursive equation for $\mathscr{P}[t < U | N_s : s \leq t]$.

8
Random Fields

1. INTRODUCTION

We have used the term stochastic process to denote a collection of random variables indexed by a single real parameter. In other words, the parameter space is a subset of the real line and usually an interval. In most applications, this parameter is interpreted as time. There are many applications where it is more appropriate to consider collections of random variables indexed by points in a more general parameter space. For example, in problems involving propagation of electromagnetic waves through random media, the natural parameter space is a subset of R^4, representing space and time. A similar example is the velocity field in turbulence theory. The term random field is often used to denote a collection of random variables with a parameter space which is a subset of R^n. There are other possible parameter spaces. For example, the parameter space can be taken to be a function space of some kind. Such is the case with **generalized processes**. Alternatively, we can also take the parameter space to be a collection of subsets of R^n. Such, for example, is the situation for **random measures,** which have already been made use of in connection with second-order stochastic integrals. Generally speaking, the kind of assumptions that we make concerning mutual dependence of the collection of random variables reflects something of the parameter space. For example, if the parameter space is an interval, we usually assume continuity in probability. If the parameter space is a linear topological space, we usually assume that the collection of random variables as a function of the parameter is both linear and continuous in probability. If the parameter space is a σ algebra, then we usually assume that the collection of random variable is σ additive, and so on.

Compared to the one-parameter case, relatively little is known concerning processes with a more general parameter space. Of course, a great deal of the results concerning stochastic processes with a one-dimensional parameter space do not depend on the fact that the parameter is one dimensional. These results are easily generalized to more general collections of random variables. Such generalizations require little elaboration. In this chapter, we shall focus our attention on problems of the following two kinds: (1) problems which arise only when the parameter space is more complex than one-dimensional, and (2) important properties of one-dimensional processes which are not easily extended, because they depend on the parameter space being one dimensional. As an example of category (1), we have the rich interplay between the probabilistic properties of a random field and the geometry of its parameter space. Although this interplay already appears in the one-dimensional case in the form of stationarity, the geometry of the real line is obviously both degenerate and rather trivial by comparison with the geometry of higher dimensions. As an example of category (2), consider Markov processes. The definition of a Markov process makes explicit use of the well orderedness of the real line. It is difficult to see how it can be generalized to a multidimensional parameter space. The way that it is done is one of the most interesting problems that we shall discuss in this chapter.

To avoid confusion, we shall adopt the following terminology: A collection of random variables defined on a common probability space will be called a **stochastic process** or a **random field** accordingly as its parameter space is one-dimensional or multidimensional. We should note that while this terminology is widely used, it is by no means universal. For example, a random field is often called a stochastic process with a several-dimensional time.

2. HOMOGENEOUS RANDOM FIELDS

The simplest generalization is to take the parameter space to be an n-dimensional interval, that is, $A = \{(z_1, z_2, \ldots, z_n) : z_i \in I_i\}$ where

2. HOMOGENEOUS RANDOM FIELDS

each I_i is an interval, closed, open, or semiclosed. For a random field $\{X(\omega,z), z \in A\}$ most of our discussion in Chaps. 2 and 3 carry over with no difficulty, e.g., continuity questions, separability, Karhunen-Loeve expansion. Results on wide-sense stationary process can also be generalized to random fields with parameter space R^n. Here, the generalization is interesting and not entirely trivial.

Let $\{X_z, z \in R^n\}$ be a family of second-order random variables defined on a probability space $(\Omega, \mathcal{C}, \mathcal{P})$. We assume that X_z is continuous in probability, i.e., for every $\epsilon > 0$,

$$\mathcal{P}(|X_z - X_{z'}| \geq \epsilon) \xrightarrow[\|z=z'\|\to 0]{} 0 \tag{2.1}$$

where $\|z\|$ denotes the Euclidean norm

$$\|z\| = \left(\sum_{i=1}^{n} z_i^2\right)^{\frac{1}{2}} \tag{2.2}$$

As in Chap. 3, we allow X_z to be complex valued. The most straightforward generalizations to wide-sense stationary processes are **homogeneous random fields** defined as follows: We say that $\{X_z, z \in R^n\}$ is homogeneous if $EX_z = \mu$ does not depend on z and

$$EX_z \overline{X}_{z'} = EX_{z+z_0} \overline{X}_{z'+z_0} \tag{2.3}$$

for all z_0, z, z' in R^n. Setting $z_0 = -z'$ in (2.3), we see that the covariance function

$$E(X_z - EX_z)\overline{(X_{z'} - EX_{z'})} = R(z - z') \tag{2.4}$$

depends only on $z - z'$. Of course, $R(z - z')$ is also nonnegative definite; i.e., for any finite number of points $z_1, z_2, \ldots, z_N$ in R^n, and any collection of complex constants $a_1, a_2, \ldots, a_N$, we have

$$\sum_{i,j=1}^{N} a_i \bar{a}_j R(z_i - z_j) \geq 0 \tag{2.5}$$

Bochner's theorem (Proposition 3.5.1) can now be generalized as follows.

Proposition 2.1. A function $R(z)$, $z \in R^n$, is the covariance function of a homogeneous q.m. continuous random field if and only if it is of the form

$$R(z) = \int_{R^n} e^{i2\pi(\nu,z)} F(d\nu) \tag{2.6}$$

Where F is a finite Borel measure on R^n, and (ν,z) denotes the inner product

$$(\nu,z) = \sum_{i=1}^{n} \nu_i z_i \tag{2.7}$$

Proposition 2.1 can be proved in exactly the same way as Proposition 3.5.1, and we won't repeat it here. Similarly, we can obtain a spectral representation for $\{X_z, z \in R^n\}$ of the form

$$X_z = \int_{R^n} e^{i2\pi(\nu,z)} \hat{X}(d\nu) \tag{2.8}$$

where $\hat{X}(\cdot)$ is a random set function defined on Borel sets of R^n such that

$$E\hat{X}(A)\hat{X}(B) = F(A \cap B) \tag{2.9}$$

F being defined by (2.6). An integral of the form

$$\int_{R^n} f(\nu) \hat{X}(d\nu)$$

can be defined for any $f \in L^2(F)$ as the q.m. limit of a sequence of random variables resulting from approximating f by a sequence $\{f_k\}$ where each f_k is a linear combination of indicator functions of Borel sets and

$$\int_{R^n} |f(\nu) - f_k(\nu)| F(d\nu) \xrightarrow[k \to \infty]{} 0$$

The details are nearly identical to those given in Sec. 3.6.

Translations: $z \to z + z_0$ are transformations of R^n onto R^n which leave the Euclidean distance $\|z - z'\|$ between any pair of points unchanged. However, these are not the only transformations of R^n onto R^n which have this property. Transformations $t: R^n \to R^n$ of the form

$$\begin{aligned} t(z) &= (t_1(z), \ldots, t_n(z)) \\ t_i(z) &= \sum_{j=1}^{n} a_{ij} z_j \end{aligned} \tag{2.10}$$

where $\mathbf{A} = (a_{ij})$ is an orthogonal matrix, also preserve Euclidean distances. Such a transformation is called a **rotation,** and is often further designated as being **proper** or **improper** according as the determinant of $\mathbf{A}$ is $+1$ or -1. It follows that any succession of rotations and translations also preserves Euclidean distances. It also turns out that every transformation t of R^n onto R^n preserving Euclidean distances can be represented as a translation followed by a rotation. Such distance preserving transformations are called **rigid-body motions.** The collection of all rigid-body motions forms a group G which is called the **Euclidean group.**

Now consider a second-order random field $\{X_z, z \in R^n\}$ such that $EX_z = \mu$ is a constant which we assume to be zero and

$$EX_z \bar{X}_{z'} = EX_{t(z)} \bar{X}_{t(z')} \tag{2.11}$$

for all $z, z' \in R^n$ and all rigid-body motions t. Suppose that we move z'

2. HOMOGENEOUS RANDOM FIELDS

to the origin $\mathbf{0}$ by translation, then z is moved to $z - z'$. If we follow this by a rotation which takes $z - z'$ into $(\|z - z'\|, 0, 0, \ldots, 0)$, the origin $\mathbf{0}$ is left invariant. Therefore, from (2.11) we have

$$EX_z \bar{X}_{z'} = EX_{(\|z-z'\|, 0, \ldots, 0)} \bar{X}_{(0,0,\ldots,0)}$$
$$= R(\|z - z'\|) \qquad (2.12)$$

which is a function of a single, real positive variable $\|z - z'\|$. A random field satisfying (2.11) will be called a **homogeneous and isotropic random field**. It is obviously homogeneous.

For an isotropic and homogeneous random field, it is advantageous to adopt a polar-coordinate system. We define the $(n-1)$-dimensional unit sphere S^{n-1} as the set of all points z in R^n such that $\|z\| = 1$. An arbitrary point z in R^n can be uniquely identified by its distance from the origin $\mathbf{0}$ which is $\|z\|$ and the point on S^{n-1} intersected by a line connecting z and $\mathbf{0}$ which is

$$\frac{z}{\|z\|} = \left(\frac{z_1}{\|z\|}, \frac{z_2}{\|z\|}, \ldots, \frac{z_n}{\|z\|} \right)$$

Therefore any $z \in R^n$ is uniquely identified by $\|z\|$ and any $n-1$ coordinates identifying a point of S^{n-1}. We choose a coordinate system for S^{n-1}, recursively, as follows: S^1 is the unit circle in R^2, and every point on S^1 has the form

$$z = (\cos \theta_1, \sin \theta_1) \qquad 0 \leq \theta_1 < 2\pi \qquad (2.13)$$

Hence, S^1 has a single coordinate θ_1. For S^n, consider the point $z = (1, 0, \ldots, 0) \in R^{n+1}$ and call it the **north pole**. The set of all points on S^n with $z_1 = \cos \theta_n$ forms an $(n-1)$-dimensional sphere with radius $\sin \theta_n$. If a coordinate system $(\theta_1, \theta_2, \ldots, \theta_{n-1})$ has already been chosen for S^{n-1}, then this automatically produces a coordinate system $(\theta_1, \theta_2, \ldots, \theta_n)$ for S^n. Starting from (2.13), we can thus generate a coordinate system $(\theta_1, \ldots, \theta_n)$ for S^n with the property that $\theta_n =$ constant represents a sphere of dimension $n-1$. We have also generated a coordinate system for R^n where a point z is represented by $(r, \theta_1, \ldots, \theta_{n-1})$ with $r = \|z\|$, and $(\theta_1, \ldots, \theta_{n-1})$ is the intersection between S^{n-1} and a line connecting z and the origin $\mathbf{0}$. The origin $\mathbf{0}$ has a degenerate representation in this system given by $r = 0$ with $\theta_1, \ldots, \theta_{n-1}$ undefined.

Starting from (2.6), we can now find the form of Bochner's theorem for an isotropic and homogeneous random field. Since an isotropic and homogeneous random field is necessarily homogeneous, we have

$$R(\|z\|) = \int_{R^n} e^{i2\pi(\nu, z)} F(d\nu) \qquad (2.14)$$

For any *rotation* t,

$$\begin{aligned}
R(\|t(z)\|) &= R(\|z\|) \\
&= \int_{R^n} e^{i2\pi(\nu, t(z))} F(d\nu) \\
&= \int_{R^n} e^{i2\pi(\nu, z)} F(d\nu)
\end{aligned} \quad (2.15)$$

Since $(\nu, t(z)) = (t(t^{-1}(\nu)), t(z)) = (t^{-1}(\nu), z)$, we have, upon a change in the variable of integration,

$$\begin{aligned}
\int_{R^n} e^{i2\pi(\nu, t(z))} F(d\nu) &= \int_{t(R^n)} e^{i2\pi(\lambda, z)} F(t(d\lambda)) \\
&= \int_{R^n} e^{i2\pi(\nu, z)} F(t(d\nu))
\end{aligned} \quad (2.16)$$

It follows from (2.15) and (2.16) that for every $z \in R^n$ and every rotation t,

$$\int_{R^n} e^{i2\pi(\nu, z)} F(d\nu) = \int_{R^n} e^{i2\pi(\nu, z)} F(t(d\nu)) \quad (2.17)$$

Therefore, the Borel measure F must be isotropic, that is,

$$F(A) = F(t(A)) \quad (2.18)$$

for every Borel set A and every rotation t. We can now take z to be directed along the north pole (that is, $z_1 = \|z\|$, $z_2 = 0$, $z_3 = 0$, . . .), and if we make use of the isotropy of F, then (2.14) becomes

$$R(\|z\|) = \int_0^\infty \int_{S^{n-1}} e^{i2\pi\lambda\|z\|\cos\varphi_{n-1}} d\Omega_{\varphi_1, \ldots, \varphi_{n-1}} F_0(d\lambda) \quad (2.19)$$

where $d\Omega$ is the differential surface element of S^{n-1}.

The set of all points on S^{n-1} with a fixed φ_{n-1} (that is, at a fixed distance from the north pole) is a sphere of dimension $n-2$ with radius $\sin \varphi_{n-1}$, the surface area of which must be $\sin^{n-2} \varphi_{n-1}$ area (S^{n-2}). Since the integrand in (2.19) depends only on φ_{n-1}, we have

$$R(\|z\|) = K \int_0^\infty \int_0^\pi e^{i2\pi\lambda\|z\|\cos\varphi_{n-1}} \sin^{n-2}\varphi_{n-1} \, d\varphi_{n-1} F_0(d\lambda) \quad (2.20)$$

The constant K is just the total area of S^{n-2} and can be absorbed into F_0. The inside integral in (2.20) can be evaluated to be

$$\int_0^\pi e^{i2\pi\lambda\|z\|\cos\varphi} \sin^{n-2}\varphi \, d\varphi = \frac{J_{(n-2)/2}(2\pi\lambda\|z\|)}{(2\pi\lambda\|z\|)^{(n-2)/2}} \quad (2.21)$$

where J is the Bessel function. Hence, the isotropic version of Bochner's theorem is given as follows [see, e.g., Yaglom, 1962, pp. 81–86].

Proposition 2.2. A function $R(r)$, $0 \leq r < \infty$, is the covariance function of an isotropic and homogeneous q.m. continuous random field if and only if

$$R(r) = \int_0^\infty \frac{J_{(n-2)/2}(\lambda r)}{(\lambda r)^{(n-2)/2}} F_0(d\lambda) \qquad (2.22)$$

where F_0 is a finite Borel measure on $[0, \infty)$.

We shall call F_0 the **spectral measure** for the random field. We note that the constant 2π in (2.21) is absorbed into λ in (2.22) to result in a simpler formula.

For an isotropic and homogeneous random field, the spectral representation (2.8) assumes a special form which involves a countable family of random set functions defined on the Borel sets of $[0, \infty)$ in place of the n-dimensional set function $\hat{X}$. This spectral-representation formula is expressible in terms of spherical harmonics which will be introduced in the next section. In the process, we shall also introduce isotropic random fields which are not necessarily homogeneous.

3. SPHERICAL HARMONICS AND ISOTROPIC RANDOM FIELDS

We have already defined a unit n-dimensional sphere S^n as the set of all points in R^{n+1} at unit distance from the origin. Starting from (2.13), we can find a natural coordinate system for S^n so that an arbitrary point $\theta \in S^n$ has a representation $\theta = (\theta_1, \theta_2, \ldots, \theta_n)$ in which θ_n is the angle of θ from the north pole. The north pole is thus defined by having $\theta_n = 0$, and for the north pole $\theta_1, \theta_2, \ldots, \theta_{n-1}$ are not specified.

The sphere S^n is obviously invariant under any rotation (proper or improper) of R^{n+1}, and the set of all rigid-body motions of R^{n+1} which leave S^n invariant is just the set of rotations. Suppose that we define a metric d on S^n by

$$d(\theta, \theta') = \psi(\theta, \theta') \qquad (3.1)$$

where $\psi(\theta, \theta')$ is just the angle between the two straight lines connecting the origin in R^{n+1} to θ and θ'. It is easy to verify that (3.1) defines a metric. Now, suppose that we consider the set of all transformations $t: S^n \to S^n$ such that

$$d(\theta, \theta') = d(t(\theta), t(\theta')) \qquad (3.2)$$

It turns out that this set is just the set of rotations in R^{n+1}. This set is denoted by $G(S^n)$ and is easily verified to be a group. If we denote by $G(R^{n+1})$, the group of rigid-body motions in R^{n+1}, and by $T(R^{n+1})$, the group of all translations in R^{n+1}, then we have

$$G(R^{n+1}) = G(S^n) \times T(R^{n+1}) \qquad (3.3)$$

We also note that $T(R^{n+1})$ is isomorphic to R^{n+1}, since to each $t \in R^{n+1}$

$$t(z) = z + t \tag{3.4}$$

This observation and (3.3) allow us to identify R^{n+1} as $G(R^{n+1})/G(S^n)$ where the latter denotes the collection of equivalence classes of rigid motions, each equivalence class being made up of the set of all motions which take the origin to the same point.

Now, consider the Laplacian operator on R^{n+1} which we denote by $\Delta(R^{n+1})$. In Cartesian coordinates, we have

$$\Delta(R^{n+1}) = \frac{\partial^2}{\partial z_1^2} + \frac{\partial^2}{\partial z_2^2} + \cdots + \frac{\partial^2}{\partial z_{n+1}^2} \tag{3.5}$$

As domain of $\Delta(R^{n+1})$, we take the set of all complex-valued functions on R^{n+1} with bounded-continuous second partials with respect to each z_i, and we denote this set by $C^2(R^{n+1})$. For each rigid-body motion $g \in G(R^{n+1})$, we can define a mapping $T_g: C^2(R^{n+1})$ onto $C^2(R^{n+1})$ by

$$T_g f(z) = f(g(z)) \tag{3.6}$$

It is both well known and easily verified that for each $g \in G(R^{n+1})$, T_g commutes with $\Delta(R^{n+1})$. That is,

$$T_g(\Delta f) = \Delta(T_g f) \qquad g \in G(R^{n+1}) \tag{3.7}$$

The easiest proof of this fact is probably by representing each function in $C^2(R^{n+1})$ in terms of its Fourier integral.

In terms of polar coordinates, whereby each point in R^{n+1} is represented by a pair $(r,\theta) \in [0,\infty) \times S^n$, the Laplacian can be rewritten as

$$\Delta(R^{n+1}) = \frac{1}{r^n} \frac{\partial}{\partial r}\left(r^n \frac{\partial}{\partial r}\right) + \frac{1}{r^2} \Delta(S^n) \tag{3.8}$$

where $\Delta(S^n)$, the Laplacian of the n sphere, can be recursively defined as follows:

$$\Delta(S^1) = \frac{\partial^2}{\partial \theta_1^2}$$

$$\Delta(S^n) = \frac{1}{\sin^{n-1} \theta_n} \frac{\partial}{\partial \theta_n}\left(\sin^{n-1} \theta_n \frac{\partial}{\partial \theta_n}\right) + \frac{1}{\sin^2 \theta_n} \Delta(S^{n-1}) \tag{3.9}$$

From (3.8) it is easy to see that $\Delta(S^n)$ commutes with each T_g, $g \in G(S^n)$, since r is left invariant by any $g \in G(S^n)$, and T_g commutes with $\Delta(R^{n+1})$. It is now convenient to take as the domain of $\Delta(S^n)$ the space $C^2(S^n)$ of functions in $C^2(R^{n+1})$ which do not depend on the radial distance r. Now, consider the eigenvalues and eigenfunctions of $\Delta(S^n)$. An **eigenvalue** λ of $\Delta(S^n)$ is any complex number such that the equation

$$\Delta(S^n) f = \lambda f \qquad f \in C^2(S^n) \tag{3.10}$$

3. SPHERICAL HARMONICS AND ISOTROPIC RANDOM FIELDS

has a solution f which is not identically zero. For an eigenvalue λ, any not-identically-zero f satisfying (3.10) is called an **eigenfunction** corresponding to (3.10). Because $\Delta(S^n)$ is self-adjoint, the eigenvalues must be real. To find the eigenvalues, first we seek those eigenvalues such that (3.10) has a solution f which depends only on θ_n and not on $(\theta_1, \theta_2, \ldots, \theta_{n-1})$. For such a function we have

$$\frac{1}{\sin^{n-1}\theta_n}\frac{d}{d\theta_n}\left(\sin^{n-1}\theta_n\frac{df(\theta_n)}{d\theta_n}\right) = \lambda f(\theta_n) \tag{3.11}$$

If we set $\cos\theta_n = x$ and $f(\theta_n) = h(\cos\theta_n)$, then (3.11) becomes

$$(1 - x^2)\frac{d^2h(x)}{dx^2} - nx\frac{dh(x)}{dx} - \lambda h(x) = 0 \tag{3.12}$$

which is an equation of familiar form and is sometimes called the **Gegenbauer equation**. By seeking a power series solution, it is easy to discover that a bounded h exists if and only if λ is of the form

$$\lambda_m = -m(m + n - 1) \quad m = 0, 1, 2, \ldots \tag{3.13}$$

and for $n \geq 2$, the corresponding eigenfunctions are the Gegenbauer polynomials defined by

$$\begin{aligned} f_m(\theta_n) &= h_m(\cos\theta_n) = C_m^{(n-1)/2}(\cos\theta_n) \\ &= K_m\int_0^\pi (\cos\theta_n + i\sin\theta_n\cos\varphi)^m \sin^{n-2}\varphi\, d\varphi \end{aligned} \tag{3.14}$$

where the constant K_m is such that $f_m(0) = 1$.

We have thus found a subset of eigenvalues for (3.10), and for each $\lambda_m = -m(m + n - 1)$ we have found an eigenfunction of the form

$$f_m(\theta) = C_m^{(n-1)/2}(\cos\theta_n) \quad \theta \in S^n \tag{3.15}$$

For each $g \in G(S^n)$ we can define a linear operator $T_g: C^2(S^n) \to C^2(S^n)$ by

$$(T_g f)(\theta) = f(g(\theta)) \tag{3.16}$$

Since each T_g commutes with $\Delta(S^n)$, for each g and each f_m given by (3.15), the function $T_g f_m$ is again an eigenfunction corresponding to the same eigenvalue $\lambda_m = -m(m + n - 1)$. Now consider the Hilbert space $L^2(S^n)$ generated by complex-valued functions on S^n which are square integrable with respect to the uniform measure $d0$ $\left(\int_{S^n} d0 = 1\right)$. For each m, the set of functions $\{T_g f_m,\ g \in G(S^n)\}$ spans a finite-dimensional subspace $\mathcal{H}_m$ of $L^2(S^n)$ with dimension d_m. For $n \geq 2$, we have

$$1 = d_0 < d_1 < d_2 < \cdots \tag{3.17}$$

For $m \neq m'$, $\mathcal{H}_m$ and $\mathcal{H}_{m'}$ are orthogonal. We now choose a real orthonormal basis $\{h_{ml}^{(n)}, l = 1, \ldots, d_m\}$ for $\mathcal{H}_m$, where we always take $h_{m1}^{(n)}$ to be proportional to f_m as defined by (3.15). It turns out that the set of functions

$$\{h_{ml}^{(n)}, l = 1, \ldots, d_m, m = 0, 1, \ldots\} \tag{3.18}$$

is a basis for $L^2(S^n)$. Since $L^2(S^n)$ contains $C^2(S^n)$, this means that (3.13) exhausts all eigenvalues for (3.10), and for each eigenvalue λ_m every eigenfunction is a linear combination of $\{h_{ml}^{(n)}, l = 1, 2, \ldots, d_m\}$. We shall call the functions $\{h_{ml}^{(n)}\}$ **spherical harmonics** [Erdélyi, 1953, Chap. 11].

Let F be a function defined on $S^n \times S^n$ such that for every $g \in G(S^n)$ and every $(\theta, \theta') \in S^n \times S^n$,

$$F(g(\theta), g(\theta')) = F(\theta, \theta') \tag{3.19}$$

Let **0** denote the north pole as before. Then for every rotation τ which leaves **0** fixed, we have

$$F(\theta, 0) = F(\tau(\theta), 0) \tag{3.20}$$

Therefore, $F(\theta, 0)$ can depend only on θ_n, that is, the last component of θ. Suppose that $F(\cdot, 0) \in \mathcal{H}_m$. That is, suppose that

$$\Delta(S^n)F(\cdot, 0) = -m(m + n - 1)F(\cdot, 0) \tag{3.21}$$

Then it follows that $F(\theta, 0)$ as a function θ_n must satisfy (3.11) corresponding to $\lambda = -m(m + n - 1)$. This means that we must have

$$F(\theta, 0) = Kf_m(\theta) = KC_m^{(n-1)/2}(\cos \theta_n) \tag{3.22}$$

where K is a constant. Now, for any fixed pair (θ, θ') there always exists a motion g which simultaneously takes θ' into **0** and θ into $(0, 0, \ldots, \psi(\theta, \theta'))$ where $\psi(\theta, \theta')$ denotes the angle between θ and θ'. This means that

$$F(\theta, \theta') = KC_m^{(n-1)/2}(\cos \psi(\theta, \theta')) \tag{3.23}$$

What we have shown is that if F satisfies (3.19) and (3.21) then it must be of the form (3.23).

Let $\{h_{ml}^{(n)}\}$ denote the spherical harmonics that we defined earlier. Consider the bilinear sum

$$F(\theta, \theta') = \sum_{l=1}^{d_m} h_{ml}^{(n)}(\theta) h_{ml}^{(n)}(\theta') \tag{3.24}$$

We shall show that the F in (3.24) satisfies both (3.19) and (3.21). For an arbitrary $g \in G(S^n)$, $\mathcal{H}_m$ is invariant under T_g so that we can write

$$(T_g h_{ml}^{(n)})(\theta) = h_{ml}^{(n)}(g(\theta)) = \sum_{l'=1}^{d_m} \alpha_{ll'}(g) h_{ml'}^{(n)}(\theta) \tag{3.25}$$

3. SPHERICAL HARMONICS AND ISOTROPIC RANDOM FIELDS

Because

$$\int_{S^n} h_{mk}{}^{(n)}(g(\theta))h_{ml}{}^{(n)}(g(\theta))\, d0 = \int_{S^n} h_{mk}{}^{(n)}(\theta)h_{ml}{}^{(n)}(\theta)\, d0 = \delta_{kl} \quad (3.26)$$

we find that for each $g \in G(S^n)$,

$$\sum_{l'=1}^{d_m} \alpha_{ll'}(g)\alpha_{kl'}(g) = \delta_{kl} \quad (3.27)$$

This means that $\mathbf{A}(g) = [\alpha_{ij}(g)]$ is an orthogonal matrix so that we also have

$$\sum_{l'=1}^{d_m} \alpha_{l'l}(g)\alpha_{l'k}(g) = \delta_{kl} \quad (3.28)$$

It follows that

$$\sum_{l=1}^{d_m} h_{ml}{}^{(n)}(g(\theta))h_{ml}{}^{(n)}(g(\theta')) = \sum_{l'=1}^{d_m}\sum_{k'=1}^{d_m} h_{ml'}^{(n)}(\theta)h_{mk'}^{(n)}(\theta) \sum_{l=1}^{d_m} \alpha_{ll'}(g)\alpha_{lk'}(g)$$

$$= \sum_{l=1}^{d_m} h_{ml}{}^{(n)}(\theta)h_{ml}{}^{(n)}(\theta') \quad (3.29)$$

Hence, the function F defined by (3.24) satisfies (3.19). It is obvious that it also satisfies (3.21) so that

$$\sum_{l=1}^{d_m} h_{ml}{}^{(n)}(\theta)h_{ml}{}^{(n)}(\theta') = KC_m{}^{(n-1)/2}(\cos \psi(\theta,\theta')) \quad (3.30)$$

The constant K can be evaluated as follows:

$$\int_{S^n} \sum_{l=1}^{d_m} h_{ml}{}^{(n)}(\theta)h_{ml}{}^{(n)}(\theta)\, d0 = d_m$$

$$= KC_m{}^{(n-1)/2}(1) = K \quad (3.31)$$

Therefore,

$$C_m{}^{(n-1)/2}(\cos \psi(\theta,\theta')) = \frac{1}{d_m}\sum_{l=1}^{d_m} h_{ml}{}^{(n)}(\theta)h_{ml}{}^{(n)}(\theta') \quad (3.32)$$

We can now apply the results on spherical harmonics to the problem of representing isotropic random fields. A second-order random field $\{X_z, z \in R^n\}$ is said to be isotropic if for every rotation ρ

$$EX_{\rho(z)} = EX_z \quad z \in R^n \quad (3.33)$$

and

$$EX_{\rho(z)}\bar{X}_{\rho(z')} = EX_z\bar{X}_{z'} \quad z, z' \in R^n \quad (3.34)$$

If we adopt a polar-coordinate system $z = (r,\theta)$, $r \in [0, \infty)$, $\theta \in S^{n-1}$, then

(3.33) and (3.34) imply that
$$EX(r,\theta) = \mu(r) \tag{3.35}$$
and
$$EX(r,\theta)\bar{X}(r',\theta) = R(r, r', \cos\psi(\theta,\theta')) \tag{3.36}$$
where $\psi(\theta,\theta')$ is the angle between θ and θ'. Suppose that we now seek a representation of the form
$$X(r,\theta) = \sum_{m=0}^{\infty} \sum_{l=1}^{d_m} X_{ml}(r) h_{ml}^{(n-1)}(\theta) \tag{3.37}$$
where $h_{ml}^{(n-1)}$ are the spherical harmonics. The orthonormality of the spherical harmonics yields
$$X_{ml}(r) = \int_{S^{n-1}} X(r,\theta) h_{ml}^{(n-1)}(\theta)\, d\theta \tag{3.38}$$
so that
$$\begin{aligned} EX_{ml}&(r)\bar{X}_{m'l'}(r') \\ &= \int_{S^{n-1} \times S^{n-1}} EX(r,\theta)\bar{X}(r',\theta') h_{ml}^{(n-1)}(\theta) h_{m'l'}^{(n-1)}(\theta')\, d\theta\, d\theta' \\ &= \int_{S^{n-1} \times S^{n-1}} R[r, r', \cos\psi(\theta,\theta')] h_{ml}^{(n-1)}(\theta) h_{m'l'}^{(n-1)}(\theta')\, d\theta\, d\theta' \end{aligned} \tag{3.39}$$

If $\{X_z, z \in R^n\}$ is q.m. continuous, then $R(r,r',\cos\psi)$ can be expanded in terms of $C_m^{(n-2)/2}(\cos\psi)$ as
$$R(r,r',\cos\psi) = \sum_{m=0}^{\infty} d_m R_m(r,r') C_m^{(n-2)/2}(\cos\psi) \tag{3.40}$$

The bilinear form (3.32) and (3.40) can now be used in (3.39) to yield
$$EX_{ml}(r)\bar{X}_{m'l'}(r') = \delta_{mm'}\delta_{ll'} R_m(r,r') \tag{3.41}$$
This means that $\{X_{ml}(r)\}$ is a countable family of orthogonal one-dimensional stochastic processes.

Suppose that $\{X_z, z \in R^n\}$ is not only isotropic, but also homogeneous; then we know from (2.8) that we can write
$$X_z = \int_{R^n} e^{i2\pi(\nu,z)} \hat{X}(d\nu) \tag{3.42}$$
where $\hat{X}$ is a random set function defined on the Borel sets of R^n. Now, adopt a polar-coordinate system for both ν and z in (3.42) so that
$$\begin{aligned} 2\pi\nu &= (\lambda,\hat{\phi}) & 0 \leq \lambda &< \infty \\ & & \hat{\phi} &\in S^{n-1} \\ z &= (r,\theta) & 0 \leq r &< \infty \\ & & \theta &\in S^{n-1} \end{aligned}$$

3. SPHERICAL HARMONICS AND ISOTROPIC RANDOM FIELDS

It is obvious that
$$\Delta(R^n) e^{i2\pi(\nu,\cdot)} = -\lambda^2 e^{i2\pi(\nu,\cdot)}$$

It follows that we must be able to write
$$e^{i\lambda r \cos \psi(\theta,0)} = e^{i\lambda r \cos \theta_{n-1}} = \sum_{m=0}^{\infty} C_m^{(n-2)/2}(\cos \theta_{n-1}) f_m(\lambda r) \tag{3.43}$$

where f_m satisfies
$$\frac{1}{r^{n-1}} \frac{d}{dr}\left(r^{n-1} \frac{d}{dr} f_m(\lambda r) \right) - \frac{m(m+n-2)}{r^2} f_m(\lambda r)$$
$$= -\lambda^2 f_m(\lambda r) \tag{3.44}$$

$$f_m(\lambda r) = K_m \frac{J_{(n-2)/2+m}(\lambda r)}{(\lambda r)^{(n-2)/2}} \tag{3.45}$$

Therefore, from (3.43) and (3.32), we have
$$e^{i\lambda r \cos \psi(\theta,\varphi)} = \sum_{m=0}^{\infty} K_m \frac{J_{(n-2)/2+m}(\lambda r)}{(\lambda r)^{(n-2)/2}} C_m^{(n-2)/2}(\cos \psi(\theta,\hat{\varphi}))$$
$$= \sum_{m=0}^{\infty} \sum_{l=1}^{d_m} \frac{K_m}{d_m} \frac{J_{(n-2)/2+m}(\lambda r)}{(\lambda r)^{(n-2)/2}} h_{ml}^{(n-1)}(\theta) h_{ml}^{(n-1)}(\hat{\varphi}) \tag{3.46}$$

We can now use (3.46) in (3.42) and get
$$X(r,\theta) = \sum_{m=0}^{\infty} \sum_{l=1}^{d_m} h_{ml}^{(n-1)}(\theta) \int_0^{\infty} \frac{J_{(n-2)/2+m}(\lambda r)}{(\lambda r)^{(n-2)/2}} \hat{X}_{ml}(d\lambda) \tag{3.47}$$

where $\{\hat{X}_{ml}\}$ is a family of random set functions defined on the Borel sets of $[0,\infty)$ by the formula
$$\hat{X}_{ml}(A) = \frac{K_m}{d_m} \int_{A \times S^{n-1}} h_{ml}^{(n-1)}(\hat{\varphi}) \hat{X}\left(\frac{d\lambda}{2\pi} d\hat{\varphi} \right) \tag{3.48}$$

It is easy to verify that
$$E\hat{X}_{ml}(A) \hat{X}_{m'l'}(B) = \frac{K_m^2}{d_m^2} \delta_{mm'} \delta_{ll'} \int_{A \cap B} F_0(d\lambda) \tag{3.49}$$

where F_0 is the Borel measures appearing in (2.22).

A comparison of (3.37) and (3.47) shows that for an isotropic and homogeneous random field, we have
$$X_{ml}(r) = \int_0^{\infty} \frac{J_{(n-2)/2+m}(\lambda r)}{(\lambda r)^{(n-2)/2}} \hat{X}_{ml}(d\lambda) \tag{3.50}$$

with
$$R_m(r,r') = \frac{K_m^2}{d_m^2} \int_0^{\infty} \frac{J_{(n-2)/2+m}(\lambda r)}{(\lambda r)^{(n-2)/2}} \frac{J_{(n-2)/2+m}(\lambda r')}{(\lambda r')^{(n-2)/2}} F_0(d\lambda) \tag{3.51}$$

Thus we see that an isotropic random field can be decomposed into a countable number of mutually uncorrelated stochastic processes with a one-dimensional parameter. If the random field is also homogeneous, then the correlation functions of these component processes can all be expressed in terms of a single Borel measure, as is done in (3.51). Results very similar to these can be obtained for random fields with parameter spaces which are certain differentiable manifolds with a Riemannian metric, when these random fields are homogeneous with respect to the motions which preserve this metric. More generally, representation results for random fields defined on homogeneous spaces can be obtained [Yaglom, 1961; Gangoli, 1968; Wong, 1969].

4. MARKOVIAN RANDOM FIELDS

One characterization of Markov processes is that the future and the past should be independent given the present (see 2.5.17). Lévy [1956] generalized the concept of Markov property to random fields with parameter space R^n by identifying the present with any smooth, closed $(n-1)$ surface separating the parameter space into a bounded part (past) and an unbounded part (future). We should immediately note that this correspondence does not quite reduce to the usual identification when the parameter space is R^1, since a "closed" surface in R^1 corresponds to a pair of points rather than to a single point. If we had identified the present with any $(n-1)$ surface separating the parameter space into two parts, and not just with closed surfaces, the definition for a Markov random field would correspond more closely to that of Markov processes. However, this would lead to other difficulties.

Let ∂D be a smooth (infinitely differentiable), closed surface separating R^n into a bounded part D^- and an unbounded part D^+. A random field $\{X_z, z \in R^n\}$ is said to be Markov if for any such ∂D, X_z and $X_{z'}$, $z \in D^-$, $z' \in D^+$, are independent, given $\{X_z, z \in \partial D\}$. Now suppose that $\{X_z, z \in R^n\}$ is a q.m. continuous zero-mean Gaussian isotropic and homogeneous random field. From (2.22), we know that the probability law of such a random field is completely specified by a finite Borel measure F_0 on $[0, \infty)$. A question of obvious interest is what must F_0 be so that the random field $\{X_z, z \in R^n\}$ is Markov? We shall answer this question in this section. The answer, unfortunately, is not very satisfactory in that the only such Markov fields turn out to be degenerate ones.

Suppose that $\{X_z, z \in R^n\}$ is a q.m. continuous Gaussian isotropic random field. From Sec. 3, we know that if we rewrite

$$X(r,\theta) = \sum_{m=0}^{\infty} \sum_{l=1}^{d_m} X_{ml}(r) h_{ml}{}^{(n-1)}(\theta) \tag{4.1}$$

4. MARKOVIAN RANDOM FIELDS

then $\{X_{ml}(r)\}$ are mutually uncorrelated processes. Since $\{X_z, z \in R^n\}$ is Gaussian, $\{X_{ml}(\cdot)\}$ are in fact independent Gaussian processes with parameter space $[0, \infty)$. Now, suppose that $\{X_z, z \in R^n\}$ is Markov. Then each $X_{ml}(r)$, $0 \leq r < \infty$, is a Gaussian Markov process from the following reasoning: Because each $X_{ml}(\cdot)$ is Gaussian, we only need to prove that for each m and l,

$$E[X_{ml}(r)|X_{ml}(\rho), \rho \leq r_0 < r] = E[X_{ml}(r)|X_{ml}(r_0)] \tag{4.2}$$

Because $\{X_{ml}(\cdot)\}$ are mutually independent, we have

$$E[X_{ml}(r)|X_{ml}(\rho), \rho \leq r_0 < r]$$
$$= E[X_{ml}(r)|X_{m'l'}(\rho), \text{ for all } m', l' \text{ and } \rho \leq r_0 < r]$$
$$= E[X_{ml}(r)|X(\rho,\theta), \rho \leq r_0 < r, \theta \in S^{n-1}]$$

Since $X_{ml}(r)$ is given by

$$X_{ml}(r) = \int_{S^{n-1}} h_{ml}{}^{(n-1)}(\theta) X(r,\theta) \, d\theta$$

we can apply the Markov property of $\{X_z, z \in R^n\}$ with $\partial D =$ sphere with radius r_0, and find

$$E[X_{ml}(r)|X(\rho,\theta), \rho \leq r_0 < r, \theta \in S^{n-1}] = E[X_{ml}(r)|X(r_0,\theta), \theta \in S^{n-1}]$$
$$= E[X_{ml}(r)|X_{m'l'}(r_0), \text{ all } m', l']$$
$$= E[X_{ml}(r)|X_{ml}(r_0)]$$

which is just (4.2). We can summarize this result as follows.

Proposition 4.1. Let $\{X_z, z \in R^n\}$ be a q.m. continuous Gaussian isotropic Markov random field. Define

$$X_{ml}(r) = \int_{S^{n-1}} X(r,\theta) h_{ml}{}^{(n-1)}(\theta) \, d\theta \qquad 0 \leq r < \infty \tag{4.3}$$

Then $\{X_{ml}(r), 0 \leq r < \infty, m = 0, 1, 2, \ldots; l \leq d_m\}$ is a family of independent Gaussian Markov processes.

From (3.41) we know that the correlation function of $X_{ml}(\cdot)$ only depends on m, and from (2.5.14) we know that it must have the form

$$EX_{ml}(r)X_{ml}(r') = R_m(r,r') = f_m(\max(r,r'))g_m(\min(r,r')) \tag{4.4}$$

If $\{X_z, z \in R^n\}$ is also homogeneous, then we know further that R_m must be of the form [cf. (3.51)]

$$R_m(r,r') = f_m(\max(r,r'))g_m(\min(r,r'))$$
$$= A_m \int_0^\infty \frac{J_{(n-2)/2+m}(\lambda r)}{(\lambda r)^{(n-2)/2}} \frac{J_{(n-2)/2+m}(\lambda r')}{(\lambda r')^{(n-2)/2}} F_0(d\lambda) \tag{4.5}$$

where A_m are positive constants. Let Δ_m denote the operator

$$(\Delta_m f)(r) = \frac{1}{r^{n-1}} \frac{d}{dr}\left(r^{n-1} \frac{df(r)}{dr}\right) - \frac{m(m+n-2)}{r^2} f(r) \tag{4.6}$$

For $r > r'$ consider

$$[\Delta_m R_m(\cdot, r')](r) = g_m(r')(\Delta_m f_m)(r) \tag{4.7}$$

and

$$[\Delta_m R_m(r, \cdot)](r') = f_m(r)(\Delta_m g_m)(r') \tag{4.8}$$

Because of (4.5), (3.44), and (3.45), we have

$$\begin{aligned}[] [\Delta_m R_m(\cdot, r')](r) &= A_m \int_0^\infty (-\lambda^2) \frac{J_{(n-2)/2+m}(\lambda r)}{(\lambda r)^{(n-2)/2}} \frac{J_{(n-2)/2+m}(\lambda r')}{(\lambda r')^{(n-2)/2}} F_0(d\lambda) \\ &= [\Delta_m R_m(r, \cdot)](r') \end{aligned} \tag{4.9}$$

so that for $r > r'$,

$$\frac{1}{f_m(r)} (\Delta_m f_m)(r) = \frac{1}{g(r')} (\Delta_m g_m)(r') \tag{4.10}$$

Since the two sides of (4.10) are functions of different variables, they must be equal to a constant, i.e.,

$$\begin{aligned} (\Delta_m f_m)(r) &= \nu_m f_m(r) \quad r > 0 & (4.11) \\ (\Delta_m g_m)(r) &= \nu_m g_m(r) \quad r > 0 & (4.12) \end{aligned}$$

Now, consider

$$\begin{aligned} R(r) &= EX(r, \theta)X(0, \cdot) = EX_z X_0 \\ &= \sum_{m,l} R_m(r, 0) h_{ml}^{(n-1)}(\theta) h_{ml}^{(n-1)}(\varphi) \end{aligned}$$

It is obvious that we must have

$$R_m(r, 0) = 0 \quad m \geq 1$$

because $h_{01}^{(n-1)}(\theta) = 1$ and all other $h_{ml}^{(n-1)}$ are orthogonal to it. Therefore,

$$R(r) = R_0(r, 0) = f_0(r) g_0(0) \tag{4.13}$$

It follows from (4.12) that $R(\cdot)$ must satisfy

$$\frac{1}{r^{n-1}} \frac{d}{dr}\left(r^{n-1} \frac{d}{dr} R(r)\right) = \nu_0 R(r) \quad r > 0 \tag{4.14}$$

The only solution of (4.14) such that $R(0) < \infty$ and $R(\|z - z'\|)$ is non-

4. MARKOVIAN RANDOM FIELDS

negative definite is given by

$$R(r) = \frac{A J_{(n-2)/2}(\lambda_0 r)}{(\lambda_0 r)^{(n-2)/2}} \tag{4.15}$$

where λ_0 and A are nonnegative constants.

If $\lambda_0 = 0$ in (4.15) then the random field is just a single Gaussian random variable. It is Markov but only trivially so. If $\lambda_0 > 0$ then

$$X_{ml}(r) = X_{ml} \frac{J_{(n-2)/2+m}(\lambda_0 r)}{(\lambda_0 r)^{(n-2)/2}} \tag{4.16}$$

where $\{X_{ml}\}$ is a collection of independent Gaussian random variables. Since $J_{(n-2)/2+m}(\lambda_0 r)$ has zeros on $0 \leq r < \infty$, the processes $X_{ml}(r)$, $0 \leq r < \infty$ are not Markov.[1] By virtue of Proposition 4.1 the corresponding random field is not Markov.

Equation (4.14) has a second solution of the form

$$R(r) = A \frac{K_{(n-2)/2}(\lambda_0 r)}{(\lambda_0 r)^{(n-2)/2}} \tag{4.17}$$

for which $R(\|z - z'\|)$ is positive definite, but $R(0) = \infty$. Equation (4.17) is counterpart of the exponential-correlation function $R(\tau) = e^{-\alpha|\tau|}$ of a stationary Gaussian Markov process (Ornstein-Uhlenbeck process), and the correspondence becomes more evident when we write

$$\frac{K_{(n-2)/2}(\lambda_0 r)}{(\lambda_0 r)^{(n-2)/2}} = \int_0^\infty \frac{J_{(n-2)/2}(\lambda r)}{(\lambda r)^{(n-2)/2}} \frac{1}{(\lambda^2 + \lambda_0^2)} \lambda^{n-1} \, d\lambda \tag{4.18}$$

which suggests that the spectral density is of the form

$$\varphi(\lambda) = \frac{1}{\lambda^2 + \lambda_0^2} \tag{4.19}$$

Of course, (4.18) corresponds to an unbounded spectral measure

$$F_0(d\lambda) = \frac{\lambda^{n-1} \, d\lambda}{(\lambda^2 + \lambda_0^2)} \tag{4.20}$$

and there is no q.m. continuous random field with a correlation function given by (4.17). It is possible to define a generalized random field with a correlation function given by (4.17). It turns out that this generalized random field, if Gaussian, is Markovian in a well-defined sense. To extend the definition of a Markov random field to generalized random fields requires some care, but it can be done [Wong, 1969].

---

[1] This point was clarified for me by Professor Frank Spitzer.

Lévy [1948] defined a **Brownian motion** with parameter space R^n as a Gaussian random field $\{X_z, z \in R^n\}$ with zero mean and covariance function

$$EX_z X_{z'} = \tfrac{1}{2}(\|z\| + \|z'\| - \|z - z'\|) \tag{4.21}$$

For $n = 1$, it reduces to the ordinary Brownian motion on $(-\infty, \infty)$ as defined by (2.3.17). For odd-dimensional parameter spaces ($n = 2p + 1$), Lévy [1956] conjectured that $\{X_z, z \in R^{2p+1}\}$ is Markovian of order $p + 1$ in the following sense. A random field $\{X_z, z \in R^n\}$ is said to be Markov of order $\leq p + 1$ if, for any smooth, closed $(n - 1)$ surface ∂D, every approximation $\tilde{X}_z$ to X_z in a neighborhood of ∂D which has the property

$$\lim_{\delta \downarrow 0} \frac{1}{\delta^p} |\tilde{X}_z - X_z| = 0 \qquad \delta = \text{distance } (z, \partial D) \tag{4.22}$$

also has the property that given $\tilde{X}$, X_z and $X_{z'}$ are independent whenever $z \in D^-$ and $z' \in D^+$. If X_z is Markov of order $\leq p + 1$ but not $\leq p$, then it is said to be Markov of order $p + 1$. This conjecture in a somewhat different formulation was proved by McKean [1963]. McKean showed that for a Brownian motion on R^{2p+1}, given X and its "normal derivatives" $\partial^k X$, $k = 1, 2, \ldots, p$, on ∂D, X_z, $z \in D^-$ is independent of $X_{z'}$, $z' \in D^+$. However, a Brownian motion is not even once differentiable, so that the normal derivatives need to be defined, which McKean has done. Brownian motions with an even-dimensional parameter space have no Markov property at all.

One way in which Markovian random fields (of some order) arise naturally is through differential equations driven by white noise, very much in the same way that diffusion processes are generated by white noise. It is not difficult to define white noise, Gaussian or not, with a multidimensional parameter. However, stochastic partial-differential equations as extensions of Ito equations have not been studied, except for linear equations of the form

$$\Delta X_z = k X_z + \eta_z$$

where η_z is a white noise. Several examples of this type have been given by Whittle [1963].

5. MULTIPARAMETER MARTINGALES

In view of the important role that martingales have played in the development of a theory of filtering and detection, one is motivated to generalize the martingale concept to multiparameter processes. This can be done in a number of ways. One of the simplest and most natural is to make use of the

5. MULTIPARAMETER MARTINGALES

natural partial ordering defined in terms of Cartesian coordinates. For simplicity of exposition we limit the discussion for now to the two dimensional case.

Let $\mathbb{R}_+^2$ denote the positive quadrant $\{(t_1, t_2): 0 \leq t_1, t_1 < \infty\}$ and define a partial ordering $(>)$ as follows: For $t, s \in \mathbb{R}_+^2$, $t > s$ if and only if $t_1 \geq s_1$ and $t_2 \geq s_2$. Now martingales can be defined. Let $(\Omega, \mathcal{A}, \mathcal{P})$ be a probability space and $\{\mathcal{A}_t, t \in \mathbb{R}_+^2\}$ be a family of σ-algebras such that

$$t > s \Rightarrow \mathcal{A}_t \supseteq \mathcal{A}_s \quad \text{(increasing family)}$$

A two-parameter process $\{M_t, t \in \mathbb{R}_+^2\}$ is said to be a **martingale** relative to $\{\mathcal{A}_t\}$ if

$$t > s \Rightarrow E(M_t|\mathcal{A}_s) = M_s \quad \text{a.s.} \tag{5.1}$$

An example of two-parameter martingales is the Wiener process W_t defined as follows:

(a) $\{W_t, t \in \mathbb{R}_+^2\}$ is a Gaussian process
(b) $EW_t = 0$ for every t
(c) $EW_t W_s = \min(t_1, s_1)\min(t_2, s_2)$

Now, let $\mathcal{A}_{wt}$ be the σ-algebra generated by $\{W_s, s < t\}$. Then, W_t is a martingale relative to $\{\mathcal{A}_{wt}\}$.

The fact that the Wiener process is a martingale can be easily seen by relating it to a "white noise process" as follows: Let $\mathcal{B}$ denote the collection of Borel sets in $\mathbb{R}_+^2$. Let $\{\eta(A), A \in \mathcal{B}\}$ be a Gaussian family of random variables parametrized by Borel sets such that $E\eta(A) = 0$ and

$$E\eta(A)\eta(B) = \text{area}(A \cap B)$$

It is clear that η is independent on disjoint sets. Now, let A_t denote the rectangle $\{s \in \mathbb{R}_+^2 : s < t\}$. Then, a Wiener process W_t can be expressed as

$$W_t = \eta(A_t)$$

The martingale property of W can now be derived as follows:

$$\begin{aligned}
t > s \Rightarrow E(W_t|\mathcal{A}_{ws}) &= E[\eta(A_t)|\mathcal{A}_{ws}] \\
&= E[\eta(A_s) + \eta(A_t - A_s)|\mathcal{A}_{ws}] \\
&= \eta(A_s) \quad \text{a.s.} \\
&= W_s
\end{aligned}$$

where we have used the fact that $\eta(A_s)$ is $\mathcal{A}_{ws}$ measurable while $\eta(A_t - A_s)$ ($A_t - A_s$ being disjoint from A_s) is $\mathcal{A}_{ws}$ independent.

Wong and Zakai (1974) considered the problem of representing the functionals of a Wiener process in $\mathbb{R}_+^2$ by suitably defined stochastic integrals. This work was followed shortly thereafter by the paper of Cairoli and Walsh (1975), and these papers inaugurated a body of research that has grown to substantial proportion. In this section we give a brief account of the basic results on stochastic integration on $\mathbb{R}_+^2$.

Let $\{\eta(A), A \in \mathcal{B}\}$ be a white noise process and $\{W_t = \eta(A_t)\}$ be a Wiener process as before. Let $\{\mathcal{C}_t, t \in \mathbb{R}_+^2\}$ be an increasing family of σ-algebras such that $B \cap A_t = \varnothing$ implies $\eta(B)$ is $\mathcal{C}_t$ independent. Suppose $\{\phi_t, t \in T\}$ is a random field on the unit square $T = [0,1]^2$, such that for each t, ϕ_t is $\mathcal{C}_t$ measurable and

$$\int_T E\phi_t^2 \, dt < \infty$$

Denote the collection of all such random fields by $\mathcal{H}_1$. Then, $\mathcal{H}_1$ is a Hilbert space with inner product

$$(\phi, \psi) = \int_T E(\phi_t \psi_t) \, dt$$

We note define type-1 stochastic integrals

$$\phi \cdot W = \int_T \phi_s \, dW_s = \int_T \phi_s \eta(ds)$$

for $\phi \in \mathcal{H}_1$ as follows:

(a) If $\phi_t(\omega) = Z(\omega) I_A(t)$ where I_A is the indicator function of a rectangle A, then we set

$$\phi \cdot W = Z\eta(A)$$

(Note: Z is necessarily $\mathcal{C}_{t_0}$ measurable where t_0 is the minimum point on A.)

(b) If $\phi = \sum_{i=1}^m \phi_i$ and each ϕ_i is of the form given in (a), then we set

$$\phi \cdot W = \sum_{i=1}^m \phi_i \cdot W$$

and call such ϕ's **simple** functions. For simple ϕ's, we have

$$E(\phi \cdot W)^2 = (\phi, \phi) = \|\phi\|^2$$

(c) A Cauchy sequence of simple functions $\{\phi_k\}$ in $\mathcal{H}_1$ produces a Cauchy sequence $\{\phi_k \cdot W\}$ of square-integrable random variables.

5. MULTIPARAMETER MARTINGALES

Hence, we can define

$$(\lim \phi_k) \cdot W = \lim \text{ in q.m. } \phi_k \cdot W$$

(d) It can be shown in a way similar to that in Section 4.2 that every $\phi \in \mathcal{H}_1$ is the limit of a sequence of simple functions. Hence, $\phi \cdot W$ is well defined for all $\phi \in \mathcal{H}_1$.

It is clear that $\phi \cdot W$ is a straightforward generalization of the Itô integral (defined in Section 4.2) and its properties are similar. These include:

(a) linearity: $(\alpha\phi + \beta\phi') \cdot W = \alpha(\phi \cdot W) + \beta(\phi' \cdot W)$
(b) isometry: $E(\phi \cdot W)(\phi' \cdot W) = (\phi, \phi')$
(c) martingale: $E(\phi \cdot W | \mathcal{A}_t) = \phi I_{A_t} \cdot W$ almost surely for every $t \in T$
(i.e., $E[\int_T \phi_s \, dW_s | \mathcal{A}_t] = \int_{s<t} \phi_s \, dW_s$)

Type-1 stochastic integrals are not enough to represent all Wiener functionals. That is, there are square-integrable random variables Z measurable with respect to $\mathcal{A}_{(1,1)}$ that cannot be written as

$$Z = EZ + \phi \cdot W$$

To see this, consider a random variable Z constructed as follows: Let t and s be two unordered points and let Δ_t and Δ_s be two forward incremental rectangles at t and s, respectively. Define $Z = \eta(\Delta_t)\eta(\Delta_s)$. For infinitesimal Δ_t and Δ_s, to represent Z as a type-1 stochastic integral would require it to be of the form

$$Z = X\eta(\Delta_t) + Y\eta(\Delta_s)$$

where X is $\mathcal{A}_t$ measurable and Y is $\mathcal{A}_s$ measurable. That would mean

$$EZ^2 = E\big[X\eta^2(\Delta_t)\eta(\Delta_s) + Y\eta^2(\Delta_s)\eta(\Delta_t)\big]$$
$$= E\{X\eta^2(\Delta_t)E[\eta(\Delta_s)|\mathcal{A}_t] + Y\eta^2(\Delta_s)E[\eta(\Delta_t)|\mathcal{A}_s]\} = 0$$

which contradicts

$$EZ^2 = \text{area}(\Delta_t)\,\text{area}(\Delta_s)$$

What is needed is a type-2 stochastic integral which will be defined next.

Let G denote the set of all unordered pairs (t, s) in $T \times T$ such that t is lower and to the right and s is higher and to the left. In other words $t_1 > s_1$ and $t_2 < s_2$. We denote this by $s \wedge t$. Let $\mathcal{H}_2$ denote the collection of all random functions $\psi_{t,s}$ defined on G such that:

(a) for each (t,s), $\psi_{t,s}$ is a random variable measurable with respect to $\mathcal{Q}_{t_1, s_2}$;

(b) $\psi(t,s,\omega)$ is jointly measurable with respect to $\mathcal{L} \times \mathcal{L} \times \mathcal{Q}$ where $\mathcal{L}$ denotes the σ-algebra of Lebesgue measurable sets in T;

(c) $\int_G E\psi_{t,s}^2 \, dt\, ds < \infty$.

We can now define a type-2 stochastic integral, variously denoted by

$$\int_{T^2} \psi_{t,s}\, dW_t\, dW_s = \int_{T^2} \psi_{t,s}\eta(dt)\eta(ds) = \psi \cdot W^2$$

as follows:

(a) Suppose that there exist rectangles Δ_1 and Δ_2 such that $\Delta_1 \times \Delta_2 \subset G$ and

$$\psi_{t,s}(\omega) = \alpha(\omega) \quad \text{for } (t,s) \in \Delta_1 \times \Delta_2$$
$$= 0 \quad \text{otherwise}$$

Then, we set

$$\psi \cdot W^2 = \eta(\Delta_1)\eta(\Delta_2)$$

(b) We say ψ is *simple* if

$$\psi = \sum_{k=1}^m \psi_k$$

and each ψ_k is of the form given in (a). If ψ is simple, we set

$$\psi \cdot W^2 = \sum_{k=1}^m \psi_k \cdot W^2$$

(c) It can be shown (Wong and Zakai, 1974) that simple ψ's are dense in $\mathcal{H}_2$ with respect to the norm

$$\|\psi\| = \left(\int_G E\psi_{t,s}^2 \, dt\, ds \right)^{1/2}$$

and that for a simple ψ,

$$E(\psi \cdot W^2)^2 = \|\psi\|^2$$

It follows that $\psi \cdot W^2$ can be extended to all $\psi \in \mathcal{H}_2$ by approximating ψ with a sequence of simple functions ψ_n such that

5. MULTIPARAMETER MARTINGALES

$\|\psi - \psi_n\|\xrightarrow[n\to\infty]{} 0$ and by defining

$$\psi \cdot W^2 = \lim_{n\to\infty} \text{ in q.m. } \psi_n \cdot W^2$$

Type-2 stochastic integrals have the following important properties:

(a) linearity: $(\alpha\psi + \beta\psi') \cdot W^2 = \alpha(\psi \cdot W^2) + \beta(\psi' \cdot W^2)$
(b) isometry: $E(\psi \cdot W^2)(\psi' \cdot W^2) = \int_G E\psi_{t,s}\psi'_{t,s}\, dt\, ds$
(c) orthogonality: $E(\psi \cdot W^2)(\phi \cdot W) = 0$, $\psi \in \mathcal{K}_2$, $\phi \in \mathcal{K}_1$
(d) martingale: $E(\psi \cdot W^2 | \mathcal{C}_t) = \psi I_{A_t \times A_t} \cdot W^2$.

With stochastic integrals of type-1 and type-2, all square-integrable Wiener functionals can now be represented. For example, $W_{(1,1)}^2 = \eta^2(T)$ can be written as

$$W_{(1,1)}^2 = EW_{(1,1)}^2 + 2(W \cdot W) + 2(1 \cdot W^2)$$

$$= 1 + 2\int_T W_s\, dW_s + 2\int_G dW_t\, dW_s$$

Several proofs of the representation theorem have now been given (Wong and Zakai, 1974; Cairoli and Walsh, 1975; Hajek, 1979). Rather than reproducing one here, we show a heuristic argument for it.

Consider a functional

$$f(u, W) = \exp \int_T u(s)\, dW_s$$

where u is a deterministic function in $L^2(T)$. Intuitively, any square-integrable Wiener functional can be approximated by sums of the form

$$\sum_u \alpha(u) f(u, W)$$

so that it would be sufficient to obtain a representation for $f(u, W)$. Now, define

$$M_t = \int_{A_t} u(s)\, dW_s$$

and

$$X_t = \exp\left\{M_t - \frac{1}{2}\int_{A_t} u^2(s)\, ds\right\}$$

Then, M_{t_1, t_2} is a one-parameter martingale in $t_1(t_2)$ for a fixed $t_2(t_1)$. It follows from the differentiation rule for a one-parameter martingale that

$$d_{t_2} X_t = X_t \, d_{t_2} M_t$$

and

$$d_{t_1} d_{t_2} X_t = X_t \, d_{t_1} M \, d_{t_2} M + X_t \, d_{t_1} d_{t_2} M$$

Now, $d_{t_1} d_{t_2} M = u(t)\eta(dt)$ and

$$d_{t_1} M \, d_{t_2} M = \sum_{\substack{s' \wedge s \\ s \vee s' = t}} u(t) \eta(ds) \eta(ds')$$

Hence, if we denote $A(u) = \exp \frac{1}{2} \int_T u^2(s)\, ds$ then

$$f(u, W) = A(u) X_{1,1} = A(u)\left\{ 1 + \int_t X_t (d_{t_1} M d_{t_2} M + d_{t_1} d_{t_2} M) \right\}$$

$$= A(u)\left\{ 1 + \int_T u(t) X_t \, dW_t + \int_G u(s) u(s') X_{s \vee s'} \, dW_s \, dW_{s'} \right\}$$

which is the representation result that we have been seeking.

From the martingale property of the stochastic integrals of both types, it is clear that the representation theorem implies that every square-integrable Wiener martingale can be represented in the form

$$M_t = EM_0 + \phi I_A \cdot W + \psi I_{A_t \times A_t} \cdot W^2$$

However, for the stochastic integrals to be useful as an analytical and computational tool, we need more than a representation result, we need a calculus. It turns out (Cairoli and Walsh, 1975) that a host of additional concepts is needed. These include: 1- and 2- martingales and strong and weak martingales. Even with these additional concepts the resulting calculus is complicated. To remove this obstacle, a theory of stochastic differential forms is needed. Some of the basic elements of such a theory are presented in the next section.

Before taking up stochastic differential forms, we should mention that stochastic integration has been further generalized in a number of ways. Yor (1976) generalized the representation result to n-parameter Wiener functionals and in the process introduced multiple stochastic integrals of various orders of which our type-2 integral is a special case corresponding to order 2. Hajek (1979) generalized the representation result by viewing martingales as set-parametrized processes. In the process he not only extended stochastic integration to a much more general setting but also illuminated the connection between multiple stochastic integrals and multiple Wiener integrals (Itô, 1951c).

6. STOCHASTIC DIFFERENTIAL FORMS

Intuitively, differential forms are objects to be integrated on reasonably smooth k-dimensional surfaces in an n-dimensional space ($n \geq k$). For example, a 1-form in two-dimensional space can be written as

$$x = a(t)\, dt_1 + b(t)\, dt_2, \qquad t = (t_1, t_2) \in \mathbb{R}^2 \tag{6.1}$$

If a curve γ in $\mathbb{R}^2$ is represented by

$$\gamma = \{t_1(\alpha), t_2(\alpha), 0 \leq \alpha \leq 1\}$$

then $\int_\gamma x$ is given by

$$\int_\gamma x = \int_0^1 \left[a(t(\alpha)) \frac{dt_1(\alpha)}{d\alpha} + b(t(\alpha)) \frac{dt_2(\alpha)}{d\alpha} \right] d\alpha$$

Our goal is to generalize the concept to include not only cases where $a(t)$ and $b(t)$ in (6.1) are multiparameter processes but also situations where the processes involved are not sufficiently smooth to allow representation such as (6.1).

We begin with the observation (Whitney, 1957) that if we denote

$$X(\gamma) = \int_\gamma x$$

then knowing $X(\gamma)$ on a sufficiently large set of γ's completely specifies x. In fact, it is enough to know $X(\gamma)$ on horizontal and vertical lines, which would allow $a(t)$ and $b(t)$, respectively, to be determined. Thus, a differential 1-form in n-space can be viewed as a function of lines parallel to the axes, and 2-forms, can be viewed as functions of rectangles, etc. This is the approach we shall use in defining stochastic differential forms. For simplicity of exposition we shall limit ourselves to differential forms on two-dimensional spaces.

Consider the collection of all oriented horizontal and vertical line segments σ. We give σ a positive orientation if it is increasing (in the coordinate that is changing) and negative otherwise. Then, σ and $-\sigma$ represent the same set which is denoted by $|\sigma|$. If σ is subdivided into two subsegments σ' and σ'' having the same orientation as σ, then we write

$$\sigma = \sigma' + \sigma''$$

A stochastic differential 1-form X is a family of random variables parametrized by

$$\Sigma = \{\text{oriented horizontal and vertical line segments}\}$$

such that it is additive, i.e.,

$$X(\sigma' + \sigma'') = X(\sigma') + X(\sigma'')$$

To allow X to be extended to curves, we need some continuity conditions to be explained next.

Construct a linear space Γ_1 consisting of formal sums

$$\sum_{i=1}^{m} \alpha_i \sigma_i$$

where α_i are real constants and σ_i are oriented line segments with the requirements that: (a) elements of Γ_1 equal under subdivision are not distinguished, and (b) $\alpha(-\sigma) = (-\alpha)\sigma$. A differential 1-form X is easily extended to Γ_1 by linearity, i.e.,

$$X\left(\sum_{i=1}^{m} \alpha_i \sigma_i\right) = \sum_{i=1}^{m} \alpha_i X(\sigma_i)$$

Elements of Γ_1 are called 1-chains.

Now, give each rectangle ρ in $\mathbb{R}^2$ an orientation (say $-$ for "in" and $+$ for "out"). The boundary $\partial \rho$ of each oriented rectangle ρ is a 1-chain, and we adopt a right-hand-screw convention so that the boundary of a positively oriented rectangle is counter-clockwise. Let Γ_2 denote the linear space consisting of linear combinations of oriented rectangles $\sum_i \alpha_i \rho_i$ with the previously cited convention concerning equivalence under subdivision and multiplication by -1. Elements of Γ_2 are called 2-chains. We are now ready to define stochastic differential forms.

Let $(\Omega, \mathcal{C}, \mathcal{P})$ be a fixed probability space and let $\mathcal{S}$ denote the space of all random variables on the space. We define stochastic k-forms on $\mathbb{R}^2$ as follows:

A 0-form is a function $X: \mathbb{R}^2 \to \mathcal{S}$
A 1-form is a linear function $X: \Gamma_1 \to \mathcal{S}$
A 2-form is a linear function $X: \Gamma_2 \to \mathcal{S}$

In addition, we shall assume the following conditions to be satisfied by all stochastic differential forms:

(a) All 0-forms $X(\omega, t)$ are bimeasurable in (ω, t).
(b) All 1-forms satisfy two continuity conditions:

$$\forall \sigma \in \Gamma_1 \quad \lim_{\|\sigma\| \to 0} \text{in p.} \ X(\sigma) = 0 \quad \text{where } \|\sigma\| = \text{length of } \sigma \quad (6.2)$$

6. STOCHASTIC DIFFERENTIAL FORMS

$$\forall \rho \in \Gamma_2 \quad \lim_{\|\rho\| \to 0} \text{in p.} \; X(\partial \rho) = 0 \quad \text{where } \|\rho\| = \text{area of } \rho \quad (6.3)$$

(c) All 2-forms satisfy

$$\forall \rho \in \Gamma_2 \quad \lim_{\|\rho\| \to 0} \text{in p.} \; X(\rho) = 0 \quad (6.4)$$

With the continuity conditions (6.2)–(6.4), the 1 and 2 forms can now be further extended. For example, a sequence of approximating 1-chains can be constructed for any smooth curve γ by successively subdividing γ and constructing a staircase approximation using the subdivision. If the subdivisions are nested then the difference between two staircase approximations is the boundary of a 2-chain. Continuity (6.3) then allows a 1-form to be extended to γ. Similarly, (6.4) allows a 2-form to be extended to a two-dimensional set that can be approximated by 2-chains.

Before proceeding further, consider the following example. Let $\{\eta(A), A \in \mathcal{R}^2\}$ be a Gaussian collection of random variables parametrized by the collection $\mathcal{R}^2$ of Borel sets in $\mathbb{R}^2$ such that $E\eta(A) = 0$ and

$$E\eta(A)\eta(B) = \text{area}(A \cap B)$$

The set-parameter process η will be called a Gaussian white noise. Now, for any oriented rectangle, set

$$Z(\rho) = \eta(|\rho|) \quad \text{if } \rho \text{ has a } + \text{ orientation}$$
$$= -\eta(|\rho|) \quad \text{if } \rho \text{ has a } - \text{ orientation}$$

Z can be extended to Γ_2 by linearity and so extended satisfies condition (6.4). Thus, Z is a stochastic 2-form.

A 0-form W can also be defined in terms of η in a natural way. For each $t \in \mathbb{R}^2$ set

$$W_t = \pm \eta(A_t)$$

when A_t is the rectangle (unoriented) bounded by the two axes and t. The sign is $+$ if t is in the first or third quadrant and $-$ otherwise.

A 1-form $\mathcal{U}$ can be defined in terms of W as follows. Let **ab** denote an oriented horizontal or vertical line segment from point a to point b, and set

$$\mathcal{U}(\mathbf{ab}) = W(b) - W(a)$$

Then $\mathcal{U}$ can be extended by linearity to Γ_1 and to smooth curves by continuity. The above example suggests a close relationship among the three forms Z, W, and $\mathcal{U}$. To expose the precise relationship, however, requires the introduction of some additional concepts.

Assign an orientation (+ or −) to each point t in $\mathbb{R}^2$ and denote the space of all linear combinations $\sum_i \alpha_i t_i$ by Γ_0. A 0-form can be extended to Γ_0 by linearity. Just as the boundary $\partial \rho$ of a ρ in Γ_2 is in Γ_1, the boundary $\partial \rho$ of a σ in Γ_1 is in Γ_0. We can now define the **exterior derivative** dX of an r-form ($r = 0, 1$) as an $r + 1$ form such that

$$dX(\sigma) = X(\partial \sigma) \qquad \text{for all } \sigma \in \Gamma_{r+1} \tag{6.5}$$

It follows that for a zero form Y,

$$dY(\mathbf{ab}) = Y(b) - Y(a)$$

and for a 1-form $\mathcal{V}$,

$$d\mathcal{V}\left(\begin{array}{cc} d & c \\ \boxed{+} & \\ a & b \end{array} \right) = \mathcal{V}(\mathbf{ab}) + \mathcal{V}(\mathbf{bc}) + \mathcal{V}(\mathbf{cd}) + \mathcal{V}(\mathbf{da})$$

Observe that $\partial\partial(\rho) = 0$ for all $\rho \in \Gamma_2$, hence ddY is always 0.

For our previous example, it is clear that

$$\mathcal{U} = dW$$

but the relationship between Z and $\mathcal{U}$ (or W) remains obscure. To expose that relationship requires one more concept. Given a 1-form, say X, we can express it in terms of a "coordinate system" similar to (6.1) as follows. Define X_i ($i = 1, 2$) as one forms such that

$$\begin{aligned} X_1(\sigma) &= X(\sigma) \quad \text{and} \quad X_2(\sigma) = 0 &&\text{if } \sigma \text{ is horizontal} \\ &= 0 && \qquad\qquad = X(\sigma) &&\text{if } \sigma \text{ is vertical} \end{aligned}$$

Then, we can write $X = X_1 + X_2$ and the representation is good everywhere on Γ_1. For our example, we have

$$Z = d(\mathcal{U}_2) = -d(\mathcal{U}_1)$$

To see this consider a rectangle ρ given by

$$\rho = \begin{array}{cc} d & c \\ \boxed{+} & \\ a & b \end{array}$$

Then,

$$Z(\rho) = \eta(|\rho|) = \eta\left(\begin{array}{cc} d & c \\ \square & \\ a & b \end{array} \right)$$

6. STOCHASTIC DIFFERENTIAL FORMS

Now,

$$d(\mathcal{U}_2)(\rho) = \mathcal{U}_2(\partial\rho) = \mathcal{U}_2(\mathbf{ab} + \mathbf{bc} + \mathbf{cd} + \mathbf{da})$$
$$= \mathcal{U}(\mathbf{bc}) + \mathcal{U}(\mathbf{da})$$
$$= W(c) - W(b) - W(d) + W(a)$$
$$= \eta(|\rho|)$$

Note that the exterior derivative d is independent of any coordinate system. Hence, the relationship between $\mathcal{U}$ and W is coordinate independent. However, the relationship between Z and $\mathcal{U}$ (or W) depends explicitly on the coordinate system.

The concept of "martingales" can be extended easily to differential forms, and the results that follow illuminate the situation discussed in Section 5. Let $\{\mathcal{Q}_t, t \in \mathbb{R}^2\}$ be a family of σ-algebras satisfying the following conditions:

$$H_1: t > s \Rightarrow \mathcal{Q}_t \supseteq \mathcal{Q}_s$$
$$H_2: \mathcal{Q}_t = \mathcal{Q}_{t+} = \bigcap_{\substack{s_1 > t_1 \\ s_2 > t_2}} \mathcal{Q}_s$$

If σ is a point, vertical or horizontal line segment, or closed rectangle, denote by $\underline{t}(\sigma)$ and $\bar{t}(\sigma)$ the minimum and maximum points of σ, respectively. An r-form ($r = 0, 1, 2$) X is said to be **adapted** if $X(\sigma)$ is $\mathcal{Q}_{\bar{t}(\sigma)}$ measurable. Now, define $\mathcal{Q}_t^i$, $i = 1, 2$, as follows: For $t = (t_1, t_2)$,

$$\mathcal{Q}_t^1 = \bigvee_{t_2} \mathcal{Q}_t \equiv \mathcal{Q}_{t_1, \infty}$$

$$\mathcal{Q}_t^2 = \bigvee_{t_1} \mathcal{Q}_t \equiv \mathcal{Q}_{\infty, t_2}$$

A 1-form X is said to ba a **1-martingale** (2-martingale) if for an oriented horizontal (vertical) line segment σ,

$$E\big[X(\sigma)|\mathcal{Q}_{\underline{t}(\sigma)}^1\big] = 0 \quad \left(E\big[X(\sigma)|\mathcal{Q}_{\underline{t}(\sigma)}^2\big] = 0\right) \tag{6.6}$$

A 2-form can also be defined as an i-martingale ($i = 1, 2$) in exactly the same way. We merely need to take σ to be an oriented rectangle rather than a line segment in (6.6). We say an r-form ($r = 1, 2$) X is a *martingale* if it is an i-martingale for both $i = 1$ and 2.

For 0-forms we retain the definition of martingale given earlier, namely,

$$t > s \Rightarrow E(X_t|\mathcal{Q}_s) = X_s \quad \text{a.s.}$$

Now, the question is: What is the relationship among the martingale r-forms for different r? The following result due in its original form to Cairoli and Walsh (1975) relates 0-form martingales to 1-form martingales.

Proposition 6.1. Let $\{\mathcal{Q}_t\}$ satisfy the additional condition

$H_3: \mathcal{Q}_t^1$ and $\mathcal{Q}_t^2$ are conditionally independent given $\mathcal{Q}_t$

Let M be a 0-form such that M is a one-parameter martingale on each of the two axes. Then, M is a martingale if and only if dM is a martingale.

The relationship between martingale 1-forms and 2-forms is less interesting.

Proposition 6.2. If a 1-form X is an i-martingale ($i = 1, 2$), then dX_i is an i-martingale.

Now, if X is a martingale 1-form then X is a one-parameter martingale on every horizontal line and every vertical line. If X is square-integrable then X has a quadratic variation on each horizontal or vertical line. Hence, there exists a positive adapted 1-form $\langle X \rangle$ such that $\langle X \rangle_i$ ($i = 1, 2$) are the horizontal and vertical quadratic variations. Furthermore, if Y is a 0-form then we can define a 1-form $Y \wedge X$ by evaluating $Y \wedge X$ on line segments σ as follows.

$$(Y \wedge X) = \int_\sigma Y_t \, d_{t_1} X_t \quad \text{if } \sigma \text{ is horizontal}$$
$$= \int_\sigma Y_t \, d_{t_2} X_t \quad \text{if } \sigma \text{ is vertical}$$

In either case the integral is well defined as an integral with respect to a one-parameter martingale (c.f. Chapter 6). Indeed, $Y \wedge X$ remains well defined if X is only a semimartingale on horizontal and vertical lines.

We can now proceed to state the differentiation formula for two-parameter martingale 0-forms as follows:

Proposition 6.3. Let M be a square-integrable martingale 0-form. Let f be a twice continuously differentiable function. Then

$$dF(M) = f'(M) \wedge dM + \tfrac{1}{2} f''(M) \wedge \langle dM \rangle \tag{6.7}$$

where prime denotes differentiation.

Consider the 0-form W defined earlier. It is easier to see that dW is in martingale and

$$\langle dW \rangle = d\mu(A)$$

where μ denotes the area (i.e., the Lebesgue measure).

6. STOCHASTIC DIFFERENTIAL FORMS

The differentiation formula (6.7) can be generalized to $\mathbb{R}^n$ without change in form. It is also a global formula without dependence on any coordinate systems, and as such is far simpler than the differentiation formulas that had been derived earlier for two-dimensional martingales.

The operation $Y \wedge X$ (exterior product) can be defined for suitable p and r forms and results in a $p + r$ form. The new cases for $\mathbb{R}^2$ are $p = 0$ and $r = 2$ and $p = 1 = r$. If Y is a 0-form and X a martingale 2-form then $Y \wedge X$ is basically an area stochastic integral such as defined in Section 5. If Y and X are both martingale 1-forms then $Y \wedge X$ is a martingale 2-form. We shall not deal with the topic further except to consider some examples.

Consider the differential forms Z, W, and $\mathcal{U}$ ($= dW$) introduced in an earlier example. $f(W) \wedge Z$ is a 2-form defined by

$$(f(W) \wedge Z)(\sigma) = \int_\sigma f(W_t) \, dW_t$$

where the integral is a type-1 integral introduced in Section 5. Now, $\mathcal{U} \wedge \mathcal{U} \equiv 0$, but $\mathcal{U}_1 \wedge \mathcal{U}_2$ is given by the type-2 integral

$$(\mathcal{U}_1 \wedge \mathcal{U}_2)\left(\begin{array}{c} t \\ \boxed{+} \\ t' \end{array} \right) = I_{A_1 \times A_2} \cdot W^2$$

where $A_1 = [t'_1, t_1] \times [0, t_2]$ and $A_2 = [0, t_1] \times [t'_2, t_2]$. If $\{\mathcal{Q}_t\}$ is generated by the process W, then one can proceed to obtain representation theorems not only for martingale 0-forms but also for martingale 1- and 2-forms. We will not pursue the topic further here.

As a final topic, we shall present a brief discussion on the relationship between differential forms and two-parameter Markov processes. Consider the Wiener process W_t, $t \in \mathbb{R}^2_+$, defined in Section 5. One might think that if any process is Markov it should be W_t. Surprisingly, W_t is not Markov! For example, consider the triangular region S in $\mathbb{R}^2_+$ bounded by the line $\gamma = \{t: t_1 + t_2 = 1\}$. For a point t outside of S, one can show that

$$E(W_t | W_{t'}, t' \in \gamma) \neq E(W_t | W_{t'}, t' \in S)$$

demonstrating that W is not Markov. The situation is greatly illuminated by the use of differential forms. Consider the two 1-forms d_1W and d_2W. It is easy to show that each d_iW is a Markov process in the following sense. For any curve ∂D that separates $\mathbb{R}^2_+$ into a bounded region D that includes the origin and D^+,

$$\sigma \subset D \text{ and } \sigma^+ \subset D^+ \Rightarrow \begin{array}{l} d_iW(\sigma) \text{ and } d_iW(\sigma^+) \text{ are independent given} \\ \{d_iW(\sigma'), \sigma' \subset \partial D\} \end{array}$$

To prove this, take a vertical segment σ^+ in D^+ and construct a set ρ by projecting σ^+ horizontally onto ∂D. Then, $d(d_2 W) = Z$ so that

$$d(d_2 W)(\rho) = Z(\rho) = \eta(|\rho|)$$

where η is the Gaussian white noise defined earlier. Since $d(d_2 W)(\rho) = d_2 W(\partial \rho)$ and $d_2 W$ is zero on horizontal segments, we have

$$d_2 W(\sigma^+) - d_2 W(\rho \cap \partial D) = \eta(|\rho|)$$

which is independent of $d_2 W(\sigma)$ for any $\sigma \in D$, and we have proved $d_2 W$ is Markov. A similar proof works for $d_1 W$. Hence, we conclude that although W is not Markov, the components of dW are Markov.

As a second example of the relationship between differential forms and two-parameter Markov processes, consider a pair of independent zero-mean Gaussian generalized random fields ξ_1 and ξ_2 each with a covariance function given by (4.19) for $n = 2$ and $\lambda_0 = 1$, i.e.,

$$E\xi_i(t)\xi_j(t') = \delta_{ij} K_0(\|t - t'\|), \qquad t, t' \in \mathbb{R}^2$$

The pair (ξ_1, ξ_2) defines a differential 1-form X via the relationship

$$X(\sigma) = \int_\sigma \xi_{1t}\, dl_t \qquad \text{if } \sigma \text{ is horizontal}$$

$$= \int_\sigma \xi_{2t}\, dl_t \qquad \text{if } \sigma \text{ is vertical}$$

where dl_t denotes the path differential. Now, dX is a 2-form and it can be easily shown that

$$dX + \alpha \wedge X = \eta \tag{6.8}$$

is a Gaussian white noise 2-form for any 1-form

$$\alpha = \alpha_1\, dt_1 + \alpha_2\, dt_2$$

such that α_1 and α_2 are constants satisfying

$$\alpha_1^2 + \alpha_2^2 = 1$$

Equation (6.8) relates an isotropic and homogeneous Gauss-Markov field to Gaussian white noise and can be used to prove the Markovian property. It also strengths our intuition that a Markov process with parameter of any dimension is "one derivative away" from white noise.

References

Beneš, V. E. (1981): Exact finite-dimensional filters for certain diffusions with nonlinear drift, *Stochastics*, **5**:65–92.

Birkoff, G. and S. MacLane (1953): "A Survey of Modern Algebra," Macmillan, New York.

Breiman, L. (1968): "Probability," Addison-Wesley, Reading, Mass.

Brémaud, P. (1981): "Point Processes and Queues, Martingale Dynamics," Springer-Verlag, New York.

Bucy, R. S. and P. D. Joseph (1968): "Filtering for Stochastic Processes with Applications to Guidance," Interscience, Wiley, New York.

Cairoli, R. and J. B. Walsh (1975): Stochastic integrals in the plane, *Acta Math.* **134**:111–183.

Cramér, H. (1966): On stochastic processes whose trajectories have no discontinuities of the second kind, *Ann. di Matematica (iv)*, **71**:85–92.

Davenport, W. B., Jr. and W. L. Root (1958): "An Introduction to the Theory of Random Signals and Noise," McGraw-Hill, New York.

Davis, M. H. A. (1980): On a multiplicative functional transformation arising in nonlinear filtering theory, *Z. Wahrscheinlichkeitstheorie verw. Geb.*, **54**:125–139.

Dellacherie, C. and P. A. Meyer (1978): "Probabilities and Potential," North-Holland, New York.

Dellacherie, C. and P. A. Meyer (1982): "Probabilities and Potential B, Theory of Martingales," North-Holland, New York.

Doléans, C. (= Doleans-Dade, C.) (1967): Processus croissant naturels et processus tres-bien-measurables, *C.R. Acad. Sci. Paris*, **264**:874–876.

Doléans, C. (1968): Existence du processus croissant naturel associé à un potential de la classe (D) *Z. Wahrscheinlichkeitstheorie verw. Geb.*, **9**:309–314.

Doléans, C. (1970): Quelque applications de la formule de changement de variable pour les semimartingales *Z. Wahrscheinlichkeitstheorie verw. Geb.*, **16**:181–194.

Doob, J. L. (1953): "Stochastic Processes," Wiley, New York.

Doob, J. L. (1984): "Classical Potential Theory and Its Probabilistic Counterpart," Springer-Verlag, New York.

Duncan, T. E. (1968): Evaluation of likelihood functions, *Information and Control*, **13**:62–74.

Duncan, T. E. (1970): On the absolute continuity of measures, *Annals Math. Stat.*, **41**:30–38.

Dynkin, E. B. (1965): "Markov Processes," (2 vols.), Academic, New York, Springer-Verlag, Berlin.

Erdelyi, A. (1953): "Higher Transcendental Functions," vol. II, Bateman Manuscript Proj., McGraw-Hill, New York.

Fujisaki, M., G. Kallianpur and H. Kunita (1972): Stochastic differential equations for the non-linear filtering problem, *Osaka J. Math.* **9**:19–40.

Gangoli, R. (1967): Abstract harmonic analysis and Lévy's Brownian motion of several parameters, *Proc. 5th Berkeley Symp. Math. Stat. and Prob.*, **II-1**:13–30.

Girsanov, I. V. (1960): On transforming a certain class of stochastic processes by absolutely continuous substitution of measures, *Theory of Prob. and Appl.*, **5**:285–301.

Hajek, B. (1979): "Stochastic Integration, Markov Property and Measure Transformation of Random Fields," Ph.D. dissertation, University of California, Berkeley.

Halmos, P. R. (1950): "Measure Theory," Van Nostrand, Princeton, N.J.

Hazewinkel, M. and J. C. Willems (Eds.) (1981): "The mathematics of filtering and identification, Proceedings of the NATO Advanced Study Institute which took place in June–July 1980 at Les Arcs, France," Reidel, Boston.

Itô, K. (1944): Stochastic integrals, *Proc. Imp. Acad. Tokyo*, **20**:519–524.

Itô, K. (1951a): On stochastic differential equations, *Mem. Amer. Math. Soc.*, **4**:1–51.

Itô, K. (1951b): On a formula concerning stochastic differentials, *Nagoya Math. J.*, **3**:55–65.

Itô, K. (1951c): Multiple Wiener integral, *J. Math. Soc. Japan*, **3**:157–169.

Jacod, J. (1979): "Calcul Stochastique et Problèmes de Martingales," Springer-Verlag, New York.

Jazwinski, A. H. (1970): "Stochastic Processes and Filtering Theory," Academic, New York.

Kac, M. (1951): On some connections between probability theory and differential and integral equations, *Proc. 2nd Berkeley Symp. on Math. Stat. and Prob.*, 189–215.

Kailath, T. (1969): A general likelihood-ratio formula for random signals in Gaussian noise, *IEEE Trans. Inf. Th.*, **IT-5**:350–361.

Kakutani, S. (1948): On equivalence of infinite product measures, *Ann. Math.*, **47**:214–224.

REFERENCES

Kalman, R. E. and R. S. Bucy (1961): New results in linear filtering and prediction theory, *Trans. Am. Soc. Mech. Engn. Series D, J. Basic Eng.*, **83**:95–108.

Karhunen, K. (1947): Über linear methoden in der wahrscheinlichkeitsrechnung, *Ann. Acad. Sci. Fenn.*, **37**.

Kohlmann, M. and W. Vogel (Eds.) (1979): "Stochastic control theory and stochastic differential systems, Proceedings of a Workshop of the 'Sonderforschungsbereich 72 der Deutschen Forschungsgemeinschaft an der Universität Bonn,' which took place in January 1979 at Bad Honnef," Lecture Notes in Control and Information Sciences, **16**, Springer-Verlag, New York.

Kolmogorov, A. N. (1931): Über die analytische methoden in der wahrscheinlichkeitsrechnung, *Math. Ann.*, **104**:415–458.

Kunita, H. and S. Watanabe (1967): On square integrable martingales, *Nagoya Math. J.*, **30**:209–245.

Kushner, H. J. (1967): Dynamical equations for optimal nonlinear filtering, *J. Diff. Equat.*, **3**:179–190.

Lévy, P. (1956): A special problem of Brownian motion, and a general theory of Gaussian random function, *Proc. 3rd Berkeley Symp. Math. Stat. and Prob.*, **2**:133–175.

Liptser, R. S. and A. N. Shiryayev (1977): "Statistics of Random Processes, I and II," Springer-Verlag, New York.

Loève, M. (1963): "Probability Theory," 3d ed., Van Nostrand, Princeton, N.J.

McKean, H. P., Jr. (1960): The Bessel motion and a singular integral equation, *Mem. Coll. Sci. Univ. Kyota, Series A*, **33**:317–322.

McKean, H. P., Jr. (1963): Brownian motion with a several dimensional time, *Theory of Prob. and Appl.*, **8**:335–365.

McKean, H. P., Jr. (1969): "Stochastic Integrals," Academic, New York.

McShane, E. J. (1969): Toward a stochastic calculus, II, *Proc. National Academy of Sciences*, **63**:1084–1087.

McShane, E. J. (1970): Stochastic differential equations and models of random processes, *Proc. 6th Berkeley Symp. Math. Stat. and Prob.*, **3**:263–294.

Meyer, P. A. (1966): "Probability and Potentials," Blaisdell, Waltham, Mass.

Mortensen, R. E. (1966): "Optimal Control of Continuous Time Stochastic Systems," Ph.D. dissertation, Dept. of Electrical Engineering, University of California, Berkeley.

Neveu, J. (1965): "Mathematical Foundations of the Calculus of Probability," Amiel Feinstein (trans.), Holden-Day, San Francisco.

Paley, R. E. A. C. and N. Wiener (1934): "Fourier Transforms in the Complex Domain," *Amer. Math. Soc. Coll. Pub., Am. Math. Soc.*, **19**.

Prokhorov, Yu V. (1956): Convergence of random processes and limit theorems in probability theory, *Theory of Prob. and Appl.*, **1**:157–214.

Rao, K. M. (1969): On decomposition theorems of Meyer, *Math. Scand.*, **24**:66–78.

Riesz, F. and B. Sz.-Nagy (1955): "Functional analysis," Ungar, New York.

Root, W. L. (1962): Singular Gaussian measures in detection theory, *Proc. Symp.*

Time Series Analysis, Brown University, 1962, Wiley, New York, 1963, pp. 292–316.

Rudin, Walter (1966): "Real and Complex Analysis," McGraw-Hill, New York.

Skorokhod, A. V. (1965): "Studies in the Theory of Random Processes," (trans. from Russian), Addison-Wesley, Reading, Mass.

Slepian, D. (1958): Some comments on the detection of Gaussian signals in Gaussian noise, *IRE Trans. Inf. Th.*, 4:65–68.

Sosulin, Yu and R. L. Stratonovich (1965): Optimum detection of a diffusion process in white noise, *Radio Engrg. Electron. Phys.*, 10:704–714.

Stratonovich, R. L. (1960): Conditional Markov processes, *Theory Prob. Appl.*, 5:156–178.

Stratonovich, R. L. (1966): A new form of representation of stochastic integrals and equations, *SIAM J. Control*, 4:362–371.

Taylor, A. E. (1961): "Introduction to Functional Analysis," Wiley, New York.

Thomasian, A. J. (1969): "The Structure of Probability Theory with Applications," McGraw-Hill, New York.

Van Schuppen, J. H. and E. Wong (1974): Transformation of local martingales under a change of law, *Annals of Prob.*, 2:878–888.

Whitney, H. (1957): "Geometric Integration Theory", Princeton University Press, Princeton, N.J.

Whittle, P. (1963): Stochastic processes in several dimensions, *Bull. Inst. Int. Statist.*, 40:974–994.

Wiener, N. (1949): "Extrapolation, Interpolation, and Smoothing of Stationary Time Series," Wiley, New York.

Wiener, N. and P. Masani (1958): The prediction theory of multivariate stochastic processes-II, the linear predictor, *Acta Mathematica*, 99:93–137.

Wong, E. (1964): The construction of a class of stationary Markoff processes, *Proc. Symp. in Appl. Math., Am. Math. Soc.*, 16:264–276.

Wong, E. (1969): Homogeneous Gauss-Markov random fields, *Ann. Math. Stat.*, 40:1625–1634.

Wong, E. and J. B. Thomas (1961): On the multidimensional prediction and filtering problem and the factorization of spectral matrices, *J. Franklin Institute*, 272:87–99.

Wong, E. and M. Zakai (1965a): On the relationship between ordinary and stochastic differential equations, *Int. J. Engng. Sci.*, 3:213–229.

Wong, E. and M. Zakai (1965b): On the convergence of ordinary integrals to stochastic integrals, *Ann. Math. Stat.*, 36:1560–1564.

Wong, E. and M. Zakai (1966): On the relationship between ordinary and stochastic differential equations and applications to stochastic problems in control theory, *Proc. 3rd IFAC Congress*, paper 3B.

Wong, E. and M. Zakai (1969): Riemann-Stieltjes approximations of stochastic integrals, *Z. Wahrscheinlichkeitstheorie verw. Geb.*, 12:87–97.

Wong, E. and M. Zakai (1974): Martingales and stochastic integrals for processes with a multi-dimensional parameter, *Z. Wahrscheinlichkeitstheorie verw. Geb.*, 29:109–122.

Wonham, W. M. (1964): Some applications of stochastic differential equations to optimal nonlinear filtering, *Tech. Rept.* 64-3, Feb. 1964, *RIAS*, Baltimore.

Wonham, W. M. (1970): Random differential equations in control theory, in "Probabilistic Methods in Applied Math," vol. 2, Academic, New York.

Yaglom, A. M. (1961): Second-order homogeneous random fields, *Proc. 4th Berkeley Symp. Math. Stat. and Prob.*, **2**:593–620.

Yaglom, A. M. (1962): "An Introduction to the Theory of Stationary Random Functions," R. A. Silverman (trans.), Prentice-Hall, Englewood Cliffs, N.J.

Yor, M. (1976): Representation de martingale de carre integrable relative aux processus de Wiener et de Poisson a n parameters, *Z. Wahrscheinlichkeitstheorie verw. Geb.*, **35**:121–129.

Youla, D. C. (1961): On the factorization of rational matrices, *IRE Trans. Inf. Th.*, **IT-7**:172–189.

Zakai, M. (1969): On the optimal filtering of diffusion processes, *Z. Wahrscheinlichkeitstheorie verw. Geb.*, **11**:230–243.

Solutions to Exercises

CHAPTER 1

1. (a) We need only note that $[0,a) \cap [0, a+1) = [a, a+1) \notin C_1$.

(b) Let $A = \bigcup_{i=1}^{m} [a_i,b_i)$ and $B = \bigcup_{i=m+1}^{m+n} [a_i,b_i)$

Then

$$A \cup B = \bigcup_{i=1}^{m+n} [a_i,b_i) \quad \text{and} \quad A \cap B = \bigcup_{i=1}^{m} \bigcup_{j=m+1}^{m+n} [a_i,b_i) \cap [a_j,b_j)$$

But $[a_i,b_i) \cap [a_j,b_j)$ is either empty or of the form $[\min(a_i,b_j), \max(b_i,b_j))$. Hence, $A \cap B$ is again a finite union of intervals of the form $[a,b)$.

(c) Let C be any Boolean algebra containing C_1. Because $[a,b) = [0,b) \cap [0,a)$, C must contain all sets of the form $[a,b)$ and, hence, all finite unions of such sets. Hence, every Boolean algebra containing C_1 must also contain C_2. Because C_2 is a Boolean algebra, it must also be the smallest.

(d) $[a,b] = \bigcap_{n=1}^{\infty} \left[a, b + \frac{1}{n}\right)$

$(a,b) = \bigcup_{n=1}^{\infty} \left[a + \frac{1}{n}, b\right)$

$(a,b] = \bigcap_{n=1}^{\infty} \left(a, b + \frac{1}{n}\right)$

CHAPTER 1

2. (a) See Exercise 1.

(b) If $\mathcal{P}$ is σ additive, then it is also sequentially continuous so that from Solution 1.1d, we have

$$\mathcal{P}([a,b]) = \mathcal{P}\left(\lim_{n\to\infty}\left[a, b + \frac{1}{n}\right)\right)$$
$$= \lim_{n\to\infty} \mathcal{P}\left(\left[a, b + \frac{1}{n}\right)\right)$$
$$= \lim_{n\to\infty} P\left(b + \frac{1}{n}\right) - P(a)$$
$$= P(b^+) - P(a)$$
$$\mathcal{P}((a,b)) = \lim_{n\to\infty}\left\{P(b) - P\left(a + \frac{1}{n}\right)\right\}$$
$$= P(b) - P(a^+)$$
$$\mathcal{P}((a,b]) = P(b^+) - P(a^+)$$

3. (a) Set $f^{-1}(A) = \{\mathbf{x}, f(\mathbf{x}) \in A\}$. If $A \in \mathcal{R}$, then $f^{-1}(A) \in \mathcal{R}^n$, because f is a Borel function. Hence,

$$P(f^{-1}(A)) = \mathcal{P}(\{\omega: \mathbf{X}(\omega) \in f^{-1}(A)\})$$
$$= \mathcal{P}(\{\omega: f(\mathbf{X}(\omega)) \in A\})$$

(b) We can assume that f is nonnegative, because otherwise we can write $f = f^+ - f^-$ where f^+ and f^- are both nonnegative. If, in addition, f is also simple, i.e., there exist disjoint Borel sets $A_1, \ldots, A_m$ such that

$$\bigcup_{k=1}^{m} A_m = R^n$$

and

$$f(\mathbf{x}) = f_k \quad \begin{array}{l}\mathbf{x} \in A_k, \\ k = 1, \ldots, m,\end{array}$$

then, by definition, we have

$$\int_{R^n} f(\mathbf{x}) P(d\mathbf{x}) = \sum_{k=1}^{m} f_k P(A_k)$$

If we set $X^{-1}(A) = \{\omega: \mathbf{X}(\omega) \in A\}$, then we can write

$$\int_{R^n} f(\mathbf{x}) P(d\mathbf{x}) = \sum_{k=1}^{m} \int_{X^{-1}(A_k)} f(\mathbf{X}(\omega)) \mathcal{P}(d\omega)$$
$$= \int_{\Omega} f(\mathbf{X}(\omega)) \mathcal{P}(d\omega)$$

If f is not a simple function, then by definition

$$\int_{R^n} f(\mathbf{x}) P(d\mathbf{x}) = \lim_{m\to\infty} \int_{R^n} f_m(\mathbf{x}) P(d\mathbf{x})$$

when $\{f_m\}$ is a nondecreasing sequence of simple functions converging pointwise to f. We now have

$$\int_{R^n} f(\mathbf{x})P(d\mathbf{x}) = \lim_{m\to\infty} \int_{R^n} f_m(\mathbf{x})P(d\mathbf{x})$$
$$= \lim_{m\to\infty} \int_\Omega f_m(\mathbf{X}(\omega))\mathcal{P}(d\omega)$$
$$= \int_\Omega f(\mathbf{X}(\omega))\mathcal{P}(d\omega)$$

where the last equality follows from monotone convergence.

6. Let $X_1 = Y\cos\theta$, $X_2 = Y\sin\theta\cos\Phi$, and $X_3 = Y\sin\theta\sin\Phi$. If we denote the joint density function of Y, θ, and Φ by p, then

$$p(y,\theta,\varphi) = \begin{vmatrix} \cos\theta & -y\sin\theta & 0 \\ \sin\theta\cos\varphi & y\cos\theta\cos\varphi & -y\sin\theta\sin\varphi \\ \sin\theta\sin\varphi & y\cos\theta\sin\varphi & y\sin\theta\cos\varphi \end{vmatrix}$$

$$p_X(y\cos\theta, y\sin\theta\cos\varphi, y\sin\theta\sin\varphi) = \frac{y^2\sin\theta}{(2\pi)^{\frac{3}{2}}} e^{-\frac{1}{2}y^2}$$

Therefore,

$$p_Y(y) = \int_0^{2\pi} d\varphi \int_0^\pi d\theta \frac{1}{(2\pi)^{\frac{3}{2}}} y^2 \sin\theta \, e^{-\frac{1}{2}y^2}$$
$$= \sqrt{\frac{2}{\pi}} y^2 e^{-\frac{1}{2}y^2} \qquad y \geq 0$$

7. Since $Y_k = \sum_{j=1}^k X_j$, we have $X_1 = Y_1$ and

$$Y_k - Y_{k-1} = X_k \qquad k = 2, \ldots, n$$

Therefore,

$$p_Y(y_1, \ldots, y_n) = \begin{vmatrix} 1 & 0 & 0 & \cdots & 0 \\ -1 & 1 & 0 & \cdots & \cdot \\ 0 & -1 & 1 & \cdots & \cdot \\ \cdot & \cdot & \cdot & \cdot & \cdot \\ 0 & \cdots & \cdots & -1 & 1 \end{vmatrix} p_X(y_1, y_2 - y_1, \ldots, y_n - y_{n-1})$$

$$= \frac{1}{(2\pi)^{n/2}} \exp\left[-\frac{1}{2}\sum_{k=1}^n (y_k - y_{k-1})^2\right]$$

where $y_0 = 0$.

8. $E\dfrac{|X|}{1+|X|} = \int_{|X|\geq\epsilon} \dfrac{|X|}{1+|X|} d\mathcal{P} + \int_{|X|<\epsilon} \dfrac{|X|}{1+|X|} d\mathcal{P}$

This implies that

$$E\frac{|X|}{1+|X|} \geq \frac{\epsilon}{1+\epsilon}\mathcal{P}(|X|\geq\epsilon)$$

and

$$E \frac{|X|}{1+|X|} \le \mathcal{P}(|X| \ge \epsilon) + \frac{\epsilon}{1+\epsilon} \mathcal{P}(|X| < \epsilon)$$

14. (a) Because Y is independent of X_2, we have

$$E(Y|X_2) = EY = EX_1 + \alpha EX_2 = 0$$

On the other hand

$$X_1 = Y_1 - \alpha X_2$$

so that

$$E(X_1|X_2) = E(Y_1|X_2) - \alpha X_2 = -\alpha X_2$$

15. Let $X_1 = Y \cos \Phi$. The joint density function of Y and Φ is given by

$$p(y,\varphi) = \begin{vmatrix} \cos\varphi & -y\sin\varphi \\ \sin\varphi & y\cos\varphi \end{vmatrix} \frac{1}{2\pi} e^{-\frac{1}{2}(y^2\cos^2\varphi + y^2\sin^2\varphi)}$$
$$= \frac{1}{2\pi} y e^{-\frac{1}{2}y^2}$$

In other words, Y and Φ are independent, with Φ uniformly distributed on $[0,2\pi)$. Therefore,

$$E(X_1|Y) = E(Y\cos\Phi|Y) = YE\cos\Phi$$
$$= Y \frac{1}{2\pi} \int_0^{2\pi} \cos\varphi \, d\varphi = 0$$

16. By definition $\mathfrak{A}_X$ contains every set of the form $\{\omega: X_i(\omega) \in A\}$, $A \in \mathfrak{R}^1$, $i = 1, \ldots, n$. It follows that if $A_1, \ldots, A_n$ are one-dimensional Borel sets then

$$X^{-1}\left(\prod_{i=1}^n A_i\right) = \bigcap_{i=1}^n \{\omega: X_i(\omega) \in A_i\} \in \mathfrak{A}_X$$

Since $\mathfrak{R}^n$ is the smallest σ algebra containing all n products of one-dimensional Borel sets, it follows that for every $b \in \mathfrak{R}^n$,

$X^{-1}(B) \in \mathfrak{A}_X$ so that $\mathfrak{A}_X \supset \{X^{-1}(B), B \in \mathfrak{R}^n\}$

Conversely, consider the collection $\{X^{-1}(B), B \in \mathfrak{R}^n\}$. It is a σ algebra, and every X_i is clearly measurable with respect to $\{X^{-1}(B), B \in \mathfrak{R}^n\}$. Hence, $\mathfrak{A}_X \subset \{X^{-1}(B), B \in \mathfrak{R}^n\}$, and our assertion is proved.

CHAPTER 2

1. For any real number a,

$$\mathcal{P}(|X_t - X_s| > \epsilon) \ge \mathcal{P}(X_t > a + \epsilon, X_s < a - \epsilon)$$
$$= [1 - P_t(a + \epsilon)]P_s(a - \epsilon)$$
$$\xrightarrow[s \to t]{} [1 - P_t(a + \epsilon)]P_t(a - \epsilon)$$

and continuity in probability means that

$$[1 - P_t(a^+)]P_t(a) = 0$$

for all t and a. Therefore, at all continuity points P_t is either 0 or 1. It follows that P_t is a function with a single jump of size 1, say, at $f(t)$, and that for each t, $X_t(\omega) = f(t)$ with probability 1. Because the X process is continuous in probability, the function f must be continuous.

2. (a) $\mathcal{P}(\{\omega: X_t(\omega) = 0 \text{ for at least one } t \text{ in } T_n\}) = \mathcal{P}\left(\bigcup_{t \in T_n} \{\omega: X(\omega) = -t\}\right)$

$$= \sum_{t \in T_n} \mathcal{P}(X = -t) = 0$$

(b) $\mathcal{P}(X + t = 0 \text{ for at least one } t \text{ in } [0,1]) = \mathcal{P}(X \in [-1,0]) = \int_{-1}^{0} \frac{1}{\sqrt{2\pi}} e^{-x^2} dx > 0$

3. $P_t(x) = \mathcal{P}(\{\omega: X_t(\omega) < x\}) = \mathcal{P}(\{\omega: t\omega < x\})$

$= $ Lebesgue measure of $\left[0, \frac{x}{t}\right) \cap [0,1]$

$= \min\left(1, \frac{x}{t}\right)$

$P_{t,s}(x_1, x_2) = \mathcal{P}\left(\left\{\omega: \omega < \frac{x_1}{t}, \omega < \frac{x_2}{s}\right\}\right)$

$= $ Lebesgue measure of $\left[0, \frac{x_1}{t}\right) \cap \left[0, \frac{x_2}{s}\right) \cap [0,1]$

$= \min\left(1, \frac{x_1}{t}, \frac{x_2}{s}\right).$

4. (a) $\mu(t) = EX_t = \int_0^1 \omega t \, d\omega = \frac{1}{2}t$

$R(t,s) = \int_0^1 (\omega t - \frac{1}{2}t)(\omega s - \frac{1}{2}s) \, d\omega = ts \int_0^1 (\omega - \frac{1}{2})^2 \, d\omega$

$= \frac{ts}{3} \cdot \frac{1}{4} = \frac{ts}{12}$

(b) We note that
$X_t = Z \cos \theta \cos 2\pi t - Z \sin \theta \sin 2\pi t$
$= A \cos 2\pi t - B \sin 2\pi t$

where $A = Z \cos \theta$ and $B = Z \sin \theta$. If we denote the joint density function of A and B by $\tilde{p}$, then

$$\tilde{p}(z \cos \theta, z \sin \theta) \begin{vmatrix} \cos \theta & -z \sin \theta \\ \sin \theta & z \cos \theta \end{vmatrix} = \frac{1}{2\pi} z e^{-\frac{1}{2}z^2}$$

or

$$\tilde{p}(z \cos \theta, z \sin \theta) = \frac{1}{2\pi} e^{-\frac{1}{2}z^2}$$

and

$$\tilde{p}(a,b) = \frac{1}{2\pi} e^{-\frac{1}{2}(a^2+b^2)}$$

It now follows that every linear combination

$$\sum_{i=1}^{n} \alpha_i X_{t_i} = \Big(\sum_{i=1}^{n} \alpha_i \cos 2\pi t_i\Big) A - \Big(\sum_{i=1}^{n} \alpha_i \sin 2\pi t_i\Big) B$$

is a Gaussian random variable. By definition, $\{X_t, -\infty < t < \infty\}$ is a Gaussian process.

8. Let $\mathcal{B}_{rs}$ denote the smallest algebra (not σ algebra) such that for every $\tau \leq s$, all sets of the form $\{\omega: X_\tau(\omega) < a\}$ are in $\mathcal{B}_{rs}$. It is clear that $\mathcal{C}_{rs}$ is generated by $\mathcal{B}_{rs}$. Now, every set A in $\mathcal{B}_{rs}$ depends on only a finite collection $X_{t_1}, X_{t_2}, \ldots, X_{t_n}, t_i \leq s$. Therefore, for every $A \in \mathcal{B}_{rs}$

$$EI_A X_t = E\{E[I_A X_t | X_{t_1}, X_{t_2}, \ldots, X_{t_n}, X_s]\}$$

Writing $X_t = X_t^+ - X_t^-$, where both X_t^+ and X_t^- are nonnegative, we have

$$EI_A X_t^+ = EI_A X_s^+ \qquad A \in \mathcal{B}_{rs}$$
$$EI_A X_t^- = EI_A X_s^- \qquad A \in \mathcal{B}_{rs}$$

Each of these four terms defines a finite measure on $\mathcal{B}_{rs}$ which has a unique extension to $\mathcal{C}_{rs}$. It follows that

$$EI_A X_t^+ = EI_A X_s^+ \qquad A \in \mathcal{C}_{rs}$$

for otherwise, we would have two different extensions of the same measure. Similarly,

$$EI_A X_t^- = EI_A X_s^- \qquad A \in \mathcal{C}_{rs}$$

and

$$EI_A X_t = EI_A X_s \qquad A \in \mathcal{C}_{rs}$$

Since X_s is $\mathcal{C}_{rs}$ measurable, we have

$$E^{\mathcal{C}_{rs}} X_t = X_s$$

with probability 1.

11. Suppose that $X_t = f(t) W_{g(t)/f(t)}$, then

$$E X_t X_s = f(t) f(s) \min\left(\frac{g(t)}{f(t)}, \frac{g(s)}{f(s)}\right)$$

Because $g(t)/f(t)$ is nondecreasing, we have

$$\begin{aligned} E X_t X_s &= f(t) f(s) \frac{g(\min(t,s))}{f(\min(t,s))} \\ &= f(\max(t,s)) g(\min(t,s)) \\ &= e^{-|t-s|} = e^{-[\max(t,s) - \min(t,s)]} \end{aligned}$$

It follows that $f(t)g(t) = 1$ so that $g(t) = 1/f(t)$ and

$$\begin{aligned} f(\max(t,s))g(\min(t,s)) &= \frac{f(\max(t,s))}{f(\min(t,s))} \\ &= e^{-\max(t,s)} e^{-\min(t,s)} \end{aligned}$$

Hence, we can take $f(t) = ke^{-t}$ where k is any nonzero constant. Thus,
$$X_t = ke^{-t}W_{[(1/k)e^t]^2}$$

12. (a) Since $\{X_t, -\infty < t < \infty\}$ is Markov, the Chapman-Kolmogorov equation yields
$$\mathcal{P}(X_{t+s} = x_i | X_0 = x_j) = \sum_{k=1}^{n} \mathcal{P}(X_{t+s} = x_i | X_s = x_k)\mathcal{P}(X_s = x_k | X_0 = x_j)$$
or, equivalently,
$$p_{ij}(t+s) = \sum_{k=1}^{n} p_{ik}(t)p_{kj}(s) \qquad t,s > 0$$
In matrix form, this can be rewritten as
$$\mathbf{p}(t+s) = \mathbf{p}(t)\mathbf{p}(s)$$
so that
$$\lim_{t \downarrow 0} \frac{1}{t}[\mathbf{p}(t+s) - \mathbf{p}(s)] = \lim_{t \downarrow 0} \frac{1}{t}[\mathbf{p}(t) - \mathbf{I}]\mathbf{p}(s) = \mathbf{A}\mathbf{p}(s)$$
Hence,
$$\dot{\mathbf{p}}(s) = \mathbf{A}\mathbf{p}(s) \qquad s > 0$$
the unique solution of which corresponding to $\mathbf{p}(0) = \mathbf{I}$ is $\mathbf{p}(s) = e^{s\mathbf{A}}$.

(b) $\sum_{j=1}^{n} \mathcal{P}(X_{t+\tau} = x_i | X_t = x_j)\mathcal{P}(X_t = x_j) = \mathcal{P}(X_{t+\tau} = x_i)$

$\sum_{i=1}^{n} \mathcal{P}(X_{t+\tau} = x_i | X_t = x_j) = 1$

Hence, $\mathbf{p}(\tau)\mathbf{q} = \mathbf{q}$ and $\mathbf{p}^T(\tau)\mathbf{1} = 1$. $\left(\mathbf{1} = \begin{bmatrix} 1 \\ 1 \end{bmatrix}\right)$

If $\mathbf{q} = \begin{bmatrix} \frac{1}{2} \\ \frac{1}{2} \end{bmatrix}$, then we have $\mathbf{p}(\tau)\mathbf{1} = 1$ and $\mathbf{p}^T(\tau)\mathbf{1} = 1$, and $\mathbf{p}(\tau)$ must have the form
$$\mathbf{p}(\tau) = \begin{bmatrix} f(\tau) & 1 - f(\tau) \\ 1 - f(\tau) & f(\tau) \end{bmatrix}$$
and $\dot{\mathbf{p}}(0) = \dot{f}(0)\begin{bmatrix} 1 & -1 \\ -1 & 1 \end{bmatrix} = \mathbf{A}$. Because every entry in $p(\tau)$ is nonnegative, $f(0)$ must be less than or equal to zero. Setting $\dot{f}(0) = -\lambda$, we have from part (a)
$$\mathbf{p}(\tau) = e^{\tau \mathbf{A}} = \exp\left(-\lambda\tau \begin{bmatrix} 1 & -1 \\ -1 & 1 \end{bmatrix}\right) = \begin{bmatrix} \dfrac{1 + e^{-\lambda\tau}}{2} & \dfrac{1 - e^{-\lambda\tau}}{2} \\ \dfrac{1 - e^{-\lambda\tau}}{2} & \dfrac{1 + e^{-\lambda\tau}}{2} \end{bmatrix}$$

CHAPTER 2

14. Because $\{X_t, -\infty < t < \infty\}$ is stationary, its covariance function depends only on the time difference. Set

$$\rho(t-s) = \frac{E[(X_t - EX_t)(X_s - EX_s)]}{E(X_t - EX_t)^2}$$

then from Solution 2.13, ρ must satisfy

$$\rho(t+s) = \rho(t)\rho(s) \qquad t, s \geq 0$$

It follows that for any positive integers m, n, we have

$$\rho\left(\frac{m}{n}\right) = \rho^m\left(\frac{1}{n}\right)$$

$$\rho(1) = \rho^n\left(\frac{1}{n}\right)$$

so that $\rho\left(\dfrac{m}{n}\right) = \rho^{m/n}(1) = e^{(m/n)\ln \rho(1)}$. By continuity we must have

$$\rho(t) = e^{t \ln \rho(1)} \qquad t \geq 0$$

and by symmetry

$$\rho(t) = e^{|t| \ln \rho(1)}$$
$$= e^{-\lambda |t|}$$

where we have set $-\lambda = \ln \rho(1)$.

15. (a) Consider the characteristic function

$$E \exp\left(i \sum_{k=1}^n u_k X_{t_k+\tau}\right) = E\left\{\frac{1}{2\pi}\int_0^{2\pi} \exp\left[iA \sum_{k=1}^n u_k \cos(2\pi t_k + 2\pi\tau + \theta)\right] d\theta\right\}$$

$$= E\left\{\frac{1}{2\pi}\int_{2\pi\tau}^{2\pi(\tau+1)} \exp\left[iA \sum_{k=1}^n u_k \cos(2\pi t_k + \psi)\right] d\psi\right\}$$

$$= E\left\{\frac{1}{2\pi}\int_0^{2\pi} \exp\left[iA \sum_{k=1}^n u_k \cos(2\pi t_k + \psi)\right] d\psi\right\}$$

$$= E \exp\left(i \sum_{k=1}^n u_k X_{t_k}\right)$$

(b) $EX_t = EA\left(\dfrac{1}{2\pi}\displaystyle\int_0^{2\pi} \cos(2\pi t + \theta)\, d\theta\right) = 0$

$$M_T(\omega) = \frac{1}{2T}\int_{-T}^T X_t(\omega)\, dt = A(\omega)\frac{1}{2T}\int_{-T}^T \cos[2\pi t + \theta(\omega)]\, dt \xrightarrow[T\to\infty]{} 0$$

(c) Let $Y = A \cos \theta$ and $Z = A \sin \theta$, then

$$X_t = Y \cos 2\pi t - Z \sin 2\pi t$$

and $Y = X_0$, $Z = -X_{\frac{1}{4}}$, so that $\{X_t, t \in (-\infty, \infty)\}$ is a Gaussian process if and only if Y and Z are jointly Gaussian. Since

$$EYZ = EA^2 \frac{1}{2\pi} \int_0^{2\pi} \sin \theta \cos \theta \, d\theta = 0$$

$$EY^2 = EZ^2 = \tfrac{1}{2} EA^2 = \sigma^2$$

Y and Z are jointly Gaussian if and only if

$$p_{YZ}(y,z) = \frac{1}{2\pi\sigma^2} \exp\left[-\frac{1}{2\sigma^2}(y^2 + z^2)\right]$$

Hence, by the transformation rule for random variables (see, e.g., Exercise 1.5), we have

$$\frac{1}{2\pi} p_A(r) = p_{YZ}(r \cos \theta, r \sin \theta) \begin{vmatrix} \cos \theta & -r \sin \theta \\ \sin \theta & r \cos \theta \end{vmatrix}$$

$$= \frac{r}{2\pi\sigma^2} \exp\left(-\frac{r^2}{2\sigma^2}\right)$$

and

$$p_A(r) = \frac{r}{\sigma^2} \exp\left(-\frac{1}{2}\frac{r^2}{\sigma^2}\right) \qquad r \geq 0$$

$\{X_t - \infty < t < \infty\}$ is not Markov because for $t \geq \frac{1}{4}$

$$E(X_t | X_0 = y, X_{\frac{1}{4}} = z) = y \cos 2\pi t - z \sin 2\pi t$$

which depends on y, contrary to the Markov property.

16. (a) Since X_t, Y_t are independent and Markov

$$\begin{aligned}
E(Z_4^2 | X_1, X_2, X_3, Y_1, Y_2, Y_3) &= E(X_4^2 | X_3) + E(Y_4^2 | Y_3) \\
&= E[(X_4 - X_3)^2 + 2X_3(X_4 - X_3) + X_3^2 | X_3] \\
&\quad + E[(Y_4 - Y_3)^2 + 2Y_3(Y_4 - Y_3) + Y_3^2 | Y_3] \\
&= 1 + X_3^2 + 1 + Y_3^2
\end{aligned}$$

Therefore, $E(Z_4^2 | \text{observed data}) = 7$.

Now, by the Schwarz inequality

$$E(Z_4 | \text{data}) \leq \sqrt{E(Z_4^2 | \text{data})} = \sqrt{7}$$

On the other hand, the Schwarz inequality applied to summations yields

$$(xx_0 + yy_0)^2 \leq (x^2 + y^2)(x_0^2 + y_0^2)$$

so that

$$\sqrt{x^2 + y^2} \geq \frac{|xx_0 + yy_0|}{\sqrt{x_0^2 + y_0^2}} \geq \frac{(xx_0 + yy_0)}{\sqrt{x_0^2 + y_0^2}}$$

$$= \sqrt{x_0^2 + y_0^2} + \frac{x_0}{\sqrt{x_0^2 + y_0^2}} (x - x_0) + \frac{y_0}{\sqrt{x_0^2 + y_0^2}} (y - y_0)$$

It follows that

$$E(\sqrt{X_4^2 + Y_4^2} | X_3, Y_3) \geq \sqrt{X_3^2 + Y_3^2} + \frac{X_3}{\sqrt{X_3^2 + Y_3^2}} E[(X_4 - X_3)|X_3]$$

$$+ \frac{Y_3}{\sqrt{X_3^2 + Y_3^2}} E[(Y_4 - Y_3)|Y_3]$$

$$= \sqrt{X_3^2 + Y_3^2}$$

Hence, $E(Z_4|\text{data}) \geq \sqrt{1 + (2)^2} = \sqrt{5}$.

(b) Introduce $\{\Theta_t, t \geq 0\}$ so that

$$X_t = Z_t \cos \Theta_t \qquad Y_t = Z_t \sin \Theta_t$$

For $t_1 < t_2 < \cdots < t_n$

$$E[f(Z_{t_n})|Z_{t_1}, Z_{t_2}, \ldots, Z_{t_{n-1}}] = E\{E[f(Z_{t_n})|Z_{t_j}, \Theta_{t_j}, j = 1, \ldots, n-1]\}$$
$$= E\{E[f(Z_{t_n})|Z_{t_{n-1}}, \Theta_{t_{n-1}}]|Z_{t_1}, \ldots, Z_{t_{n-1}}\}$$

Now,

$$E[f(Z_{t_n})|Z_{t_{n-1}}, \Theta_{t_{n-1}}] = E[f(\sqrt{X_{t_n}^2 + Y_{t_n}^2})|X_{t_{n-1}}, Y_{t_{n-1}}]$$

and

$$E[f(Z_{t_n})|Z_{t_{n-1}} = r_0, \Theta_{t_{n-1}} = \theta_0] = E[f(\sqrt{X_{t_n}^2 + Y_{t_n}^2})|$$
$$X_{t_{n-1}} = r_0 \cos \theta_0, Y_{t_{n-1}} = r_0 \sin \theta_0]$$
$$= \int_0^\infty dr \int_0^{2\pi} d\theta \frac{rf(r)}{2\pi(t_n - t_{n-1})}$$
$$\exp\left\{-\frac{1}{2(t_n - t_{n-1})}[r^2 + r_0^2 - 2rr_0 \cos(\theta - \theta_0)]\right\}$$

Because $\cos \theta$ is periodic with period 2π, a change in variable yields

$$E[f(Z_{t_n})|Z_{t_{n-1}} = r_0, \Theta_{t_{n-1}} = \theta_0] = \int_0^\infty dr \int_0^{2\pi} d\theta' \frac{rf(r)}{2\pi(t_n - t_{n-1})}$$
$$\exp\left[-\frac{1}{2(t_n - t_{n-1})}(r^2 + r_0^2 - 2rr_0 \cos \theta')\right]$$

which is independent of θ_0. It follows that

$$E[f(Z_{t_n})|Z_{t_{n-1}}, \Theta_{t_{n-1}}] = E[f(Z_{t_n})|Z_{t_{n-1}}]$$

and

$$E[f(Z_{t_n})|Z_{t_1}, \ldots, Z_{t_{n-1}}] = E[f(Z_{t_n})|Z_{t_{n-1}}]$$

so that $\{Z_t, t \geq 0\}$ is Markov (see Proposition 4.6.4 for an easy means of proving this fact).

CHAPTER 3

1. (a) Compute

$$S(\nu) = \int_{-\infty}^{\infty} R(\tau)\, d\tau = \int_{-1}^{1} (1 - |\tau|)e^{-i2\pi\nu\tau}\, d\tau$$
$$= 2\int_{0}^{1} (1 - \tau)\cos 2\pi\nu\tau\, d\tau$$
$$= \frac{2}{2\pi\nu}\int_{0}^{1} \sin 2\pi\nu\tau\, d\tau = \frac{2(1 - \cos 2\pi\nu)}{(2\pi\nu)^2}$$
$$= \left(\frac{\sin \pi\nu}{\pi\nu}\right)^2 \geq 0$$

It follows from one-half of Bochner's theorem that R is nonnegative definite.

(b) Repeat the same procedure as in (a).

(c) If $R(t,s) = e^{|t-s|}$ then $R(t,t) = 1$, and it violates (1.8), viz.

$$|R(t,s)| \leq \sqrt{R(t,t)R(s,s)}$$

(d) R is continuous at all diagonal points (t,t), but not continuous everywhere, so it cannot be nonnegative definite.

3. (a) Since R is continuous and period c, we can write it in a Fourier series as

$$R(\tau) = \sum_{n=-\infty}^{\infty} R_n e^{in(2\pi/T)t}$$

where $R_n \geq 0$. It follows that

$$E Z_m \bar{Z}_n = \iint_{0}^{T} R(t - s)e^{-i(2\pi/T)(mt-ns)}\, dt\, ds$$
$$= \sum_{k=-\infty}^{\infty} R_n \delta_{mk}\delta_{nk} = R_n \delta_{mn}$$

(b) $E\left| X_t - \sum_{n=-N}^{N} Z_n e^{in(2\pi/T)t} \right|^2 = R(0) - \sum_{n=-N}^{N} E|Z_n|^2$
$$= R(0) - \sum_{n=-N}^{N} R_n \xrightarrow[N\to\infty]{} 0$$

(c) We can write

$$R(\tau) = \sum_{n=-\infty}^{\infty} R_n e^{in(2\pi/T)\tau}$$

where $R_n = 1/2(1 + n^2)$, $n \neq 0$ and $R_0 = 1$. Since the family $\{e^{in(2\pi/T)t}, 0 \leq t \leq T, n = 0, \pm 1, \ldots\}$ is orthogonal, we can clearly take $\varphi_n(t) = (1/\sqrt{T})e^{in(2\pi/T)t}$ to be the orthonormal eigenfunction. The eigenvalues are $\lambda_n = R_n T$.

CHAPTER 3								327

4. First, we write

$$\lambda\varphi(t) = \int_0^t (1 - t + s)\varphi(s)\, ds + \int_t^{\frac{1}{2}} (1 - s + t)\varphi(s)\, ds$$

Differentiating it once, we get

$$\lambda\varphi'(t) = -\int_0^t \varphi(s)\, ds + \int_t^{\frac{1}{2}} \varphi(s)\, ds$$

Differentiating once more, we find

$$\lambda\varphi''(t) = -2\varphi(t)$$

as was suggested by the hint. The second of the above equations yields the boundary condition $-\varphi'(0) = \varphi'(\frac{1}{2})$. The first and second equations yield $\varphi(0) + \varphi(\frac{1}{2}) = \frac{3}{2}\varphi'(0)$.

From the equation $\lambda\varphi''(t) = -2\varphi(t)$, we get

$$\varphi(t) = A \cos\sqrt{\frac{2}{\lambda}}\, t + B \sin\sqrt{\frac{2}{\lambda}}\, t$$

Applying the condition $-\varphi'(0) = \varphi'(\frac{1}{2})$ yields

$$\varphi(t) = C \cos\sqrt{\frac{2}{\lambda}}\, (t - \tfrac{1}{4})$$

Applying the second boundary condition yields the transcendental equation

$$\tfrac{3}{4}\sqrt{2/\lambda} = \frac{\cos\tfrac{1}{4}\sqrt{2/\lambda}}{\sin\tfrac{1}{4}\sqrt{2/\lambda}} = \cot\tfrac{1}{4}\sqrt{2/\lambda}$$

which is to be solved for the eigenvalues λ. Finally, for normalization we choose $C = 2$ so that

$$\varphi(t) = 2 \cos\sqrt{\frac{2}{\lambda}}\, (t - \tfrac{1}{4})$$

6. Since the W process is a Brownian motion, we can write

$$W_{\tau(t)} = \sum_{n=0}^{\infty} \sqrt{\lambda_n}\, \varphi_n(\tau(t))Z_n \qquad 0 \leq t \leq T$$

where λ_n and φ_n are given (4.32) and (4.33), respectively, and $\{Z_n\}$ are Gaussian and orthonormal. Hence, the desired expansion is obtained by setting

$$\alpha_n(t) = f(t)\sqrt{\lambda_n}\, \varphi_n(t)$$

Suppose that we define $\tau^{-1}(t) = \min\{s\colon \tau(s) = t\}$. Then,

$$X_{\tau^{-1}(t)} = f(\tau^{-1}(t))W_t \qquad 0 \leq t \leq \tau(T)$$

Since a Brownian motion is q.m. continuous, the Hilbert space $\mathcal{H}_W$ generated by $\{W_t, 0 \leq t \leq \tau(T)\}$ is spanned by $\{W_t, t \in S\}$ where S is any dense subset of $[0,\tau(T)]$. It follows that every $Z_n \in \mathcal{H}_X$ if and only if there exists a dense subset S of $[0,\tau(T)]$ such that

$$f(\tau^{-1}(t)) \neq 0 \qquad \text{for every } t \in S$$

7. $R(\tau) = \frac{1}{8}e^{-|\tau|}(3\cos\tau + \sin|\tau|)$

$$S(\nu) = \int_{-\infty}^{\infty} e^{-i2\pi\nu\tau} R(\tau)\, d\tau = 2\int_{0}^{\infty} \cos 2\pi\nu\tau R(\tau)\, d\tau$$

$$= \frac{1}{4}\int_{0}^{\infty} e^{-\tau}\cos 2\pi\nu\tau(3\cos\tau + \sin\tau)\, d\tau$$

$$= \frac{1}{8}\int_{0}^{\infty} e^{-\tau}[3\cos(1+2\pi\nu)\tau + 3\cos(1-2\pi\nu\tau) + \sin(1+2\pi\nu)\tau$$
$$+ \sin(1-2\pi\nu)\tau]\, d\tau$$

$$= \frac{1}{8}\left[\frac{3}{1+(1+2\pi\nu)^2} + \frac{3}{1+(1-2\pi\nu)^2} + \frac{1+2\pi\nu}{1+(1+2\pi\nu)^2} + \frac{1-2\pi\nu}{1+(1-2\pi\nu)^2}\right]$$

$$= \frac{1}{8}\frac{16+4(2\pi\nu)^2}{4+(2\pi\nu)^4} = \frac{1}{2}\frac{4+(2\pi\nu)^2}{4+(2\pi\nu)^4}$$

8. Let $\rho(t-s) = EY_t\bar{X}_s$. Then if $Y_t = \int_{-\infty}^{\infty} e^{i2\pi\nu t}H(\nu)\, d\hat{X}_\nu$

$$\rho(t-s) = \int_{-\infty}^{\infty} e^{i2\pi\nu(t-s)}H(\nu)S(\nu)\, d\nu$$

Therefore,

$$H(\nu) = \frac{1}{S(\nu)}\int_{-\infty}^{\infty} e^{-i2\pi\nu\tau}\rho(\tau)\, d\tau$$

$$= \frac{1}{S(\nu)}\frac{1}{4+(2\pi\nu)^4}$$

$$= \frac{2}{4+(2\pi\nu)^2}$$

The covariance function of the Y process is given by

$$R_y(\tau) = \int_{-\infty}^{\infty} e^{i2\pi\nu\tau}|H(\nu)|^2\, d\nu$$

$$= \int_{-\infty}^{\infty} \frac{4}{[4+(2\pi\nu)^2]^2} e^{i2\pi\nu\tau}\, d\nu$$

$$= \frac{1}{8}e^{-2|\tau|}(1+2|\tau|)$$

$\cdot\ E(e^{i2\pi Wt}X_t)(\overline{e^{i2\pi Ws}X_s}) = e^{i2\pi W(t-s)}EX_t\bar{X}_s$
$$= e^{i2\pi W(t-s)}R_x(t-s)$$

$\cos 2\pi WtX_t$ is not wide-sense stationary.

2. $EY_t\bar{Y}_s = \int_{-\infty}^{\infty} |\eta(\nu)|^2 e^{i\varphi(\nu)(t-s)}\, dF(\nu)$

1. (a) More generally, $Y_t = \int_{-\infty}^{\infty} H(\nu)e^{i2\pi\nu t}\, d\hat{X}_\nu$ is real valued i

$H(\nu) = \bar{H}(-\nu)$

To prove this, first assume $H \in L^2(-\infty, \infty)$, then H is the fourier transform of a

CHAPTER 3

real-valued function h and

$$Y_t = \int_{-\infty}^{\infty} h(t-s) X_s \, ds$$

If $H \notin L^2(-\infty, \infty)$ we truncate $H(\nu)$ to $\nu = \pm n$. It follows then that Y_t is the q.m. limit of a sequence of real-valued processes.

(b) Since $EX_t X_s = e^{-|t-s|}$,

$$E|d\hat{X}_\nu|^2 = \frac{2}{1 + (2\pi\nu)^2} \, d\nu$$

It follows that

$$E\tilde{X}_t \tilde{X}_s = E\tilde{X}_t \bar{\tilde{X}}_s = \int_{-\infty}^{\infty} |-i \operatorname{sgn} \nu|^2 e^{i 2\pi\nu(t-s)} \frac{2}{1+(2\pi\nu)^2} \, d\nu$$
$$= e^{-|t-s|}$$

$$E\tilde{X}_t X_s = E\tilde{X}_t \bar{X}_s = \int_{-\infty}^{\infty} (-i \operatorname{sgn} \nu) e^{i 2\pi\nu(t-s)} \frac{2}{1+(2\pi\nu)^2} \, d\nu$$
$$= 4 \int_0^{\infty} \frac{\sin 2\pi\nu(t-s)}{1+(2\pi\nu)^2} \, d\nu$$

(c) $\displaystyle Z_t = \int_{-\infty}^{\infty} (\cos 2\pi\nu_0 t - i \operatorname{sgn} \nu \sin 2\pi\nu_0 t) e^{i 2\pi\nu t} \, d\hat{X}_\nu$
$$= \int_{-\infty}^{\infty} e^{-i \operatorname{sgn} \nu (2\pi_0) t + i 2\pi\nu t} \, d\hat{X}_\nu$$

14. (a) Check the three cases $\min(t,s) > 0$, $\max(t,s) < 0$, and $\max(t,s) > 0 > \min(t,s)$.

(b) $E\zeta_{nt}\bar{\zeta}_{ns} = n^2(Z_{t+1/n} - Z_t)\overline{(Z_{s+1/n} - Z_s)}$

$$= \frac{n^2}{2}\left(\left|t+\frac{1}{n}\right| + \left|s+\frac{1}{n}\right| - |t-s| + |t| + |s| - |t-s| - |t| \right.$$
$$\left. - \left|s+\frac{1}{n}\right| + \left|t-s-\frac{1}{n}\right| - \left|t+\frac{1}{n}\right| - |s| + \left|t+\frac{1}{n}-s\right| \right)$$

$$= \frac{n^2}{2}\left(\left|t-s-\frac{1}{n}\right| + \left|t-s+\frac{1}{n}\right| - 2|t-s| \right)$$

$$= \begin{cases} n(1 - n|t-s|) & 0 \le |t-s| \le \frac{1}{n} \\ 0 & \text{elsewhere} \end{cases}$$

$$S_n(\nu) = \int_{-1/n}^{1/n} n(1 - n\tau) e^{-i 2\pi\nu\tau} \, d\tau = 2n \int_0^{1/n} (1 - n\tau) \cos 2\pi\nu\tau \, d\tau$$
$$= \frac{2n^2}{2\pi\nu} \int_0^{1/n} \sin 2\pi\nu\tau \, d\tau = \frac{2n^2}{(2\pi\nu)^2}\left(1 - \cos\frac{2\pi\nu}{n}\right)$$
$$= \frac{4n^2}{(2\pi\nu)^2} \sin^2 \frac{\pi\nu}{n} = \left(\frac{\sin \pi\nu/n}{\pi\nu/n}\right)^2$$

(c) $\displaystyle E\left[\int_{-\infty}^{\infty} \zeta_{nt} f(t) \, dt\right]\left[\overline{\int_{-\infty}^{\infty} \zeta_{nt} g(t) \, dt}\right] = \iint_{-\infty}^{\infty} f(t)\bar{g}(s) R_\zeta^{(n)}(t-s) \, ds$

$$= \int_{-\infty}^{\infty} \hat{f}(\nu)\bar{\hat{g}}(\nu)\left(\frac{\sin \pi\nu/n}{\pi\nu/n}\right)^2 d\nu \xrightarrow[n \to \infty]{\text{(dominated convergence)}} \int_{-\infty}^{\infty} \hat{f}(\nu)\bar{\hat{g}}(\nu) \, d\nu$$
$$= \int_{-\infty}^{\infty} f(t)\bar{g}(t) \, dt$$

15. Set $Z_{nt} = \int_0^t \zeta_{ns}\, ds$. Then, we have

$$Z_{nt} = n \int_0^t (Z_{s+1/n} - Z_s)\, ds$$
$$= n \left(\int_{1/n}^{t+1/n} Z_s\, ds - \int_0^t Z_s\, ds \right)$$
$$= n \left(\int_t^{t+1/n} Z_s\, ds - \int_0^{1/n} Z_s\, ds \right)$$

Therefore,

$$E|Z_t - Z_{nt}|^2 = n^2 E \left| \int_t^{t+1/n} (Z_t - Z_s)\, ds + \int_0^{1/n} Z_s\, ds \right|$$
$$\leq 2n^2 \left[E \left| \int_t^{t+1/n} (Z_t - Z_s)\, ds \right|^2 + E \left| \int_0^{1/n} Z_s\, ds \right|^2 \right]$$
$$\leq 2n^2 \left[\frac{1}{n} \int_t^{t+1/n} (s-t)\, ds + \frac{1}{n} \int_0^{1/n} s\, ds \right]$$
$$= \frac{2}{n} \xrightarrow[n \to \infty]{} 0 \quad \text{uniformly in } t$$

It follows that

$$\int_a^b \dot{f}(t) Z_{nt}\, dt + \int_a^b (t)\zeta_{nt}\, dt = f(b) Z_{nb} - (a) Z_{na}$$

We have already shown that $Z_{nt} \xrightarrow[n \to \infty]{\text{q.m.}} Z_t$ uniformly in t, and by definition,

$$\int_a^b f(t)\, dZ_t = \lim_{n \to \infty} \text{in q.m.} \int_a^b f(t) \zeta_{nt}\, dt$$

Hence, the desired result follows.

16. Starting from $Y_t = Y_a - \int_a^t Y_s\, ds + Z_t - Z_a$, we set $U_t = \int_a^t Y_s\, ds$ and write

$$\dot{U}_t + U_t = Y_a + Z_t - Z_a$$

Hence, we have

$$U_t = e^{-t} \left[\int_a^t Y_a e^s\, ds + \int_a^t (Z_s - Z_a) e^s\, ds \right]$$

and

$$Y_t = \dot{U}_t = -U_t + Y_a + Z_t - Z_a$$

If we replace $\int_a^t Z_s e^s\, ds$ by $-\int_0^t e^s\, dZ_s + Z_t e^t - Z_a e^a$, we get

$$Y_t = Y_a e^{-(t-a)} + \int_a^t e^{-(t-s)}\, dZ_s$$

which was to be shown.

17. (a) Define $X(I_{cd}) = X_d - X_c$ as required. If $f = \sum_{i=1}^n \alpha_i f_i$ and each f_i is an indicator function of a half-open interval, call f a step function and set $X(f) = \sum_{i=1}^n \alpha_i X(f_i)$.

CHAPTER 3

If f and g are step functions, then

$$EX(f)\bar{X}(g) = \int_a^b f(t)\bar{g}(t)\,dt + \iint_a^b \rho(t,s)f(t)\bar{g}(s)\,ds$$

It follows that

$$\int_a^b |f(t)|^2\,dt \le E|X(f)|^2 \le K\int_a^b |f(t)|^2\,dt$$

By taking q.m. limits, $X(f)$ can be defined for any f which is the L^2 limit of a sequence of step functions. But the set of step functions is dense in $L^2(a,b)$, hence our definition is complete.

(b) $E\bar{X}(f)X(g) = \int_a^b f(t)\bar{g}(t)\,dt + \iint_a^b \rho(t,s)f(t)\bar{g}(s)\,dt\,ds$

(c) If Y is a finite sum $Y = \sum_{i=0}^n \alpha_i X_{t_i}$, then $Y = \int_a^b \eta(t)\,dX_t + kX_a$

where η is a step function. Every $Y \in \mathcal{K}_X$ is the q.m. limit of a sequence of finite sums. Since $X(\eta_n)$ converges in q.m. if and only if η_n is a L^2-convergent sequence, the result is proved.

18. We use the condition

$$E(A - \tilde{A}_t)X_s = 0 \qquad 0 \le s \le t$$

and find

$$s = EX_s \int_0^t h(t,\tau)\,dX_\tau = s\int_0^t h(t,\tau)\,d\tau + \int_0^s h(t,\tau)\,d\tau \qquad 0 \le s \le t$$

Hence, $\int_0^t h(t,\tau)\,d\tau = t/(1+t)$ and

$$\int_0^s h(t,\tau)\,d\tau = \frac{s}{1+t}$$

$$h(t,s) = \frac{1}{1+t} \qquad 0 \le s \le t$$

Therefore, $\tilde{A}_t = [1/(1+t)]X_t$.

19. (a) We use the condition

$$E(Y - \tilde{Y}_t)\bar{X}_s = 0 \qquad s \le t$$

and find

$$\int_0^t h(t,\tau)\frac{\partial}{\partial \tau}(EX_\tau \bar{X}_s)\,d\tau + EX_0\bar{X}_s = \cos 2\pi W s \qquad s \le t$$

Since $EX_t \bar{X}_s = \cos 2\pi Wt \cos 2\pi Ws + e^{-\nu_0|t-s|}$, we have

$$\cos 2\pi Ws \left[2\pi W \int_0^t h(t,\tau) \sin 2\pi W\tau \, d\tau \right]$$
$$= e^{-\nu_0 s} - \nu_0 \int_0^t h(t,\tau) \, \text{sgn} \, (\tau - s) e^{-\nu_0|\tau-s|} \, d\tau \qquad s \le t \quad (*)$$

Differentiating both sides twice with respect to s yields

$$-(2\pi W)^2 \cos 2\pi Ws \left[2\pi W \int_0^t h(t,\tau) \sin 2\pi W\tau \, d\tau \right]$$
$$= \nu_0^2 e^{-\nu_0 s} - \nu_0 \int_0^t h(t,\tau) \, \text{sgn} \, (\tau - s) e^{-\nu_0|\tau-s|} \, d\tau + 2\nu_0 \frac{\partial h(t,s)}{\partial s}$$

or

$$2\nu_0 \frac{\partial h(t,s)}{\partial s} + \left\{ [\nu_0^2 + (2\pi W)^2](2\pi W) \int_0^t h(t,\tau) \sin 2\pi W\tau \, d\tau \right\} \cos 2\pi Ws = 0$$

That means that $h(t,s)$ must have the form

$$h(t,s) = g(t) + f(t) \sin 2\pi Ws$$

and

$$f(t) = -\frac{1}{2\nu_0} [\nu_0^2 + (2\pi W)^2] \int_0^t h(t,\tau) \sin 2\pi W\tau \, d\tau$$

which yields

$$f(t) \left\{ 1 + \frac{1}{2\nu_0} [\nu_0^2 + (2\pi W)^2] \right\} \int_0^t \sin^2 2\pi W\tau \, d\tau$$
$$+ \left\{ \frac{1}{2\nu_0} [\nu_0^2 + (2\pi W)^2] \int_0^t \sin 2\pi W\tau \, d\tau \right\} g(t) = 0 \quad (**)$$

From $(*)$ we find (upon setting $s = 0$)

$$2\pi W \int_0^t h(t,\tau) \sin 2\pi W\tau \, d\tau = 1 - \nu_0 \int_0^t h(t,\tau) e^{-\nu_0 \tau} \, d\tau$$

which yields

$$f(t) \left[\nu_0 \int_0^t e^{-\nu_0 \tau} \sin 2\pi W\tau \, d\tau - \frac{4\pi \nu_0 W}{\nu_0^2 + (2\pi W)^2} \right] + \left(\nu_0 \int_0^t e^{-\nu_0 \tau} \, d\tau \right) g(t) = 1 \quad (***)$$

Equations $(**)$ and $(***)$ suffice to determine $f(t)$ and $g(t)$, which in turn determine $h(t,s)$ completely.

(b) We begin by noting that $\int_{-\infty}^{\infty} e^{-\nu_0|\tau|} e^{i 2\pi \nu \tau} \, d\tau = 2\nu_0/(|2\pi\nu|^2 + \nu_0^2)$ so that $\{N_t, -\infty < t < \infty\}$ can be viewed as a white noise filtered by transfer function $\sqrt{2\nu_0}/(\nu_0 + i2\pi\nu)$ or

$$\dot{N}_t + \nu_0 N_t = \sqrt{2\nu_0} \, \zeta_t$$

where ζ_t is a standard white-noise process. It follows that

$$\dot{X}_t = (-2\pi W \sin 2\pi Wt) Y + \dot{N}_t$$
$$= (-2\pi W \sin 2\pi Wt) Y - \nu_0 (X_\cdot - \cos 2\pi WtY) + \sqrt{2\nu_0} \, \zeta_t$$

CHAPTER 3

Let $Z_t = e^{\nu_0 t} X_t$. Then

$$\dot{Z}_t = Y e^{\nu_0 t}(\nu_0 \cos 2\pi W t - 2\pi W \sin 2\pi W t) + \sqrt{2\nu_0}\, e^{\nu_0 t} \zeta_t$$
$$\dot{Y} = 0$$

which is in the standard form (9.32) and (9.34), with Z_t replacing X_t. Hence, we have $A(t) = 0 = F(t)$, $B(t) = \sqrt{2\nu_0}\, e^{\nu_0 t}$ and

$$H(t) = e^{\nu_0 t}(\nu_0 \cos 2\pi W t - 2\pi W \sin 2\pi W t)$$

The estimator $\tilde{Y}_t$ must satisfy

$$\dot{\tilde{Y}}_t = k(t)[\dot{Z}_t - H(t)\tilde{Y}_t]$$

Equations (9.39) and (9.40) now become

$$\dot{\Sigma}(t) = +B^2(t)k^2(t) - 2H(t)\Sigma(t)k(t)$$
$$K(t)B^2(t) = \Sigma(t)H(t)$$

Continuing, we get

$$\dot{\Sigma}(t) = -\frac{H^2(t)}{B^2(t)}\Sigma^2(t)$$

which can be transformed into a linear equation by setting

$$\frac{\dot{u}(t)}{u(t)} = \frac{H^2(t)}{B^2(t)}\Sigma(t)$$

we then get

$$\ddot{u}(t) = \dot{u}(t)\frac{d}{dt}\frac{H^2(t)}{B^2(t)}$$

$$\dot{u}(t) = \dot{u}(0) \exp\left[\frac{H^2(t)}{B^2(t)} - \frac{H^2(0)}{B^2(0)}\right]$$

It is now clear that $K(t)$ can be found explicitly.

20. We begin by factoring $S_x(\nu)$ into the form

$$S_x(\nu) = \frac{1}{[(i2\pi\nu) + 1 + i][(i2\pi\nu) + 1 - i][(i2\pi\nu) - 1 + i][(i2\pi\nu) - 1 - i]}$$

It is clear that an $\hat{h}$ satisfying (9.10) can be taken to be

$$\hat{h}(\nu) = \frac{1}{[(i2\pi\nu) + 1 + i][(i2\pi\nu) + 1 - i]}$$

corresponding to

$$h(t) = e^{-t}\sin t \quad t \geq 0$$
$$ = 0 \quad\quad\quad\;\; t < 0$$

The function $g(t)$ can be found by using (9.24) as

$$g(t) = \int_{-\infty}^{\infty} e^{i2\pi\nu t} \left[\frac{S_{xy}(\nu)}{\hat{h}(\nu)} \right]$$

$$= \int_{-\infty}^{\infty} e^{i2\pi\nu t} e^{i2\pi\nu\alpha} \hat{h}(\nu) \, d\nu$$

$$= e^{-(t+\alpha)} \sin(t+\alpha) \qquad t \geq -\alpha$$

$$= 0 \qquad t \leq -\alpha$$

Now, $\hat{\gamma}(\nu)$ is given by

$$\hat{\gamma}(\nu) = \int_0^{\infty} g(t) e^{-i2\pi\nu t} \, dt$$

$$= e^{-\alpha} \int_0^{\infty} e^{-t} \frac{e^{i(t+\alpha)} - e^{-i(t+\alpha)}}{2i} e^{-i2\pi\nu t} \, dt$$

$$= \frac{e^{-\alpha}}{2i} \left[\frac{e^{i\alpha}}{2\pi i\nu + 1 - i} - \frac{e^{-i\alpha}}{2\pi i\nu + 1 + i} \right]$$

$$= e^{-\alpha} \left[\frac{\cos\alpha + \sin\alpha(2\pi i\nu + 1)}{(2\pi i\nu + 1 - i)(2\pi i\nu + 1 + i)} \right]$$

From (9.29) we get

$$\hat{E}(X_{t+\alpha}|\mathcal{K}_{X^t}) = \int_{-\infty}^{\infty} e^{-\alpha} [\cos\alpha + \sin\alpha(2\pi i\nu + 1)] e^{i2\pi\nu t} \, d\hat{X}_\nu$$

$$= (\cos\alpha + \sin\alpha) e^{-\alpha} X_t + e^{-\alpha} \sin\alpha \dot{X}_t$$

21. Here, $S_y(\nu)$ has the form

$$S_y(\nu) = \frac{1}{1 + (2\pi\nu)^2} + \frac{1}{4 + (2\pi\nu)^4} = \frac{5 + (2\pi\nu)^2 + (2\pi\nu)^4}{[1 + (2\pi\nu)^2][4 + (2\pi\nu)^4]} = |\hat{h}(\nu)|^2$$

The function $\hat{h}(\nu)$ has the form

$$\hat{h}(\nu) = \frac{(2\pi i\nu + z_1)(2\pi i\nu + z_2)}{(2\pi i\nu + p_1)(2\pi i\nu + p_2)(2\pi i\nu + p_3)}$$

where we take $p_1 = 1 + i$, $p_2 = 1 - i$, $p_3 = 1$, and z_1 and z_2 have positive real parts. For this problem

$$S_{xy}(\nu) = S_x(\nu) = \frac{1}{4 + (2\pi\nu)^4}$$

Therefore,

$$\overline{\left[\frac{S_{xy}(\nu)}{\hat{h}(\nu)} \right]} = \frac{-2\pi i\nu + \bar{p}_3}{(-2\pi i\nu + \bar{z}_1)(-2\pi i\nu + \bar{z}_2)(2\pi i\nu + p_1)(2\pi i\nu + p_2)}$$

Since $g(\cdot)$ is the inverse transform of $[\overline{S_{xy}(\nu)/\hat{h}(\nu)}]$ and only $g(t)$ for $t \geq 0$ contributes

CHAPTER 3

to $\hat{\gamma}(\nu)$, if we write

$$\left[\overline{\frac{S_{xy}(\nu)}{\hat{h}(\nu)}}\right] = \frac{a_1}{2\pi i\nu + p_1} + \frac{a_2}{2\pi i\nu + p_2} + \frac{a_3}{-2\pi i\nu + \bar{z}_1} + \frac{a_4}{-2\pi i\nu + \bar{z}_2}$$

then, $\hat{\gamma}(\nu) = \dfrac{a_1}{2\pi i\nu + p_1} + \dfrac{a_2}{2\pi i\nu + p_2}$. Finally,

$$H(\nu) = \frac{\hat{\gamma}(\nu)}{\hat{h}(\nu)} = \frac{[(a_1 p_2 + p_1 a_2) + (a_1 + a_2)2\pi i\nu](2\pi i\nu + p_3)}{(2\pi i\nu + z_1)(2\pi i\nu + z_2)}$$

Using calculus of residues, we get

$$a_1 = \frac{p_1 + \bar{p}_3}{(p_1 + \bar{z}_1)(p_1 + \bar{z}_2)(p_1 + p_2)}$$

$$a_2 = \frac{p_2 + \bar{p}_3}{(p_2 + \bar{z}_1)(p_2 + \bar{z}_2)(-p_2 + p_1)}$$

Now, it only remains to compute z_1 and z_2 which can be shown to be given by $z_1, z_2 = (5)^{\frac{1}{2}} \exp(\pm i/2 \tan^{-1}\sqrt{19})$.

22. For this case (9.40) becomes

$$K(t) = \begin{bmatrix} 1 \\ 0 \end{bmatrix} + \Sigma(t)\begin{bmatrix} 0 \\ 1 \end{bmatrix}$$

and (9.53) becomes

$$\dot{\Sigma}(t) = \begin{bmatrix} 1 & 0 \\ 0 & 0 \end{bmatrix} - \begin{bmatrix} 1 & 0 \\ 0 & 0 \end{bmatrix} - \Sigma(t)\begin{bmatrix} 0 & 0 \\ 1 & 0 \end{bmatrix} - \begin{bmatrix} 0 & 1 \\ 0 & 0 \end{bmatrix}\Sigma(t) + \Sigma(t)\begin{bmatrix} 0 & 0 \\ 0 & 1 \end{bmatrix}\Sigma(t)$$

$$+ \begin{bmatrix} 0 & -1 \\ 1 & 0 \end{bmatrix}\Sigma(t) + \Sigma(t)\begin{bmatrix} 0 & 1 \\ -1 & 0 \end{bmatrix}$$

Since $\Sigma(0) = 0$, $\Sigma(t) \equiv 0$, $t \geq 0$ is a solution. If the solution to (9.53) is unique, as can be shown, then $\Sigma(t) \equiv 0$ is the only solution, and we have $K(t) = \begin{bmatrix} 1 \\ 0 \end{bmatrix}$. We can verify by writing

$$\dot{Y}_t = FY_t + \begin{bmatrix} 1 \\ 0 \end{bmatrix}\zeta_t$$
$$\dot{X}_t = HY_t + \zeta_t$$

Hence,

$$\dot{Y}_t = FY_t + \begin{bmatrix} 1 \\ 0 \end{bmatrix}(\dot{X}_t - HY_t)$$

23. Let $Y_{2t} = Y_t$ and $Y_{1t} = \dot{Y}_t$. Then

$$\begin{bmatrix} \dot{Y}_{1t} \\ \dot{Y}_{2t} \end{bmatrix} = \begin{bmatrix} 0 & 1 \\ 1 & 0 \end{bmatrix}\begin{bmatrix} Y_{1t} \\ Y_{2t} \end{bmatrix} + \begin{bmatrix} 1 & 0 \\ 0 & 0 \end{bmatrix}\begin{bmatrix} \eta_t \\ \xi_t \end{bmatrix}$$

$$\dot{X}_t = [0 \ 1]\begin{bmatrix} Y_{1t} \\ Y_{2t} \end{bmatrix} + [0 \ 1]\begin{bmatrix} \eta_t \\ \xi_t \end{bmatrix}$$

24. Since $e^{-|t-s|} = \int_{-\infty}^{\infty} e^{i2\pi \nu t} 2/[1 + (2\pi\nu)^2] \, d\nu$, $(1/\sqrt{2})\left(\dfrac{d}{dt}\xi_t + \xi_t\right) = \zeta_t$ is a standard white noise. Therefore,

$$\left(\frac{d}{dt} + 1\right)\dot{X}_t = \left(\frac{d}{dt} + 1\right)Y_t + \sqrt{2}\,\zeta_t$$

Let $Z_t = [(d/dt) + 1]X_t$, then we have

$$\dot{Z}_t = [1 \quad 1]\mathbf{Y}_t + [0 \quad \sqrt{2}]\begin{bmatrix} \eta \\ \zeta_t \end{bmatrix}$$

which is in the form of (9.34).

CHAPTER 4

1. Without loss of generality we can assume $a = 0$ and $b = 1$ and set

$$W_t^{(n)} = W_{k/n} \quad \begin{array}{l} k/n \le t < k + 1/n \\ k = 0, 1, \ldots, n-1 \end{array}$$

Because $E\varphi_t\bar{\varphi}_s$ is clearly continuous on $[0,1]^2$, we have

$$\sum_{k=0}^{n-1} \varphi_{k/n}(W_{k+1/n} - W_{k/n}) \xrightarrow[n\to\infty]{} \int_0^1 \varphi_t \, dW_t$$

On the other hand

$$\varphi_1 W_1 - \varphi_0 W_0 - \int_0^1 \dot{\varphi}_t W_t^{(n)} \, dt = \varphi_1 W_1 - \varphi_0 W_0 - \sum_{k=0}^{n-1} W_{k/n} \int_{k/n}^{k+1/n} \dot{\varphi}_t \, dt$$

$$= \varphi_1 W_1 - \varphi_0 W_0 - \sum_{k=0}^{n-1} W_{k/n}(\varphi_{(k+1)/n} - \varphi_{k/n})$$

$$= \sum_{k=1}^{n} \varphi_{k/n}(W_{k/n} - W_{(k-1)/n})$$

$$= \sum_{k=0}^{n-1} \varphi_{k/n}(W_{(k+1)/n} - W_{k/n})$$

$$+ \sum_{k=1}^{n} (\varphi_{k/n} - \varphi_{(k-1)/n})(W_{k/n} - W_{(k-1)/n})$$

The first term goes to $\int_0^1 \varphi_t \, dW_t$ as $n \to \infty$, while

$$\left|\sum_{k=1}^{n} (\varphi_{k/n} - \varphi_{k-1/n})(W_{k/n} - W_{(k-1)/n})\right| \le \sqrt{\sum_{k=1}^{n} (\varphi_{k/n} - \varphi_{(k-1)/n})^2 \sum_{k=1}^{n} (W_{k/n} - W_{(k-1)/n})^2}$$

$\xrightarrow[n\to\infty]{\text{a.s.}} 0$. Since $\int_0^1 \dot{\varphi}_t W_t^{(n)} \, dt \xrightarrow[n\to\infty]{\text{a.s.}} \int_0^1 \dot{\varphi}_t W_t \, dt$, we have

$$\varphi_1 W_1 - \varphi_0 W_0 - \int_0^1 \dot{\varphi}_t W_t \, dt = \int_0^1 \varphi_t \, dW_t$$

which was to be proved.

CHAPTER 4

2. Set $X_t = \ln Z_t$ then

$$dX_t = \frac{1}{Z_t} dZ_t - \frac{1}{2} \frac{1}{Z_t^2} Z_t^2 \varphi_t^2 \, dt = \varphi_t \, dW_t - \tfrac{1}{2} \varphi_t^2 \, dt$$

Since $X_0 = \ln Z_0 = 0$ we have

$$Z_t = e^{X_t} = \exp\left(\int_0^t \varphi_s \, dW_s - \frac{1}{2} \int_0^t \varphi_s^2 \, ds \right)$$

3. Define $Y_t = \ln X_t - \ln X_0$. Then, Ito's differentiation rule yields

$$dY_t = \frac{1}{X_t} dX_t - \frac{1}{2} \frac{1}{X_t^2} \sigma^2(t) X_t^2 \, dt$$

$$= m(t) \, dt + \sigma(t) \, dW_t - \frac{\sigma^2(t)}{2} dt$$

Since $Y_0 = 0$, we have

$$Y_t = \int_0^t \sigma(s) \, dW_s + \int_0^t \left[m(s) - \frac{\sigma^2(s)}{2} \right] ds$$

or

$$X_t = X_0 e^{Y_t} = \exp\left\{ \int_0^t \sigma(s) \, dW_s + \int_0^t \left[m(s) - \frac{\sigma^2(s)}{2} \right] ds \right\}$$

4. We first write

$$X_t = X_0 + \int_0^t m(X_s, s) \, ds + \int_0^t \sigma(X_s, s) \, dW_s$$

By the Cauchy-Schwarz inequality, we have

$$\left(\sum_{k=1}^N a_k \right)^2 \leq N \sum_{k=1}^N a_k^2$$

It follows that

$$EX_t^2 \leq 3 \left\{ EX_0^2 + \int_0^t E\sigma^2(X_s, s) \, ds + E\left[\int_0^t m(X_s, s) \, ds \right]^2 \right\}$$

$$\leq 3 \left[EX_0^2 + K \int_0^t E(1 + X_s^2) \, ds + Kt \int_0^t E(1 + X_s^2) \, ds \right]$$

$$= 3 \left[EX_0^2 + K(1 + t) \int_0^t E(1 + X_s^2) \, ds \right]$$

If we set $f(t) = \int_0^t E(1 + X_s^2) \, ds$, then

$$\frac{d}{dt} f(t) - 3K(1 + t) f(t) \leq 1 + 3EX_0^2$$

or

$$\frac{d}{dt} [e^{-3K(t+t^2/2)} f(t)] \leq (1 + 3EX_0^2) e^{-3K(t+t^2/2)}$$

Since $f(0) = 0$, we have

$$f(t) \leq (1 + 3EX_0^2)e^{3K(t+t^2/2)} \int_0^\infty e^{-3K(t+t^2/2)} \, dt = Be^{3K(t+t^2/2)}$$

and

$$EX_t^2 < f'(t) \leq (1 + 3EX_0^2) + 3K(1+t)f(t) \leq Ae^{3Kt^2}$$

5. We merely have to note that

$$\varphi(W_t) - \varphi(0) = \int_0^t \varphi'(W_s) \, dW_s + \frac{1}{2} \int_0^t \varphi''(W_s) \, ds$$

Therefore, $X_t = \varphi(0) + \int_0^t \varphi'(W_s) \, dW_s$. Since φ'' is bounded (say $|\varphi''| \leq B$), we have $|\varphi'(x)| \leq |\varphi'(0)| + B|x|$. Hence, $\int_0^t E|\varphi'(W_s)|^2 \, ds < \infty$, and $\int_0^t \varphi'(W_s) \, dW_s$ is a martingale and so is X_t.

7. First, let $Z_t = \sum_{k=1}^n W_{kt}^2$. Then,

$$(Z_t - Z_s) = \sum_{k=1}^n (W_{kt} - W_{ks})^2 + 2 \sum_{k=1}^n W_{ks}(W_{kt} - W_{ks})$$

Therefore,

$$E[(Z_t - Z_s)|W_{k\tau}, \tau \leq s, k = 1, \ldots, n] = \sum_{k=1}^n E(W_{kt} - W_{ks})^2 = n(t-s)$$

and

$$E[(Z_t - Z_s)^2|W_{k\tau}, \tau \leq s, k = 1, \ldots, n] = \sum_{k=1}^n E(W_{kt} - W_{ks})^4$$

$$+ \sum_{k \neq l} E(W_{kt} - W_{ks})^2 E(W_{lt} - W_{ls})^2 + 4 \sum_{k=1}^n W_{ks}^2 E(W_{kt} - W_{ks})^2$$

$$+ 4 \sum_{k \neq l} W_{ks} W_{ls} E[(W_{kt} - W_{ks})(W_{lt} - W_{ls})] - 3n(t-s)^2 + n(n-1)(t-s)^2$$

$$+ 4(t-s)Z_s$$

It follows that

$$\lim_{\Delta \downarrow 0} E[(Z_{t+\Delta} - Z_t)|Z_\tau, \tau \leq t] = n$$

$$\lim_{\Delta \downarrow 0} E[(Z_{t+\Delta} - Z_t)^2|Z_t, \tau \leq t] = 4Z_t$$

From Proposition 6.4 we know that Z is Markov and satisfies

$$dZ_t = n \, dt + 2\sqrt{Z_t} \, dW_t$$

for some Brownian motion W. Since $X_t = \sqrt{Z_t}$, we have

$$dX_t = \frac{1}{2}\frac{1}{\sqrt{Z_t}}dZ_t - \frac{1}{8}Z_t^{-\frac{3}{2}}(4Z_t)\,dt = \frac{1}{2X_t}(n\,dt + 2X_t\,dW_t) - \frac{1}{2X_t}dt$$

$$= \left(\frac{n}{2} - \frac{1}{2}\right)\frac{1}{X_t}dt + dW_t$$

We note that neither the stochastic differential equation for Z nor the one for X satisfies the conditions of Proposition 4.1. However, since Z is Markov and $X_t = \sqrt{Z_t}$, the process X is also Markov.

8. (a) For $\mu = \mu(X_0, t)$ the white-noise equation is equivalent to a stochastic differential equation.

$$dX_t = \mu(X_0, t)[b + \sigma\,dW_t]$$

or

$$X_t = X_0 + b\int_0^t \mu(X_0, s)\,ds + \sigma\int_0^t \mu(X_0, s)\,dW_s$$

Therefore,

$$E(X_T^2 | X_0 = x) = \left[x + b\int_0^T \mu(x, s)\,ds\right]^2 + \sigma^2\int_0^T \mu^2(x, s)\,ds$$

Suppose we denote the above functional of μ by $F(\mu)$. We get a minimizing μ by setting

$$\left.\frac{\partial F(\mu + \epsilon\,\delta\mu)}{\partial\epsilon}\right|_{\epsilon=0} = 0$$

and get

$$\int_0^T \left\{2\sigma^2\mu(x,s) + 2b\left[x + b\int_0^T \mu(x,\tau)\,d\tau\right]\right\}\delta\mu(x,s)\,ds = 0$$

for all $\delta\mu$. This yields

$$\mu(x, s) = \mu(x) = \frac{bx}{(\sigma^2 + b^2 T)} \qquad 0 \leq s \leq T$$

(b) For $\mu_t = \alpha(t)X_t$, the equivalent stochastic differential equation is now

$$dX_t = b\alpha(t)X_t\,dt + \sigma\alpha(t)X_t\,dW_t + \tfrac{1}{2}\sigma^2\alpha^2(t)X_t\,dt$$
$$= X_t[m(t)\,dt + \sigma\alpha(t)\,dW_t]$$

We know from Solution 4.3 that the solution is given by

$$X_t = X_0 \exp\left\{\int_0^t \sigma\alpha(s)\,dW_s + \int_0^t \left[m(s) - \frac{\sigma^2\alpha^2(s)}{2}\right]ds\right\}$$

$$= X_0 \exp\left[\int_0^t \sigma\alpha(s)\,dW_s + \int_0^t b\alpha(s)\,ds\right]$$

Now since $\int_0^T \sigma\alpha(s)\,dW_s$ is a Gaussian random variable with zero mean and variance $\Sigma^2 = \sigma^2\int_0^T \alpha^2(s)\,ds$, we have

$$E\exp\left[2\int_0^T \sigma\alpha(s)\,dW_s\right] = \exp\left[2\sigma^2\int_0^T \alpha^2(s)\,ds\right]$$

Therefore,

$$\begin{aligned} E(X_T{}^2|X_0 = x) &= x^2 \exp\left[2\int_0^T b\alpha(s)\,ds + 2\sigma^2 \int_0^T \alpha^2(s)\,ds\right] \\ &= x^2 \exp\left\{2\int_0^T \left[\sigma\alpha(s) + \frac{1}{2\sigma}b\right]^2 ds - \frac{1}{2}\left(\frac{b}{\sigma}\right)^2 T\right\} \\ &\geq x^2 e^{-\frac{1}{2}(b/\sigma)^2 T} \end{aligned}$$

so that $\alpha(s) = b/2\sigma^2$, $0 \leq s \leq T$, is the minimizing function.

9. We use the standard technique of separation of variables. Consider the equation

$$\frac{\partial f(x,t)}{\partial t} = \frac{1}{2}\frac{\partial^2}{\partial x^2}[\sigma^2(x)f(x,t)] - \frac{\partial}{\partial x}[m(x)f(x,t)]$$

If $f(x,t)$ is a product $f(x,t) = g(t)W(x)\varphi(x)$, then we have

$$\begin{aligned} W(x)\varphi(x)\frac{dg(t)}{dt} &= g(t)\left(\frac{d}{dx}\left\{\frac{1}{2}\frac{d}{dx}[\sigma^2(x)W(x)\varphi(x)] - m(x)W(x)\varphi(x)\right\}\right) \\ &= g(t)\frac{1}{2}\frac{d}{dx}\left[\sigma^2(x)W(x)\frac{d\varphi(x)}{dx}\right] \end{aligned}$$

Therefore,

$$\frac{1}{g(t)}\frac{dg(t)}{dt} = \frac{1}{W(x)\varphi(x)}\frac{1}{2}\frac{d}{dx}\left[\sigma^2(x)W(x)\frac{d\varphi(x)}{dx}\right]$$

The two sides, being functions of different variables, must be constants. Hence, if $f(x,t)$ is a product, then it must have the form

$$f_\lambda(x,t) = e^{-\lambda t}\varphi_\lambda(x)$$

when φ_λ satisfies the Sturm-Liouville equation

$$\frac{1}{2}\frac{d}{dx}\left[\sigma^2(x)W(x)\frac{d\varphi(x)}{dx}\right] + \lambda W(x)\varphi(x) = 0$$

Under rather general conditions, it can be shown that every solution $f(x,t)$ can be represented as a linear combination of products. Since $p(x,t|x_0,t_0)$ is a function of $t - t_0$, x, and x_0, it must have the assumed form.

10. The Sturm-Liouville equation in this case is

$$\tfrac{1}{2}f''(x) + \lambda f(x) = 0$$

Let $\lambda = \tfrac{1}{2}\nu^2$, then $f_\nu(x) = e^{i\nu x}$ are the bounded solutions. Now

$$\int_{-\infty}^\infty e^{-i\nu x}\frac{1}{\sqrt{2\pi t}}\exp\left(-\frac{1}{2}\frac{x^2}{t}\right)dx = e^{-\frac{1}{2}\nu^2 t}$$

By the inversion formula of the Fourier integral, we get

$$\frac{1}{\sqrt{2\pi t}}\exp\left(-\frac{1}{2}\frac{x^2}{t}\right) = \frac{1}{2\pi}\int_{-\infty}^\infty e^{-\frac{1}{2}\nu^2 t}e^{i\nu x}\,d\nu$$

and

$$\frac{1}{\sqrt{2\pi(t-t_0)}} \exp\left[-\frac{1}{2}\frac{(x-x_0)^2}{(t-t_0)}\right] = \frac{1}{2\pi}\int_{-\infty}^{\infty} e^{i\nu(x-x_0)} e^{-\frac{1}{2}\nu^2(t-t_0)}\,d\nu$$

11. The Fokker-Planck equation for this case has the form

$$\frac{\partial p}{\partial t} = \tfrac{1}{2}\beta(t)\frac{\partial^2 p}{\partial x^2} - \alpha(t)\frac{\partial}{\partial x}(xp)$$

Assuming p to have the form

$$p(x,t|x_0,t_0) = \frac{1}{\sqrt{2\pi a^2}} \exp\left[-\frac{1}{2a^2}(x-bx_0)^2\right]$$

we get

$$\frac{1}{p}\frac{\partial p}{\partial t} = -\frac{1}{2}\frac{1}{a^2}\frac{d}{dt}(a^2) + \frac{1}{2a^4}(x-bx_0)^2\frac{da^2}{dt} + \frac{x_0}{a^2}(x-bx_0)\frac{db}{dt}$$

$$\frac{1}{p}\frac{\partial p}{\partial x} = -\frac{1}{a^2}(x-bx_0)$$

$$\frac{1}{p}\frac{\partial^2 p}{\partial x^2} = \frac{1}{a^4}(x-bx_0)^2 - \frac{1}{a^2}$$

Upon substituting into the Fokker-Planck equation, we get

$$\left[\frac{(x-bx_0)^2}{2a^4} - \frac{1}{2a^2}\right]\frac{da^2}{dt} + \frac{x_0}{a^2}(x-bx_0)\frac{db}{dt} = \frac{\beta}{2}\left[\frac{(x-bx_0)^2}{a^4} - \frac{1}{a^2}\right]$$
$$+ \frac{\alpha(t)x}{a^2}(x-bx_0) - \alpha(t)$$
$$= \left[\frac{(x-bx_0)^2}{a^2} - 1\right]\left(\frac{\beta}{2a^2} + \alpha\right)$$
$$+ \frac{\alpha}{a^2} bx_0(x-bx_0)$$

Therefore,

$$\frac{da^2}{dt} = 2a^2\left(\frac{\beta}{2a^2} + \alpha\right) = \beta + 2\alpha a^2$$

$$\frac{db}{dt} = a^2\frac{\alpha}{a^2}b = \alpha b$$

which are to be solved with $a^2(t_0,t_0) = 0$, $b(t_0,t_0) = 1$.

12. (a) We assume $\dfrac{\partial p}{\partial t} \xrightarrow[t\to\infty]{} 0$ so that

$$\frac{1}{2}\frac{d^2 p(x)}{dx^2} + \frac{d}{dx}[\operatorname{sgn} x p(x)] = 0$$

or

$$\frac{dp(x)}{dx} + 2\operatorname{sgn} x p(x) = \text{constant}$$

Since $p'(x), p(x) \xrightarrow[|x| \to \infty]{} 0$, this constant must be zero. Therefore,

$$p(x) = e^{-2|x|}$$

which is already normalized.

(b) Consider the Sturm-Liouville equation

$$\frac{1}{2}\frac{d}{dx}\left[p(x)\frac{d\varphi(x)}{dx}\right] + \lambda p(x)\varphi(x) = 0$$

This can be written as

$$\frac{d^2\varphi(x)}{dx^2} - 2\,\text{sgn}\,x\,\frac{d\varphi(x)}{dx} + \lambda\varphi(x) = 0$$

The solutions satisfying, (1) $d\varphi/dx$ continuous at $x = 0$, and (2) $p\varphi^2$ bounded, are of the form

$$\varphi(x) = A_\nu e^{|x|} \sin \nu x \qquad 0 < \nu < \infty$$

where $\nu^2 = \lambda - 1$. It is convenient to choose A_ν to obtain the normalization

$$\int_{-\infty}^{\infty} \varphi_\nu(x)\varphi_{\nu'}(x)p(x)\,dx = \delta(\nu - \nu')$$

It can be shown that A_ν should be $\sqrt{2/\pi}$. Hence,

$$\varphi(\nu,x) = \sqrt{\frac{2}{\pi}}\, e^{|x|} \sin \nu x$$

The transition-density function must have the form

$$p(x,t|x_0,t_0) = p(x) \int_0^{\infty} e^{-(\nu^2+1)(t-t_0)} \varphi(\nu,x)\psi(\nu,x_0)\,d\nu$$

Because $p(x,t_0|x_0,t_0) = \delta(x - x_0)$, we have

$$\psi(\nu,x_0) = \int_{-\infty}^{\infty} \varphi(\nu,x)\delta(x - x_0)\,dx = \varphi(\nu,x_0)$$

Therefore,

$$p(x,t|x_0,t_0) = e^{-2|x|}e^{|x|+|x_0|}\frac{2}{\pi}\int_0^{\infty} e^{-(1+\nu^2)(t-t_0)} \sin \nu x \sin \nu x_0\, d\nu$$

$$= e^{-(|x|-|x_0|)}e^{-(t-t_0)}\frac{1}{\pi}\int_0^{\infty} e^{-\nu^2(t-t_0)}[\cos \nu(x - x_0) - \cos \nu(x + x_0)]\,d\nu$$

$$= e^{-(|x|-|x_0|)-(t-t_0)}\frac{1}{\sqrt{2\pi}}\left\{\exp\left[-\frac{(x-x_0)^2}{2(t-t_0)}\right] - \exp\left[-\frac{(x+x_0)^2}{2(t-t_0)}\right]\right\}$$

CHAPTER 5

1. $(H_t f)(x) = E_x(f(X_t))$
$= \frac{1}{2}(1 + e^{-t})f(x) + \frac{1}{2}(1 - e^{-t})f(-x) \qquad x = 1, -1$

If we set $\mathbf{f} = \begin{bmatrix} f(1) \\ f(-1) \end{bmatrix}$, then the operator H_t has a representation as a matrix, and

CHAPTER 5

we have
$$\mathbf{H}_t \mathbf{f} = \begin{bmatrix} \frac{1}{2}(1+e^{-t}) & \frac{1}{2}(1-e^{-t}) \\ \frac{1}{2}(1-e^{-t}) & \frac{1}{2}(1+e^{-t}) \end{bmatrix} \mathbf{f}$$

The generator A also has a matrix representation and can be found by

$$\mathbf{A} = \lim_{t \downarrow 0} \frac{1}{t}(\mathbf{H}_t - \mathbf{I}) = \lim_{t \downarrow 0} \frac{1}{2t}(1 - e^{-t})\begin{bmatrix} -1 & 1 \\ 1 & -1 \end{bmatrix}$$
$$= \frac{1}{2}\begin{bmatrix} -1 & 1 \\ 1 & -1 \end{bmatrix}$$

We can verify that $\mathbf{H}_t = e^{t\mathbf{A}}$ in many ways. For example, we note that
$$\mathbf{A}^2 = \frac{1}{2}\begin{bmatrix} 1 & -1 \\ -1 & 1 \end{bmatrix} = -\mathbf{A}$$

so that for $n = 1, 2, \ldots$,

$$\mathbf{A}^n = (-1)^{n-1}\mathbf{A} = \frac{(-1)^n}{2}\begin{bmatrix} 1 & -1 \\ -1 & 1 \end{bmatrix}$$

Therefore,

$$e^{t\mathbf{A}} = \sum_{n=0}^{\infty} \frac{t^n}{n!} \mathbf{A}^n = \mathbf{I} + \frac{1}{2}\begin{bmatrix} 1 & -1 \\ -1 & 1 \end{bmatrix} \sum_{n=1}^{\infty} \frac{(-t)^n}{n!}$$
$$= \begin{bmatrix} 1 & 0 \\ 0 & 1 \end{bmatrix} + \frac{1}{2}\begin{bmatrix} 1 & -1 \\ -1 & 1 \end{bmatrix}(e^{-t} - 1)$$
$$= \begin{bmatrix} \frac{1}{2}(1+e^{-t}) & \frac{1}{2}(1-e^{-t}) \\ \frac{1}{2}(1-e^{-t}) & \frac{1}{2}(1+e^{-t}) \end{bmatrix} = \mathbf{H}_t$$

2. Imitating the example at the end of Sec. 5.2, we find

$$\mathcal{D}_A = \{f : f \text{ bounded continuous on } [0, \infty), f'' \text{ bounded continuous on } (0, \infty)\}$$

and

$$(Af)(a) = \frac{1}{2}\frac{d^2 f(a)}{da^2} \quad a > 0$$
$$= 0 \quad a = 0$$

3. (a) With f as defined, we have

$$u_\lambda(a) = E_a \int_0^\infty e^{-\lambda t} f(X_t)\, dt = \int_0^\infty \text{prob}(X_t > c | X_0 = a) e^{-\lambda t}\, dt$$
$$= \int_0^\infty \left[\int_c^\infty \frac{1}{\sqrt{2\pi t}} \exp\left[-\frac{1}{2t}(x-a)^2\right] dx\right] e^{-\lambda t}\, dt$$

Since $\int_0^\infty \frac{1}{\sqrt{2\pi t}} \exp\left[-\frac{1}{2t}(x-a)^2\right] e^{-\lambda t}\, dt = \exp(-\sqrt{2\lambda}|x-a|)/\sqrt{2\lambda}$, we find

$$u_\lambda(a) = \int_c^\infty \frac{\exp(-\sqrt{2\lambda}|x-a|)}{\sqrt{2\lambda}}\, dx$$

$$= \begin{cases} \dfrac{1}{2\lambda} \exp[-\sqrt{2\lambda}\,(c-a)] & a \le c \\ \dfrac{1}{\lambda} - \dfrac{1}{2\lambda} \exp[-\sqrt{2\lambda}\,(a-c)] & a \ge c \end{cases}$$

Equation (3.13) now yields (with $\tau = \tau_c$)

$$u_\lambda(a) = u_\lambda(c) E_a e^{-\lambda \tau_c}$$

Since the Brownian motion X_t starts from 0 at $t = 0$, we get

$$E_0 e^{-\lambda \tau_c} = \frac{u_\lambda(0)}{u_\lambda(c)} = \exp(-\sqrt{2\lambda}\,c)$$

(b) Assume that τ_c has a density function $q(s)$, $0 \le s \le \infty$ so that

$$\mathcal{P}(\tau_c \le t) = \int_0^t q(s)\, ds$$

Taking the Laplace transform, we get

$$\int_0^\infty e^{-\lambda t} \int_0^t q(s)\, ds\, dt = \frac{1}{\lambda} \int_0^\infty e^{-\lambda t} q(t)\, dt = \frac{1}{\lambda} E e^{-\lambda \tau_c} = \frac{1}{\lambda} \exp(-\sqrt{2\lambda}\,c)$$

On the other hand,

$$u_\lambda(0) = \int_0^\infty e^{-\lambda t} \mathcal{P}(X_t \ge c)\, dt = \frac{1}{2\lambda} \exp(-\sqrt{2\lambda}\,c)$$

Because of the uniqueness of Laplace transforms

$$2\mathcal{P}(X_t \ge c) = \mathcal{P}(\tau_c \le t)$$

4. Let $\tau_0 = \min\{t : t > 0,\, X_t = 0\}$. Then,

$$\mathcal{P}(X_t \ge 0,\, 0 \le t \le 1) = \int_0^\infty \frac{1}{\sqrt{2\pi}} e^{-a^2/2}\, \mathcal{P}(\tau_0 \ge 1 | X_0 = a)\, da$$

Now, let $p(x,t|x_0,s)$ denote the transition density and define

$$f(x) = 0 \quad x \ge 0$$
$$ = 1 \quad x < 0$$

Then, (3.12) yields

$$u_\lambda(a) = \int_0^\infty e^{-\lambda t} \mathcal{P}(X_t < 0 | X_0 = a)\, dt = \int_{-\infty}^0 \left[\int_0^\infty e^{-\lambda t} p(x,t|a,0)\, dt \right] dx$$

Using τ_0 in (3.13), we get

$$u_\lambda(a) = u_\lambda(0) E_a e^{-\lambda \tau_0}$$

Let $q_a(\cdot)$ denote the density of τ_0 given $X_0 = a$. Then we have

$$\int_0^\infty q_a(s) e^{-\lambda s}\, ds = \frac{u_\lambda(a)}{u_\lambda(0)} = 2\lambda \int_0^\infty e^{-\lambda t} \mathcal{P}(X_t < 0 | X_0 = a)\, dt$$

For $a > 0$, we get

$$q_a(t) = 2 \frac{\partial}{\partial t} \mathcal{P}(X_t < 0 | X_0 = a)$$

CHAPTER 5

Therefore, for $a > 0$,

$$\mathcal{P}(\tau_0 \geq 1 | X_0 = a) = \int_1^\infty q_a(t)\, dt = 2\{\tfrac{1}{2} - \mathcal{P}(X_1 < 0 | X_0 = a)\}$$

Finally,

$$\mathcal{P}(X_t \geq 0, 0 \leq t \leq 1) = \int_0^\infty \frac{1}{\sqrt{2\pi}} e^{-a^2/2}[1 - 2\mathcal{P}(X_1 < 0 | X_0 = a)]\, da$$

$$= \tfrac{1}{2} - 2\int_0^\infty \frac{1}{\sqrt{2\pi}} e^{-a^2/2}\mathcal{P}(X_1 < 0 | X_0 = a)\, da$$

The joint density function for X_0 and X_1 has the form

$$p(x,t; x_0, 0) = \frac{1}{2\pi\sqrt{1-\rho^2}} \exp\left[-\frac{1}{2(1-\rho^2)}(x^2 - 2\rho x x_0 + x_0^2)\right]$$

where $\rho = e^{-1}$. Therefore,

$$\int_0^\infty \frac{1}{\sqrt{2\pi}} e^{-a^2/2} \mathcal{P}(X_1 \leq 0 | X_0 = a)\, da$$

$$= \int_0^\infty da \int_{-\infty}^0 dx \frac{1}{2\pi\sqrt{1-\rho^2}} \exp\left[-\frac{1}{2(1-\rho^2)}(x^2 - 2\rho x a + a^2)\right]$$

(Letting $a = r\cos\theta$, $x = r\sin\theta$, we get)

$$= \int_0^\infty r\, dr \int_{-\pi/2}^0 d\theta \frac{1}{2\pi\sqrt{1-\rho^2}} \exp\left[-\frac{r^2}{2(1-\rho^2)}(1 - \rho\sin 2\theta)\right]$$

$$= \frac{\sqrt{1-\rho^2}}{2\pi} \int_{-\pi/2}^0 \frac{1}{1 - \rho\sin 2\theta}\, d\theta$$

$$= \frac{1}{4} - \frac{1}{2\pi} \sin^{-1}\rho$$

Thus, we finally arrive at the expression

$$\mathcal{P}(X_t \geq 0, 0 \leq t \leq 1) = \frac{1}{\pi}\sin^{-1}(e^{-1})$$

5. See Solution 4.7 where we determined that

$$m(x) = 2 \qquad \sigma^2(x) = 4x$$

6. By (5.7) we can take

$$u(x) = \int_c^x \exp\left[-\int_0^y \frac{2m(z)}{\sigma^2(z)}\, dz\right] dy$$

Since $2m(z)/\sigma^2(z) = 1/z$, it is convenient to take $c = 1$ and find

$$u(x) = \int_1^x e^{-\ln y}\, dy = \ln x$$

Since $u(0) = -\infty$, S is open at 0.

7. We rewrite (2.12) for $H_t g \in \mathcal{D}_A$ as

$$\frac{d}{dt} H_t g = A H_t g$$

If we set $f_\lambda = \int_0^\infty e^{-\lambda t} H_t g\, dt$, then

$$\int_0^\infty e^{-\lambda t} \frac{d}{dt} H_t g\, dt = -g + \lambda \int_0^\infty e^{-\lambda t} H_t g\, dt = -g + \lambda f_\lambda$$

Using (2.12), we get

$$\int_0^\infty e^{-\lambda t} \frac{d}{dt} H_t g\, dt = A \int_0^\infty e^{-\lambda t} H_t g\, dt = A f_\lambda$$

Hence,

$$A f_\lambda = \lambda f_\lambda - g$$

8. First, we write

$$h(x, t+\delta) = E_x \left\{ f(X_{t+\delta}) \exp\left[-\int_0^{t+\delta} k(X_s)\, ds\right] \right\}$$

$$= E_x \left(E \left\{ f(X_{t+\delta}) \exp\left[-\int_0^{t+\delta} k(X_s)\, ds\right] \middle| X_\tau, 0 \le \tau \le \delta \right\} \right)$$

Because X is Markov and has a stationary transition function,

$$h(x, t+\delta) = E_x \left(\exp\left[-\int_0^\delta k(X_s)\, ds\right] E\left\{ f(X_{t+\delta}) \exp\left[-\int_0^t k(X_{s+\delta})\, ds\right] \middle| X_\delta \right\} \right)$$

$$= E_x \left\{ \exp\left[-\int_0^\delta k(X_s)\, ds\right] h(X_\delta, t) \right\}$$

$$= h(x,t) + h'(x,t) E_x(X_\delta - x) + \tfrac{1}{2} h''(x,t) E_x(X_\delta - x)^2$$

$$\qquad - k(x) h(x,t)\, \delta + o(\delta)$$

Therefore,

$$\frac{\partial h(x,t)}{\partial t} = m(x) h'(x,t) + \tfrac{1}{2} \sigma^2(x) h''(x,t) - k(x) h(x,t)$$

9. Setting $k(x) = \beta(1 + \operatorname{sgn} x/2)$ and $f(x) = 1$ in Kac's theorem, we get

$$u_\lambda(x) = E_x \int_0^\infty e^{-\lambda t} \exp\left(-\beta \int_0^t \frac{1 + \operatorname{sgn} X_s}{2}\, ds\right) dt$$

and u_λ satisfies

$$\frac{1}{2} \frac{d^2 u_\lambda(x)}{dx^2} - \beta \frac{1 + \operatorname{sgn} x}{2} u_\lambda(x) - \lambda u_\lambda(x) = -1$$

Therefore,

$$\frac{1}{2} \frac{d^2 u_\lambda(x)}{dx^2} - (\beta + \lambda) u_\lambda(x) = -1 \qquad x > 0$$

$$\frac{1}{2} \frac{d^2 u_\lambda(x)}{dx^2} - \lambda u_\lambda(x) = -1 \qquad x < 0$$

and u_λ must have the form

$$u_\lambda(x) = A \exp[-\sqrt{2(\beta + \lambda)}\, x] + \frac{1}{\lambda + \beta} \qquad x > 0$$

$$= B \exp(\sqrt{2\lambda}\, x) + \frac{1}{\lambda} \qquad x < 0$$

Continuity of u_λ and u_λ' at 0 yields

$$A = \frac{1}{\sqrt{\lambda + \beta}} \frac{1}{\sqrt{\lambda}} - \frac{1}{\lambda + \beta}$$

$$B = \frac{1}{\sqrt{\lambda + \beta}} \frac{1}{\sqrt{\lambda}} - \frac{1}{\lambda}$$

Since the Brownian motion starts at 0, we only need $u_\lambda(0)$ which is given by

$$u_\lambda(0) = \frac{1}{\sqrt{\lambda(\lambda + \beta)}}$$

which is the double Laplace transform of the distribution of

$$\tau(t) = \int_0^t \left(\frac{1 + \operatorname{sgn} X_s}{2}\right) ds$$

If we define $q(s,t)$ as the density of $\tau(t)$, that is, $q(s,t)\, ds = \mathcal{P}(\tau(t) \in ds)$, then

$$u_\lambda(0) = \frac{1}{\sqrt{\lambda(\lambda + \beta)}} = \int\!\!\int_0^\infty e^{-(\lambda t - \beta s)} q(s,t)\, ds\, dt$$

Inverting the Laplace transform once (with respect to β) yields

$$\int_0^\infty e^{-\lambda t} q(s,t)\, dt = \frac{1}{\sqrt{\pi s}} \frac{1}{\sqrt{\lambda}} e^{-\lambda s}$$

Inverting once again yields

$$q(s,t) = \frac{1}{\pi} \frac{1}{\sqrt{s}} \frac{1}{\sqrt{t-s}} \qquad t > s$$

$$= 0 \qquad t < s$$

Finally

$$\mathcal{P}(\tau(1) \leq t) = \int_0^t q(s,t)\, ds = \frac{1}{\pi} \int_0^t \frac{1}{\sqrt{s(1-s)}}\, ds = \frac{2}{\pi} \sin^{-1} \sqrt{t}$$

10. Define

$$h(x,t) = E_x \exp\left(-\beta \int_0^t X_s^2\, ds\right)$$

and consider the equation

$$\frac{\partial h}{\partial t} = \frac{1}{2} \frac{\partial^2 h}{\partial x^2} - \rho x^2 h \qquad h(x,0) = 1$$

We attempt a solution of the form

$$h(x,t) = A(t) e^{-\alpha(t) x^2} \qquad \begin{array}{l} \alpha(0) = 0 \\ A(0) = 1 \end{array}$$

Substituting the trial solution in the differential equation yields
$$\frac{\dot{A}(t)}{A(t)} - x^2\alpha(t) = 2\alpha^2(t)x^2 - \alpha(t) - \beta x^2$$
Equating like terms, we get
$$\dot{\alpha} + 2\alpha^2(t) = \beta$$
$$\frac{\dot{A}(t)}{A(t)} = -\alpha(t)$$
If we let $\alpha(t) = \dot{v}(t)/2v(t)$, then
$$\dot{\alpha}(t) + 2\alpha^2(t) = \frac{\ddot{v}(t)}{2v(t)} = \beta$$
so that
$$\ddot{v}(t) = 2\beta v(t)$$
With the initial condition $\alpha(0) = 0$, we get
$$\alpha(t) = \sqrt{2\beta}\, \tanh \sqrt{2\beta}\, t$$
and
$$\begin{aligned} A(t) &= \exp\left(-\int_0^t \sqrt{2\beta}\, \tanh \sqrt{2\beta}\, s\, ds\right) \\ &= \exp(-\ln \cosh \sqrt{2\beta}\, t) \\ &= \frac{1}{\cosh \sqrt{2\beta}\, t} \end{aligned}$$
Therefore,
$$h(x,t) = \frac{1}{\cosh \sqrt{2\beta}\, t} \exp[-\sqrt{2\beta}\,(\tanh \sqrt{2\beta}\, t)x^2]$$
and
$$h(0,t) = \frac{1}{\cosh \sqrt{2\beta}\, t}$$
and the density function for $Z = \int_0^1 X_t^2\, dt$ can be found by inverting $h(0,1)$, that is,
$$p_Z(z) = \frac{1}{2\pi i} \int_{C-i\infty}^{C+i\infty} e^{\beta z} \frac{1}{\cosh \sqrt{2\beta}}\, d\beta$$

11.

	$\sigma^2(x)$	$m(x)$	$u(x)$	Closed or open	Regularity
	1	-1	$\dfrac{e^{2x}-1}{2}$	Closed	Regular
	1	$-x$	$\displaystyle\int_0^x e^{v^2}\, dy$	Closed	Regular
	x	$-x$	$\dfrac{e^{2v}-1}{2}$	Closed	Regular
	x^2	$1 - \tfrac{3}{2}x$	$\displaystyle\int_0^x y^3 e^{-2v}\, dy$	Open	

CHAPTER 6

1.
$$E\left[|M_t - M_s|^2 | \mathcal{A}_s\right] = E\left[|M_t|^2 - 2\operatorname{Re}(M_t\overline{M}_s) + |M_s|^2 | \mathcal{A}_s\right]$$
$$= E\left[|M_t|^2 | \mathcal{A}_s\right] - 2\operatorname{Re}\left(\overline{M}_s E[M_t|\mathcal{A}_s]\right) + |M_s|^2$$
$$= E\left[|M_t|^2 | \mathcal{A}_s\right] - 2|M_s|^2 + |M_s|^2$$

2. It suffices to prove that for any $t \geq 0$ the events $\{S \wedge T \leq t\}$, $\{S \vee T \leq t\}$, and $\{T^* \leq t\}$ are in $\mathcal{A}_t$. These events can be expressed in terms of events in $\mathcal{A}_t$ as $\{S \leq t\} \cup \{T \leq t\}$, $\{S \leq t\} \cap \{T \leq t\}$, and $\bigcap_n \{T_n \leq t\}$, respectively, which implies the result desired.

3. Let $s \leq t$. Then $T' = (T \wedge t) \vee s$ is a bounded stopping time with $T' \geq s$ so that $E[M_{T'}|\mathcal{A}_s] = M_s$ by the optional sampling theorem. Now
$$M_{t \wedge T} = M_{T'} + (M_T - M_s)I_{\{T \leq s\}}$$
so
$$E[M_{t \wedge T}|\mathcal{A}_s] = M_s + (M_T - M_s)I_{\{T \leq s\}} = M_{T \wedge s}$$

4. It suffices to consider the case that X is a submartingale. It is easy to check that X_S is $\mathcal{A}_S$ measurable, and by definition $E[X_R|\mathcal{A}_S]$ is $\mathcal{A}_S$ measurable. Hence it suffices to prove that for any bounded $\mathcal{A}_S$-measurable random variable U,
$$E\left[UE[X_R|\mathcal{A}_S]\right] \geq EUX_S$$
or, equivalently, $EUX_R \geq EUX_S$. Now
$$E[U(X_R - X_S)] = E\left[I_{\{R=t_2, S=t_1\}}U(X_{t_2} - X_{t_1})\right]$$

Since $\{R = t_2\} = \{R \leq t_1\}^c$ is in $\mathcal{A}_{t_1}$ and $UI_{\{S=t_1\}}$ is $\mathcal{A}_{t_1}$ measurable, this becomes
$$E[U(X_R - X_S)] = E\left[E[X_{t_2} - X_{t_1}|\mathcal{A}_{t_1}]UI_{\{R=t_2, S=t_1\}}\right]$$

Since $E[X_{t_2} - X_{t_1}|\mathcal{A}_{t_1}] \geq 0$, so is the left-hand side of this equation. This completes the proof.

5. (a) For $m \geq n$,
$$E[Z_m|Z_1,\ldots,Z_n] = E[U_m \ldots U_{n+1}Z_n|U_1,\ldots,U_n]$$
$$= (EU_m EU_{m-1}\ldots EU_{n+1})Z_n = Z_n$$

(b) By the martingale convergence theorem, Z_n converges a.s. to a random variable Z_∞. Now
$$\ln(Z_n) = \sum_{i=1}^n \ln U_i, \qquad E\ln U_i = \frac{1}{2}\int_0^2 \ln(u)\,du = (\ln 2) - 1 < 0$$

and $E[(\ln(U_i))^2] < +\infty$. Then, by the weak law of large numbers, $(1/n)\ln(Z_n) \xrightarrow[n \to \infty]{P} \ln 2 - 1$. Thus $\ln(Z_n)$ converges in probability to $-\infty$, which means that Z_n converges in probability to zero. Thus $Z_\infty = 0$—that is, Z_n converges a.s. to zero. Since $EZ_n = 1$ for all n, Z_n does not converge in ν-mean for any $\nu \geq 1$.

6. Let $A \in \mathcal{B}$. Then $M(A) = \int_A (dM^{\mathcal{B}}/dM_0^{\mathcal{B}}) \, dM_0$. On the other hand, by the definition of conditional expectation, we also have

$$M(A) = \int_A \frac{dM}{dM_0} \, dM_0 = \int_A E_0 \left[\frac{dM}{dM_0} \Big| \mathcal{B} \right] dM_0$$

Equating these two expressions for $M(A)$ and noting that each side of Eq. (1.9) is $\mathcal{B}$ measurable, the conclusion follows.

7. Let $R_n \uparrow +\infty$ be stopping times such that $(M_{R_n \wedge t}, \mathcal{Q}_t : t \geq 0)$ is a martingale for each $n \geq 1$. Then, as n tends to $+\infty$, we use a conditional version of Fatou's lemma to deduce that for $t > s$,

$$E[M_t | \mathcal{Q}_s] = E\left[\liminf_{n \to \infty} M_{R_n \wedge t} | \mathcal{Q}_s \right] \quad \text{a.s.}$$

$$\leq \liminf_{n \to \infty} E\left[M_{R_n \wedge t} | \mathcal{Q}_s \right] = \liminf_{n \to \infty} M_{R_n \wedge s} = M_s \quad \text{a.s.}$$

8.
$$E\left[\mathcal{E}(h \cdot w)_\tau^p \right] = E \exp\left(ph \cdot w_\tau - \frac{p}{2} \int_0^\tau h_s^2 \, ds \right)$$

$$= E\left[\mathcal{E}(ph \cdot w)_\tau \exp\left(\frac{p^2 - p}{2} \int_0^\tau h_s^2 \, ds \right) \right]$$

$$\leq \exp\left((p^2 - p) \frac{K}{2} \right),$$

as desired. Then, by the Markov inequality,

$$E\left[\mathcal{E}(h \cdot w)_\tau I_{\{\mathcal{E}(h \cdot w)_\tau \geq c\}} \right] \leq \frac{1}{c^{p-1}} \exp\left((p^2 - p) \frac{K}{2} \right)$$

which implies that $\mathcal{E}(h \cdot w)$ is class D. Finally, since $\mathcal{E}(h \cdot w)$ is a local martingale, there exists a sequence $\tau_n \uparrow +\infty$ of stopping times such that $E\mathcal{E}(h \cdot w)_{\tau_n} = 1$ for each n. Then $E\mathcal{E}(h \cdot w)_\infty = 1$ by Proposition 1.3.

9. If $\mathcal{P} \ll \mathcal{P}_0$ then the Radon-Nikodym derivative $\Lambda = d\mathcal{P}/d\mathcal{P}_0$ exists and $L_n = E[\Lambda | \mathcal{Q}_n]$. By Proposition 1.4, $L_\infty = \Lambda$ a.s. so that $EL_\infty = 1$. Conversely, suppose that $EL_\infty = 1$. Then for $c \geq 1$,

$$E\left[L_n I_{\{L_n \leq c\}} \right] \geq E\left[L_n I_{\{L_\infty \leq c - 1/c, |L_n - L_\infty| \leq 1/c\}} \right]$$

$$\geq E\left[(L_\infty - 1/c) I_{\{L_\infty \leq c - 1/c, |L_n - L_\infty| \leq 1/c\}} \right]$$

so that

$$\lim_{c \to \infty} \liminf_{n \to \infty} E\left[L_n I_{\{L_n \leq c\}} \right] \geq \lim_{c \to \infty} E\left[\left(L_\infty - \frac{1}{c} \right) I_{\{L_\infty \leq c - 1/c\}} \right] = 1$$

Equivalently, since $EL_n = 1$ for each n,

$$\lim_{c \to \infty} \limsup_{n \to \infty} E\left[L_n I_{\{L_n > c\}} \right] = 0$$

so that L is a uniformly integrable martingale. In particular, $L_n = E[L_\infty | \mathcal{Q}_n]$. Now if U is a bounded random variable which is $\mathcal{A}_n$ measurable for some n, then $EU = E_0 U L_n = E_0 U L_\infty$. Then, by the monotone class theorem, $EU = E_0 U L_\infty$ for all $\mathcal{A}_\infty$ bounded measurable U. Hence $\mathcal{P} \ll \mathcal{P}_0$ and L_∞ is the Radon-Nikodym derivative.

10. L_n defined in the previous problem is given by

$$L_n = \exp\left(\sum_{i=1}^n a_i X_i - \frac{1}{2} \sum_{i=1}^n a_i^2 \right)$$

Suppose that $S < +\infty$. Then

$$E_0[L_n \ln L_n] = E[\ln L_n] = \frac{1}{2} \sum_{i=1}^n a_i^2 \leq \frac{1}{2} S$$

so that for $c > 1$,

$$E_0\left[L_n I_{\{L_n \geq c\}} \right] \leq \frac{E_0[L_n \ln L_n]}{\ln(c)} \leq \frac{S}{\ln(c)} \xrightarrow[c \to \infty]{} 0$$

so that (L_n) is uniformly integrable. Thus $\mathcal{P} \ll \mathcal{P}_0$.

Suppose now that $S = +\infty$. Under measure $\mathcal{P}_0$, $\ln L_n$ is a Gaussian random variable with mean $-\frac{1}{2} \sum_{k=1}^n a_k^2$ and variance $\sum_{k=0}^n a_k^2$. Therefore, $P_0[\ln L_n \leq c] \xrightarrow[n \to \infty]{} 1$ for all constants c. Combining this with the fact that, by the martingale convergence theorem, L_n converges $\mathcal{P}_0$ a.s., conclude that $\mathcal{P}_0(A) = 1$ where A is the event that $\lim_{n \to \infty} L_n = 0$. On the other hand, L^{-1} is a nonnegative $\mathcal{P}$ martingale so that

$$\mathcal{P}\left[\lim_{n \to \infty} L_n^{-1} \text{ exists and is finite} \right] = 1.$$

Therefore $\mathcal{P}(A) = 0$. Thus $\mathcal{P} \perp \mathcal{P}_0$.

11. By Itô's formula,

$$w_t^4 = 4 \int_0^t w_s^3 \, dw_s + 6 \int_0^t w_s^2 \, ds$$

so we see that $6 \int_0^t w_s^2 \, ds$ is the predictable compensator of w_t^4. Next (simply check the jumps)

$$N_t^4 = \int_0^t (N_{s-} + 1)^4 - (N_{s-})^4 \, dN_s$$

so the predictable compensator of N_t^4 is

$$\int_0^t (N_{s-} + 1)^4 - (N_{s-})^4 \, ds \text{ which equals } \int_0^t (N_s + 1)^4 - N_s^4 \, ds.$$

12. $0 = E\left[B_{n+1} | \mathcal{A}_n^V \right] = \mathcal{P}\left[B_{n+1} = 1 | \mathcal{A}_n^V \right] - \mathcal{P}\left[B_{n+1} = -1 | \mathcal{A}_n^V \right]$

and

$$1 = \mathcal{P}\left[B_{n+1} = 1 | \mathcal{A}_n^V \right] + \mathcal{P}\left[B_{n+1} = -1 | \mathcal{A}_n^V \right]$$

Thus, $\mathcal{P}[B_{n+1} = 1|\mathcal{Q}_n^V] = \mathcal{P}[B_{n+1} = -1|\mathcal{Q}_n^V] = 0.5$ for each n, and this implies part (a).

There exists a sequence of functions F_n so that

$$M_{n+1}(\omega) = F_{n+1}(B_1(\omega), B_2(\omega), \ldots, B_{n+1}(\omega)), \qquad n \geq 0$$

Now

$$M_n = E\left[M_{n+1}|\mathcal{Q}_n^V\right] = \tfrac{1}{2}\{F_{n+1}(B_1, \ldots, B_n, 1) + F_{n+1}(B_1, \ldots, B_n, -1)\}$$

so if we define a predictable process H by $H_0 = 0$ and

$$H_{n+1} = \tfrac{1}{2}\{F_{n+1}(B_1, \ldots, B_n, 1) - F_{n+1}(B_1, \ldots, B_n, -1)\}, \qquad n \geq 0$$

then $M_{n+1} = M_n + B_{n+1}H_{n+1}$ for $n \geq 0$. This implies that the representation to be proved is valid.

CHAPTER 7

1. For each possible value θ of O_{k+1},

$$\begin{aligned}
\Pi_{k+1}(j) &= \mathcal{P}(z_{k+1} = j|O_{k+1} = \theta, \mathcal{O}_k) \\
&= \mathcal{P}(z_{k+1} = j, O_{k+1} = \theta|\mathcal{O}_k)/\mathcal{P}(O_{k+1} = \theta|\mathcal{O}_k) \\
&= \frac{\sum_i \mathcal{P}(z_{k+1} = j, O_{k+1} = \theta, z_k = i|\mathcal{O}_k)}{\sum_{i,j'} \mathcal{P}(z_{k+1} = j', O_{k+1} = \theta, z_k = i|\mathcal{O}_k)}
\end{aligned}$$

Now the ith term in the numerator of the last expression is equal to

$$\mathcal{P}(z_k = i|\mathcal{O}_k)\mathcal{P}(z_{k+1} = j, O_{k+1} = \theta|z_k = i, \mathcal{O}_k)$$

which is the same as $\Pi_k(i)R_{ij}(\theta)$. The denominator above is the sum of similar terms, and the desired conclusion follows.

2. Since H_t is not random, it is equal to $E[(\xi_t - \hat{\xi}_t)^2]$. Equation (4.12) for H can be written

$$\frac{d}{dt}(H_t - a) + (H_t - a)^2 = a^2 + b^2.$$

Setting $\dot{u}(t)/u(t) = H_t - a$, we get

$$\frac{\ddot{u}(t)}{u(t)} = a^2 + b^2, \qquad \text{so } u_t = Ae^{\rho t} + Be^{-\rho t} \qquad \text{where } \rho = (a^2 + b^2)^{1/2}$$

for some constants A and B. Using the initial condition $\dot{u}(0)/u(0) = H_0 - a = -a$, we get

$$\frac{\dot{u}(t)}{u(t)} = \rho\left\{\frac{(1+\theta)e^{\rho t} - (1-\theta)e^{-\rho t}}{(1+\theta)e^{\rho t} + (1-\theta)e^{-\rho t}}\right\} \qquad \text{where } \theta = \frac{-a}{\rho}$$

Since $H_t = a + \dot{u}(t)/u(t)$, this yields H_t. We find that H_t tends to $a + \rho$ as t tends to infinity. This limit value can also be obtained by setting $\dot{H}_t = 0$ in Eq. (4.12).

3. The goal is to find a recursive equation for $\hat{\xi}_t$ where $\xi = I_{[0,U)}$. Let $\mathcal{Q}_t = \sigma(\xi_s, N_s : 0 \le s \le t)$. Then, arguing heuristically,

$$E[d\xi_t | \mathcal{Q}_t] = -\mathcal{P}[t < U \le t + dt | \mathcal{Q}_t]$$
$$= -P[t < U | \mathcal{Q}_t] \mathcal{P}[t < U \le t + dt | t < U, \mathcal{Q}_t]$$
$$= -\xi_t f(t) \, dt / (1 - F(t))$$

This suggests the easily verified fact that if $\beta_t = -\xi_t f(t)/(1 - F(t))$; then m defined by Eq. (3.2) is an $\mathcal{Q}_\cdot$ martingale. Now $\lambda_s = (a - b)\xi_s + b$, $\xi_s^2 = \xi_s$, $\hat{\xi}_s = \hat{\xi}_{s-}$, and ψ in Eq. (4.3) is zero, so Eq. (4.3) becomes

$$K_s = \frac{a\hat{\xi}_{s-} - (a - b)(\hat{\xi}_{s-})^2 - b\hat{\xi}_{s-}}{(a - b)\hat{\xi}_{s-} + b}$$

and Eq. (4.1) yields the recursive equation ($\hat{\xi}_0 = 1$):

$$\hat{\xi}_t = \hat{\xi}_0 - \int_0^t \hat{\xi}_s f(s)/(1 - F(s)) \, ds + \int_0^t K_s \left(dN_s - \left[(a - b)\hat{\xi}_s + b \right] ds \right)$$

INDEX

α-measurable function, 8
α. Poisson process, 222
Absolute continuity, 215
   with respect to Lebesgue measure, 9
Absorbing boundary, 203
Absorbing Brownian motion, 195, 203
Adapted, 209
Algebra, 2
Almost-sure convergence, 19
   criterion for, 22
Arc-sine law, 208
Atom of a σ algebra, 26
Average, *see* Expectation

Backward equation of diffusion, 171

Bandlimited process, 105
Bandpass process, 107
Basic space, 3
Bochner's theorem, 94–96
   generalization to R^n, 281
   homogeneous and isotropic case, 284–285
Boolean algebra, 2
Borel-Cantelli lemma, 12
Borel function, 7
Borel measure, 5, 8
Borel probability measure, 5, 8
   absolute continous, 9
   singular, 9
Borel sets, 5
Borel σ algebra, 5
Boundary

Boundary (*cont.*)
  absorbing, 203
  closed or open, 200
  exit, 202
  reflecting, 203
  regular, 202
Boundary condition, 202–203
Brownian motion, 50
  absorbing, 195, 203
  approximations to, 160
  Gaussian white noise as derivative of, 156–157
  generator, 189
  Markov property, 50
  martingale property, 51
  modulus of continuity, 61
  multidimensional parameter, 296
  quadratic variation, 53
  reflecting, 195, 203
  sample continuity, 58–59
  semigroup, 188–190
  stochastic integral with respect to, 141–145, 163–164

C_0 space, 89
Cadlag, *see* corlol
Cauchy sequence, 78
Chapman-Kolmogorov equation, 63
  stationary case, 181
  in terms of conditional expectation, 66
Characteristic function, 18
Characteristic operator, Å, 196
Characterization theorem of Levy and Watanabe, 241
Chebyshev inequality, *see* Markov inequality, $\nu = 2$
Compensator, 224, 238
Completion of a probability space, 3
Conditional density function, 26
Conditional distribution function, 25, 32–33
Conditional expectation, 27
  convergence properties, 29
  smoothing properties, 29–31
Conditional independence, 31
Conditional probability, 25

Continuity
  almost-sure, 55
  almost-surely sample, 55
    of Brownian motion, 58–59
    sufficient condition, 57
  modulus of, 60–61
  in νth mean, 55
  in probability, 43, 55
  in quadratic mean, 55, 77
    characterization, 77
Convergence
  almost-sure, 19, 22
  of distributions, 23–24
  of expectations, 17
  mutual, 19–21
  in νth mean, 20
  in probability, 19
  in quadratic mean, 20
  sequential, of events, 11
  to stochastic integral, 160
  strong, in Banach space, 183
  to a white noise, 110–111
Convolution, 91
Coordinate function
  finite-dimensional, 9
  infinite-dimensional, 39
Corlol, 210
Correction term for white-noise integral, 160–162
Correlation function, 74
Counting process, 221
Covariance function, 48, 74
  properties, 75–77

Decreasing sequences
  of events, 11
  of random variables, 13
Detection, 250, 254–257
Differentiability of measures, 214–215
Differential equation
  driven by white noise, 113–115, 126
  stochastic, 149–150
Differentiation in quadratic mean, 79

INDEX

Differentiation rule of stochastic integral, 147–148, 239
Diffusion equations, 169
  backward, 171
  forward, 172, 274
    fundamental solution, 172
  multidimensional case, 176
  solution using Laplace transform, 205–206
Diffusion process, 198
Discontinuities of first kind, 59
  random telegraph process, 60
random telegraph process, 60
  Distribution function, 5–7
probability, 7–8
  joint, 8
Dominated convergence, 17
Dynkin's formula, 192

Eigenfunction
  of integral equation, 83
  of Laplacian operator, 287
Eigenvalue
  of integral equation, 83
  of Laplacian operator, 286
Envelope, 109
Ergodic process, 67
  condition for a Gaussian process, 69
Ergodic theorem, 68
Estimator
  linear least squares, 116–117
  recursive, 269
Euclidean distance, 282
Euclidean group, 282
Euclidean norm, 281
Event, 1–3
  convergence, 11
Exit boundary, 202
Exit time, 194
Expectation, 15–17
  conditional, 27, 29–31
  Convergence, 17
Exponential of a semimartingale, 241
Extension thereom, 3

Fatou's lemma, 17
Feller process, 192, 194
Filtering
  and detection, 250
  Kalman-Bucy, 125–127, 268–269
  linear time-invariant, 91, 101
  recursive, 231, 269, 274
  Wiener, 120–122
Finite-dimensional distributions of a stochastic process, 37–41
  compatibility condition, 38–39
First exit time, 194
First passage time, 190
Fokker-Planck equation, 172
Forward equation of diffusion, 172
  fundamental solution, 172
Fourier integral, 90
  inversion formula, 90
Fundamental solution, 172

Gauss-Markov process, condition for, 64–65
Gauss-Markov random field, isotropic, 292–293
  and homogeneous, 293–295
Gaussian process, 46
  characteristic function, 47
  density function, 49
  linear operations on, 47–48
  Markov, 64–65
Gaussian random variable, 46
Gaussian white noise
  convergence, 160–161
  correction term, 160–162
  as derivative of Brownian motion, 156–157
  in differential equations, 156–160
  in integrals, 157
  simulation, 162–163
Gegenbauer equation, 287
Gegenbauer polynomials, 287
Generalized process, 279
Generator (of Markov semigroup), 183
Girsanov's theorem, 244
Green's function, 199

Hilbert space:
  projection onto, 117
  second-order process, 101
  wide-sense stationary process, 101
Hilbert transform, 105, 108
Hölder condition, 173
Homogeneous random field, 281
  isotropic case, 283
Hypothesis testing, 254–255

Impulse response, 91
Increasing sequence
  of events, 11
  of random variables, 13
Independence
  of events, 24
  of random variables, 25
Independent increments, 50
Indicator function, 14
Infimum, 13
Innovations, 254
  derivation of filters, 262–267
Integrable random variable, 17
Integral
  quadratic-mean, 79–80
  stochastic, see Stochastic integral
Integral equation for orthogonal expansion, 83
Intensity of a counting process, 226
Invariant random variable, 67
Inverse Fourier transform, 90
Inverse image, 7
Inversion formula
  Fourier integral, 90
  Fourier-Stieltjes integrals, 96–97
  spectral distribution, 96–97
Isotropic random field, 289
Itô integral, 140; (see also Stochastic integral)
Itô's differentiation rule, 147–148, 239
Itô's product formula, 239

Kac's theorem, 208
Kakutani's dichotomy theorem, 249

Kalman-Bucy filtering, 125–127, 268–269
Karhunen-Loève expansion, 82–88
Kolmogorov condition (for sample continuity), 57
Kolmogorov's equations, see Diffusion equations

L^2 martingale inequality, 214
L_p space, 89
Laplacian operator, 286
Lebesgue decomposition, 215
  of Borel probability measure, 9
Lebesgue measure, 5
Lebesgue-Stieltjes integral, 18, 33, 217–219
  martingale property, 222, 232–233
  for random processes, 220–221
Likelihood ratio, see Radon-Nikodym derivative
Likelihood ratio test, 254–255
Limit, sequential, 11
Limit inferior
  of events, 11
  of random variables, 13
Limit superior:
  of events, 11
  of random variables, 13
Linear least-squares estimator, 116
Linear operations
  on Gaussian process, 47–48
  on second-order process, 78
Linear time-invariant filtering, 91
  characterization, 101
Lipschitz condition, 150
Local martingale, 165, 234
Local square integrable martingale, 234
Locally bounded, 234
Locally finite variation, 219

Markov inequality, $\nu=2$, 21
Markov process, 50, 61
  and conditional expectation, 65–66
  Gaussian, 64

representation by stochastic equation, 168
    strong, 191
Markov random field, 292
Markov semigroup, 182–183
Markov time, 190
Martingale, 51, 209
    local, 165, 234
    and stochastic integral, 145, 166, 228, 232–233
Martingale inequality, 51, 214
Martingale representation theorem, 246
Mean function (or mean), 48
Measurable function, 7
    Borel, 7
Measurable process, 45
Measurable space, 5, 215
Measure
    absolute continuity, 215
    Borel, 5, 8
    finite, 5
    Lebesgue, 5
    mutual absolute continuity, 215
    probability, 3
    Borel, 5, 8, 9
    σ finite, 5
    singular, 215
    spectral, 94, 284
Mercer's theorem, 85
Meyer's predictable compensator theorem, 225
Minimum phase condition, 118
Modulus of continuity, 60–61
    Brownian motion, 61
Monotone class theorem, 33
Monotone convergence, 17
Monotone sequence of events, 11
of random variables, 13
Monotone sequential continuity at $\emptyset$, 2
Mutual convergence, 19–21
    almost-sure, 19
    in probability, 20
    in νth mean, 21
    in quadratic mean, 21

Natural scale, 197
Nonanticipative filter, 118
Nonnegative definite function, 75
Null set, 3

Optional sampling theorem, 213
Ornstein-Uhlenbeck process, 66
Orthogonal expansion, 81
Orthogonal increments, 98
Orthonormal family, 81

Paley-Wiener condition, 118
Pathwise
    integration, 217
    solutions to filtering equations, 274–275
Polar coordinates, 285
Predictable
    compensator, 224, 238
    continuous time process, 224
    discrete time process, 223
    projection, 252
    σ algebra, 224
Prediction, linear least-squares, 116
Probability density function, 9
Probability distribution function, 7–8
Probability measure, 3
    Borel, 5, 8, 9
    elementary, 2
Probability space, 3
    completion, 3
Process
    with independent increments, 50
    with orthogonal increments, 98
Progressively measurable, 220
Projection, 117
    optional, 252
    predictable, 252

Quadratic-mean continuity, 55, 77
Quadratic-mean convergence, 20
Quadratic-mean derivative, 79
Quadratic-mean integral, 79–80

Quadratic variation, 236
Quasi-martingales, 167

Radon-Nikodym derivative, 28, 215
　connection to martingales, 216, 243–244
　finite-dimensional approximations, 216–217
　representation, 257
Radon-Nikodym theorem, 27, 215
Random field, 279–280
　homogeneous, 281
　　and isotropic, 283
　isotropic, 289
　spectral representation, 282, 291
Random measure, 280
Random telegraph process, 59–60
　discontinuities of, 60, 63
Random variable
　discrete, 8
　invariant, 67
　real, 8
　second-order, 74
　simple, 14
Realization
　of probability distribution, 9
　for stochastic process, 39
Recursive estimation, 269
Recursive filtering, 269
　equation of motion, 274
　reduction to Kalman-Bucy filtering, 268–269
Reflecting boundary, 203
Reflecting Brownian motion, 195, 203
Reflection principle of D. André, 207
Resolvent, 185
Riccati equation, 127
Rigid-body motion, 282
Rotation, 282

$\mathcal{S}$-space, 89
Sample continuity, 55
　of Brownian motion, 58–59
　condition for, 57

Sample function, 37
Sample space, 216
Sampling theorem, 106
　for bandpass process, 109
Scale function, 197
Second-order calculus, 79–80
Second-order random variable, 74
Semigroup
　of Brownian motion, 188–190
　of Markov processes, 182–183
Semigroup property, 183
Semimartingale, 234
　approximation of integral, 235
　integral, 235
Separability of stochastic process, 41–45
Separable and measurable modification, 45
Separable process, 41–45
Separable σ algebra, 26
Separating set, 42
　of process continuous in probability, 43
Sequence
　of events, 11
　of random variables, 13–14
Sequential continuity, 11
Sequential limits of events, 11
σ additivity, 3
σ algebra, 2
　atoms of, 26
　Borel, 5
　generated, 2
　minimal, 2
　predictable, 224
　separable, 26
Simple random variables, 14
Simulation of white noise, 162–163
Singularity, 215
　with respect to Lebesgue measure, 9
Spectral-density function, 92, 94
　input-output relationship, 93
　interpretation in terms of average power, 93
Spectral-distribution function, 94
　for random fields, 281, 284

INDEX

Spectral factorization, 119–120
Spectral measure, 94
 for random fields, 281, 284
Spectral representation
 of homogeneous random fields, 282
  isotropic case, 291
 of wide-sense stationary process, 101
Spherical harmonics, 288
State space (of diffusion process), 181
Stationary process, 66–67
 wide-sense, 66, 88–89, 101
  jointly, 104
Stationary transition function, 180
Step function, 141
Stieltjes integral, 18, 33
Stochastic differentiation equation, 149–150
 and diffusion equations, 173
 properties of solution, 150–155
 representation of Markov process, 168
Stochastic integral, 98, 141–145, 163–164
 differentiation rule, 147–148, 239
 local martingales defined by, 165
 martingales defined by, 145
 pathwise, 220–221
 with respect to Brownian motion, 141–145, 163–164
 with respect to martingales, 166
 with respect to process with orthogonal increments, 98
 sample continuity, 146–147
Stochastic integral equation, see Stochastic differential equation
Stochastic process
 definition, 37
 sample functions, 37
 second-order, 74
Stopping time, 212
Strong convergence, 183
Strong Markov process, 191
Submartingale, 41, 209
Supermartingale, 51, 209

Supremum, 13

Transfer function, 91
Transformation rule
 of Borel probability measures, 10
 of probability density function, 10
 of stochastic integral, 147–148
Transition density function, 169
Transition function, 62, 169
 stationary, 180
Translation, 282
Translation group
 for stationary process, 67
 for wide-sense stationary process, 88–89

Uniformly integrable, 211
 martingale, 211
Usual conditions, 210

Variation, 218
 quadratic, 236

White noise, 109–115
 convergence to, 110–111
 differential equations driven by, 113–115, 126
 Gaussian, see Gaussian white noise
 integral, as stochastic integral, 111
Wide-sense stationary process, 66, 88–89, 101
Wiener-filtering, 120–122
Wiener measure, 53, 59
Wiener martingale, 209
Wiener process, see Brownian motion

Zero-one law, 245

University Series in Modern Engineering

Elements of State Space Theory of Systems
A. V. Balakrishnan

Kalman Filtering Theory
A. V. Balakrishnan

Systems & Signals
N. Levan